CAMBRIDGE STUDIES IN ADVANCED MATHEMATICS 134

Editorial Board
B. BOLLOBÁS, W. FULTON, A. KATOK, F. KIRWAN,
P. SARNAK, B. SIMON, B. TOTARO

GEOMETRIC ANALYSIS

The aim of this graduate-level text is to equip the reader with the basic tools and techniques needed for research in various areas of geometric analysis. Throughout, the main theme is to present the interaction of partial differential equations (PDE) and differential geometry. More specifically, emphasis is placed on how the behavior of the solutions of a PDE is affected by the geometry of the underlying manifold, and vice versa. For efficiency, the author mainly restricts himself to the linear theory, and only a rudimentary background in Riemannian geometry and partial differential equations is assumed.

Originating from the author's own lectures, this book is an ideal introduction for graduate students, as well as a useful reference for experts in the field.

Peter Li is Chancellor's Professor at the University of California, Irvine.

CAMBRIDGE STUDIES IN ADVANCED MATHEMATICS

Editorial Board:
B. Bollobás, W. Fulton, A. Katok, F. Kirwan, P. Sarnak, B. Simon, B. Totaro

All the titles listed below can be obtained from good booksellers or from Cambridge University Press. For a complete series listing visit: http//www.cambridge.org/mathematics.

Already published
 86 J. J. Duistermaat & J. A. C. Kolk *Multidimensional real analysis, I*
 87 J. J. Duistermaat & J. A. C. Kolk *Multidimensional real analysis, II*
 88 M. C. Golumbic & A. N. Trenk *Tolerance graphs*
 90 L. H. Harper *Global methods for combinatorial isoperimetric problems*
 91 I. Moerdijk & J. Mrčun *Introduction to foliations and Lie groupoids*
 92 J. Kollár, K. E. Smith & A. Corti *Rational and nearly rational varieties*
 93 D. Applebaum *Lévy processes and stochastic calculus (1st Edition)*
 94 B. Conrad *Modular forms and the Ramanujan conjecture*
 95 M. Schechter *An introduction to nonlinear analysis*
 96 R. Carter *Lie algebras of finite and affine type*
 97 H. L. Montgomery & R. C. Vaughan *Multiplicative number theory, I*
 98 I. Chavel *Riemannian geometry (2nd Edition)*
 99 D. Goldfeld *Automorphic forms and L-functions for the group GL(n,R)*
100 M. B. Marcus & J. Rosen *Markov processes, Gaussian processes, and local times*
101 P. Gille & T. Szamuely *Central simple algebras and Galois cohomology*
102 J. Bertoin *Random fragmentation and coagulation processes*
103 E. Frenkel *Langlands correspondence for loop groups*
104 A. Ambrosetti & A. Malchiodi *Nonlinear analysis and semilinear elliptic problems*
105 T. Tao & V. H. Vu *Additive combinatorics*
106 E. B. Davies *Linear operators and their spectra*
107 K. Kodaira *Complex analysis*
108 T. Ceccherini-Silberstein, F. Scarabotti & F. Tolli *Harmonic analysis on finite groups*
109 H. Geiges *An introduction to contact topology*
110 J. Faraut *Analysis on Lie groups: An introduction*
111 E. Park *Complex topological K-theory*
112 D. W. Stroock *Partial differential equations for probabilists*
113 A. Kirillov, Jr *An introduction to Lie groups and Lie algebras*
114 F. Gesztesy et al. *Soliton equations and their algebro-geometric solutions, II*
115 E. de Faria & W. de Melo *Mathematical tools for one-dimensional dynamics*
116 D. Applebaum *Lévy processes and stochastic calculus (2nd Edition)*
117 T. Szamuely *Galois groups and fundamental groups*
118 G. W. Anderson, A Guionnet & O. Zeitouni *An introduction to random matrices*
119 C. Perez-Garcia & W. H. Schikhof *Locally convex spaces over non-Archimedean valued fields*
120 P. K. Friz & N. B. Victoir *Multidimensional stochastic processes as rough paths*
121 T. Ceccherini-Silberstein, F. Scarabotti & F. Tolli *Representation theory of the symmetric groups*
122 S. Kalikow & R. McCutcheon *An outline of ergodic theory*
123 G. F. Lawler & V. Limic *Random walk: A modern introduction*
124 K Lux & H. Pahlings *Representations of groups*
125 K. S. Kedlaya *p-adic differential equations*
126 R. Beals & R. Wong *Special functions*
127 E. de Faria & W. de Melo *Mathematical aspects of quantum field theory*
128 A. Terras *Zeta functions of graphs*
129 D. Goldfeld & J. Hundley *Automorphic representations and L-functions for the general linear group, I*
130 D. Goldfeld & J. Hundley *Automorphic representations and L-functions for the general linear group, II*
131 D. A. Craven *The theory of fusion systems*
132 J. Väänänen *Models and games*
133 G. Malle & D. Testerman *Linear algebraic groups and finite groups of Lie type*

Geometric Analysis

PETER LI
University of California, Irvine

CAMBRIDGE
UNIVERSITY PRESS

University Printing House, Cambridge CB2 8BS, United Kingdom

Cambridge University Press is part of the University of Cambridge.

It furthers the University's mission by disseminating knowledge in the pursuit of education, learning and research at the highest international levels of excellence.

www.cambridge.org
Information on this title: www.cambridge.org/9781107020641

© Peter Li 2012

This publication is in copyright. Subject to statutory exception and to the provisions of relevant collective licensing agreements, no reproduction of any part may take place without the written permission of Cambridge University Press.

First published 2012

A catalogue record for this publication is available from the British Library

Library of Congress Cataloguing in Publication data

Li, Peter, 1952–
Geometric analysis / Peter Li, University of California, Irvine.
pages   cm. – (Cambridge studies in advanced mathematics ; 134)
ISBN 978-1-107-02064-1 (Hardback)
1.   Geometric analysis.   I. Title.
QA360.L53 2012
515'.1–dc23
                                                                    2011051365

ISBN 978-1-107-02064-1 Hardback

Cambridge University Press has no responsibility for the persistence or accuracy of URLs for external or third-party internet websites referred to in this publication, and does not guarantee that any content on such websites is, or will remain, accurate or appropriate.

I would like to dedicate this book to my wife, Glenna, for her love and unwavering support.

Contents

	Preface	page ix
1	First and second variational formulas for area	1
2	Volume comparison theorem	10
3	Bochner–Weitzenböck formulas	19
4	Laplacian comparison theorem	32
5	Poincaré inequality and the first eigenvalue	40
6	Gradient estimate and Harnack inequality	57
7	Mean value inequality	68
8	Reilly's formula and applications	77
9	Isoperimetric inequalities and Sobolev inequalities	86
10	The heat equation	96
11	Properties and estimates of the heat kernel	109
12	Gradient estimate and Harnack inequality for the heat equation	122
13	Upper and lower bounds for the heat kernel	134
14	Sobolev inequality, Poincaré inequality and parabolic mean value inequality	149

15	Uniqueness and the maximum principle for the heat equation	169
16	Large time behavior of the heat kernel	177
17	Green's function	189
18	Measured Neumann Poincaré inequality and measured Sobolev inequality	203
19	Parabolic Harnack inequality and regularity theory	216
20	Parabolicity	241
21	Harmonic functions and ends	256
22	Manifolds with positive spectrum	267
23	Manifolds with Ricci curvature bounded from below	284
24	Manifolds with finite volume	299
25	Stability of minimal hypersurfaces in a 3-manifold	306
26	Stability of minimal hypersurfaces in a higher dimensional manifold	315
27	Linear growth harmonic functions	326
28	Polynomial growth harmonic functions	340
29	L^q harmonic functions	349
30	Mean value constant, Liouville property, and minimal submanifolds	361
31	Massive sets	370
32	The structure of harmonic maps into a Cartan–Hadamard manifold	381
Appendix A	Computation of warped product metrics	392
Appendix B	Polynomial growth harmonic functions on Euclidean space	395
References		399
Index		404

Preface

The main goal of this book is to present the basic tools that are necessary for research in geometric analysis. Though the main theme centers around linear theory, i.e., the Laplace equation, the heat equation, and eigenvalues for the Laplacian, the methods of dealing with these problems are quite often useful in the study of nonlinear partial differential equations that arise in geometry.

A small portion of this book originated from a series of lectures given by the author at a Geometry Summer Program in 1990 at the Mathematical Sciences Research Institute in Berkeley. The lecture notes were revised and expanded when the author taught a regular course in geometric analysis. During the author's visit to the Global Analysis Research Institute at Seoul National University, he was encouraged to submit these notes, though still in a rather crude form, for publication in their lecture notes series [L6].

The part of this book that concerns harmonic functions originated from the author's lecture notes for a series of courses he gave on the subject at the University of California, Irvine. A part of this material was also used in a series of lectures the author gave at the XIV Escola de Geometria Diferencial in Brazil during the summer of 2006. These notes [L9] were printed for distribution to the participants of the program.

As well as updating the Korean lecture notes with more recent developments and combining with them the harmonic function notes, the author has also inserted a treatment on the heat equation. The result takes the form of an introduction to the subject of geometric analysis on the one hand, with some application to geometric problems via linear theory on the other. Due to the vast literature in geometric analysis, it is prudent not to make any attempt to try to discuss nonlinear theory. The interested reader is encouraged to consult the excellent book of Schoen and Yau [SY2] in this direction.

The aim of this book is to address entry-level geometric analysts by introducing the basic techniques in the most economical way. The theorems discussed are chosen sometimes for their fundamental usefulness and sometimes for the purpose of demonstrating various techniques. In many cases, they do not represent the best possible or most current results.

The book is roughly divided into three main parts. The first part (Chapters 1–9) contains basic background material, including reviews of various topics that may be found in a standard Riemannian geometry book. It also provides a quick glimpse of a powerful technique, namely the maximal principle method, in obtaining estimates on a manifold.

The second part (Chapters 10–19) gives an outline of the theory of the heat equation that forms a basis for further study of nonlinear geometric flows. It also established various estimates for nonnegative solutions of the heat equation. As a consequence, estimates for the constants that appear in the Sobolev inequality, the Poincaré inequality, and the mean value inequality in terms of geometric quantities are established. Chapters 18 and 19 are quite technical and contain a presentation of Moser's argument for the parabolic Harnack inequality on a manifold. Moreover, the dependency on the background geometry is explicitly stated.

The last part (Chapters 20–32) of the book is primarily on harmonic functions and various applications to other geometric problems, such as minimal surfaces, harmonic maps, and the geometric structure of certain manifolds.

The author would like to express his gratitude to Ovidiu Munteanu, Lei Ni, and Jiaping Wang for their suggestions on how to improve this book. He is particularly in debt to Munteanu for his detailed proof-reading of the draft. Acknowledgement is also due to the author's graduate students Lihan Wang and Fei He, who were extremely helpful in pointing out necessary corrections to the manuscript. The preparation of this manuscript was partially supported by NSF grant DMS-0801988.

1
First and second variational formulas for area

In this chapter, we will derive the first and second variational formulas for the area of a submanifold. This will be useful in our later discussion on the volume and Laplacian comparison theorems. It will also be used in our studies of the stability issues of minimal submanifolds.

Let M be a Riemannian manifold of dimension m with metric denoted by ds^2. In terms of local coordinates $\{x_1, \ldots, x_m\}$ the metric is written in the form

$$ds^2 = g_{ij}\, dx_i\, dx_j,$$

where we are adopting the convention that repeated indices are being summed over. If X and Y are tangent vectors at a point $p \in M$, we will also denote their inner product by

$$ds^2(X, Y) = \langle X, Y \rangle.$$

If we let $\mathcal{S}(TM)$ be the set of smooth vector fields on M, then the Riemannian connection $\nabla : \mathcal{S}(TM) \times \mathcal{S}(TM) \to \mathcal{S}(TM)$ satisfies the following properties:

(1) $\nabla_{(f_1 X_1 + f_2 X_2)} Y = f_1 \nabla_{X_1} Y + f_2 \nabla_{X_2} Y$;
(2) $\nabla_X (f_1 Y_1 + f_2 Y_2) = X(f_1) Y_1 + f_1 \nabla_X Y_1 + X(f_2) Y_2 + f_2 \nabla_X Y_2$;
(3) $X \langle Y_1, Y_2 \rangle = \langle \nabla_X Y_1, Y_2 \rangle + \langle Y_1, \nabla_X Y_2 \rangle$; and
(4) $\nabla_X Y - \nabla_Y X = [X, Y]$, for all $X, Y \in \mathcal{S}(TM)$,

for all $X, X_1, X_2, Y, Y_1, Y_2 \in \mathcal{S}(TM)$ and for all $f_1, f_2 \in C^\infty(M)$. Property (3) says that ∇ is compatible with the Riemannian metric, while property (4) means that ∇ is torsion free. Moreover, the Riemannian connection is the only connection satisfying the above properties. The curvature tensor of the Riemannian metric is then given by

$$\mathcal{R}_{XY}Z = \nabla_X \nabla_Y Z - \nabla_Y \nabla_X Z - \nabla_{[X,Y]}Z,$$

for $X, Y, Z \in \mathcal{S}(TM)$, and it satisfies the properties:

(1) $\mathcal{R}_{XY}Z = -\mathcal{R}_{YX}Z$;
(2) $\mathcal{R}_{XY}Z + \mathcal{R}_{YZ}X + \mathcal{R}_{ZX}Y = 0$; and
(3) $\langle \mathcal{R}_{XY}Z, W \rangle = \langle \mathcal{R}_{ZW}X, Y \rangle$,

for all $X, Y, Z, W \in \mathcal{S}(TM)$. The sectional curvature of the 2-plane section spanned by a pair of orthonormal vectors X and Y is defined by

$$K(X, Y) = \langle \mathcal{R}_{XY}Y, X \rangle.$$

If we take $\{e_1, \dots, e_m\}$ to be an orthonormal basis of the tangent space of M, then the Ricci curvature is defined to be the symmetric 2-tensor given by

$$\mathcal{R}_{ij} = \sum_{k=1}^{m} \langle \mathcal{R}_{e_i, e_k} e_k, e_j \rangle.$$

Observe that the diagonal elements of the Ricci curvature are given by

$$\mathcal{R}_{ii} = \sum_{k \neq i} K(e_i, e_k).$$

Let N be an n-dimensional submanifold in M with $n < m$. The Riemannian metric ds_M^2 defined on M when restricted to N induces a Riemannian metric ds_N^2 on N. One can easily check that for vector fields $X, Y \in \mathcal{S}(TN)$, if we define

$$\nabla_X^t Y = (\nabla_X Y)^t$$

to be the tangential component of $\nabla_X Y$ to N, then ∇^t is the Riemannian connection of N with respect to ds_N^2. The normal component of ∇ yields the negative of the second fundamental form of N. In particular, one defines the second fundamental form by

$$\vec{II}(X, Y) = -(\nabla_X Y)^n,$$

and checks that it is tensorial with respect to $X, Y \in \mathcal{S}(TN)$. Taking the trace of the bilinear form $\vec{II}$ over the tangent space of N yields the mean curvature vector, given by

$$\text{tr}(\vec{II}) = \vec{H}.$$

1 First and second variational formulas for area

Let us now consider a one-parameter family of deformations of N given by $N_t = \phi(N, t)$ for $t \in (-\epsilon, \epsilon)$ with $N_0 = N$. Let $\{x_1, \ldots, x_n\}$ be a coordinate system around a point $p \in N$. We can consider $\{x_1, \ldots, x_n, t\}$ to be a coordinate system of $N \times (-\epsilon, \epsilon)$ near the point $(p, 0)$. Let us denote $e_i = d\phi\,(\partial/\partial x_i)$ for $i = 1, \ldots, n$ and $T = d\phi\,(\partial/\partial t)$. The induced metric on N_t from M is then given by $g_{ij} = \langle e_i, e_j \rangle$. We may further assume that $\{x_1, \ldots, x_n\}$ form a normal coordinate system at $p \in N$. Hence $g_{ij}(p, 0) = \delta_{ij}$ and $\nabla_{e_i} e_j(p, 0) = 0$. Let us define dA_t to be the area element of N_t with respect to the induced metric. For t sufficiently close to 0, we can write $dA_t = J(x, t)\,dA_0$. With respect to the normal coordinate system $\{x_1, \ldots, x_n\}$, the function $J(x, t)$ is given by

$$J(x,t) = \frac{\sqrt{g(x,t)}}{\sqrt{g(x,0)}}$$

with $g(x, t) = \det(g_{ij}(x, t))$. To compute the first variation for the area of N, we compute $J'(p, t) = (\partial J/\partial t)(p, t)$. By the assumption that $g_{ij}(p, 0) = \delta_{ij}$, we have $J'(p, 0) = \frac{1}{2}g'(p, 0)$. However,

$$g = \det(g_{ij})$$
$$= \sum_{j=1}^{n} g_{1j}\,c_{1j},$$

where c_{ij} are the cofactors of g_{ij}. Therefore

$$g'(p, 0) = \sum_{j=1}^{n} g'_{1j}(p, 0)\,c_{1j}(p, 0) + \sum_{j=1}^{n} g_{1j}(p, 0)\,c'_{1j}(p, 0)$$
$$= g'_{11}(p, 0) + c'_{11}(p, 0).$$

By induction on the dimension, we conclude that $g'(p, 0) = \sum_{i=1}^{n} g'_{ii}$. On the other hand,

$$g'_{ii} = T\langle e_i, e_i \rangle$$
$$= 2\langle \nabla_T e_i, e_i \rangle$$
$$= 2\langle \nabla_{e_i} T, e_i \rangle,$$

because $\{x_1, \ldots, x_n, t\}$ form a coordinate system for $N \times (-\epsilon, \epsilon)$. Let us point out that the quantity

$$\sum_{i=1}^{n} \langle \nabla_{e_i} T, e_i \rangle$$

is now well defined under an orthonormal change of basis and hence is globally defined. If we write $T = T^t + T^n$, where T^t is the tangential component of T on N and T^n is its normal component, then

$$\sum_{i=1}^{n} \langle \nabla_{e_i} T, e_i \rangle = \sum_{i=1}^{n} \langle \nabla_{e_i} T^t, e_i \rangle + \sum_{i=1}^{n} \langle \nabla_{e_i} T^n, e_i \rangle$$

$$= \mathrm{div}(T^t) + \sum_{i=1}^{n} e_i \langle T^n, e_i \rangle - \sum_{i=1}^{n} \langle T^n, \nabla_{e_i} e_i \rangle$$

$$= \mathrm{div}(T^t) + \langle T^n, \vec{H} \rangle,$$

where $\vec{H}$ is the mean curvature vector of N. Hence the first variation for the volume form at the point $(p, 0)$ is given by

$$\frac{d}{dt} dA_t|_{(p,0)} = \left(\mathrm{div} T^t + \langle T^n, \vec{H} \rangle \right) dA_0 \Big|_{(p,0)}.$$

However, the right-hand side is intrinsically defined independent of the choice of coordinates and hence this formula is valid at any arbitrary point.

If T is a compactly supported variational vector field on N, then using the divergence theorem the first variation of the area of N is given by

$$\frac{d}{dt} A(N_t) \Big|_{t=0} = \int_N \langle T^n, \vec{H} \rangle.$$

This shows that the mean curvature of N is identically 0 if and only if N is a critical point of the area functional.

Definition 1.1 An immersed submanifold $N \hookrightarrow M$ is said to be minimal if its mean curvature vector vanishes identically, i.e., $\vec{H} \equiv 0$.

When N is a curve in M that is parametrized by arc-length with unit tangent vector e, then the first variational formula for length can be written as

$$\frac{d}{dt} L \Big|_{t=0} = \langle T^t, e \rangle |_0^l - \int_0^l \langle T^n, \nabla_e e \rangle$$

$$= \langle T, e \rangle |_0^l - \int_0^l \langle T, \nabla_e e \rangle.$$

We will now proceed to derive the second variational formula for area. Let $\phi : N \times (-\epsilon, \epsilon) \times (-\epsilon, \epsilon) \longrightarrow M$ be a two-parameter family of variations of N. Using similar notation, we write $d\phi(\partial/\partial x_i) = e_i$ for $i = 1, \ldots, n$, and denote the variational vector fields by $d\phi(\partial/\partial t) = T$ and $d\phi(\partial/\partial s) = S$.

1 First and second variational formulas for area

In terms of a general coordinate system, the first partial derivative of J can be written as

$$\frac{\partial J}{\partial t}(x,t,s) = \sum_{i,j=1}^{n} g^{ij} \langle \nabla_{e_i} T, e_j \rangle J(x,t,s),$$

where (g^{ij}) denotes the inverse matrix of (g_{ij}). Differentiating this with respect to s and evaluating at $(p,0,0)$ we have

$$\frac{\partial^2 J}{\partial s \partial t} = \sum_{i,j=1}^{n} S\left(g^{ij} \langle \nabla_{e_i} T, e_j \rangle J\right)$$

$$= \sum_{i,j=1}^{n} (Sg^{ij}) \langle \nabla_{e_i} T, e_j \rangle J + \sum_{i,j=1}^{n} g^{ij} \left(S \langle \nabla_{e_i} T, e_j \rangle\right) J$$

$$+ \sum_{i,j=1}^{n} g^{ij} \langle \nabla_{e_i} T, e_j \rangle S(J)$$

$$= \sum_{i,j=1}^{n} (Sg^{ij}) \langle \nabla_{e_i} T, e_j \rangle + \sum_{i=1}^{n} S \langle \nabla_{e_i} T, e_i \rangle$$

$$+ \left(\sum_{i=1}^{n} \langle \nabla_{e_i} T, e_i \rangle\right) \left(\sum_{j=1}^{n} \langle \nabla_{e_j} S, e_j \rangle\right). \tag{1.1}$$

However, differentiating the formula $\sum_{k=1}^{n} g^{ik} g_{kj} = \delta_{ij}$, we obtain

$$\sum_{k=1}^{n} (Sg^{ik}) g_{kj} = -\sum_{k=1}^{n} g^{ik} (Sg_{kj}),$$

hence

$$Sg^{ij} = -\sum_{k,l=1}^{n} g^{ik} (Sg_{kl}) g^{lj}$$

$$= -Sg_{ij}$$

$$= -S\langle e_i, e_j \rangle$$

$$= -\langle \nabla_S e_i, e_j \rangle - \langle \nabla_S e_j, e_i \rangle$$

$$= -\langle \nabla_{e_i} S, e_j \rangle - \langle \nabla_{e_j} S, e_i \rangle.$$

The first term on the right-hand side of (1.1) now becomes

$$\sum_{i,j=1}^{n}(Sg^{ij})\langle \nabla_{e_i}T, e_j\rangle = -\sum_{i,j=1}^{n}\langle \nabla_{e_i}S, e_j\rangle\langle \nabla_{e_i}T, e_j\rangle$$

$$-\sum_{i,j=1}^{n}\langle \nabla_{e_j}S, e_i\rangle\langle \nabla_{e_i}T, e_j\rangle.$$

The second term on the right-hand side of (1.1) can be written as

$$\sum_{i=1}^{n}S\langle \nabla_{e_i}T, e_i\rangle = \sum_{i=1}^{n}\langle \nabla_S\nabla_{e_i}T, e_i\rangle + \sum_{i=1}^{n}\langle \nabla_{e_i}T, \nabla_S e_i\rangle$$

$$= \sum_{i=1}^{n}\langle \mathcal{R}_{Se_i}T, e_i\rangle + \sum_{i=1}^{n}\langle \nabla_{e_i}\nabla_S T, e_i\rangle + \sum_{i=1}^{n}\langle \nabla_{e_i}T, \nabla_{e_i}S\rangle,$$

where the term $\langle \mathcal{R}_{Se_i}T, e_i\rangle$ on the right-hand side denotes the curvature tensor of M. Therefore, we have

$$\frac{\partial^2 J}{\partial s \partial t} = -\sum_{i,j=1}^{n}\langle \nabla_{e_i}S, e_j\rangle\langle \nabla_{e_i}T, e_j\rangle - \sum_{i,j=1}^{n}\langle \nabla_{e_j}S, e_i\rangle\langle \nabla_{e_i}T, e_j\rangle$$

$$+ \sum_{i=1}^{n}\langle \mathcal{R}_{Se_i}T, e_i\rangle + \sum_{i=1}^{n}\langle \nabla_{e_i}\nabla_S T, e_i\rangle + \sum_{i=1}^{n}\langle \nabla_{e_i}T, \nabla_{e_i}S\rangle$$

$$+ \left(\sum_{i=1}^{n}\langle \nabla_{e_i}T, e_i\rangle\right)\left(\sum_{j=1}^{n}\langle \nabla_{e_j}S, e_j\rangle\right). \tag{1.2}$$

We will now consider some special cases that will simplify (1.2). Let us first assume that N is a curve parametrized by arc-length in M with unit tangent vector given by e, then the second variational formula for the length is given by

$$\frac{\partial^2 L}{\partial s \partial t}\bigg|_{(s,t)=(0,0)} = \int_0^l \{-\langle \nabla_e S, e\rangle\langle \nabla_e T, e\rangle + \langle \mathcal{R}_{Se}T, e\rangle\}$$

$$+ \int_0^l \{\langle \nabla_e \nabla_S T, e\rangle + \langle \nabla_e T, \nabla_e S\rangle\}.$$

1 First and second variational formulas for area

If we further assumed that N is a geodesic satisfying the geodesic equation $\nabla_e e \equiv 0$, then we have

$$\frac{\partial^2 L}{\partial s \partial t}\bigg|_{(s,t)=(0,0)} = \int_0^l \{-(e\langle S, e\rangle)(e\langle T, e\rangle) + \langle \mathcal{R}_{Se}T, e\rangle\}$$

$$+ \int_0^l \{e\langle \nabla_S T, e\rangle + \langle \nabla_e T, \nabla_e S\rangle\}$$

$$= \int_0^l \{\langle \nabla_e T, \nabla_e S\rangle + \langle \mathcal{R}_{Se}T, e\rangle - (e\langle S, e\rangle)(e\langle T, e\rangle)\}$$

$$+ \langle \nabla_S T, e\rangle|_0^l.$$

The second special case is when N is a general n-dimensional manifold and then if the two variational vector fields are the same and are normal to N, (1.2) becomes

$$\frac{\partial^2 J}{\partial t^2}\bigg|_{t=0} = -\sum_{i,j=1}^n \langle \nabla_{e_i} T, e_j\rangle^2 - \sum_{i,j=1}^n \langle \nabla_{e_j} T, e_i\rangle \langle \nabla_{e_i} T, e_j\rangle + \sum_{i=1}^n \langle \mathcal{R}_{Te_i} T, e_i\rangle$$

$$+ \sum_{i=1}^n \langle \nabla_{e_i} \nabla_T T, e_i\rangle + \sum_{i=1}^n |\nabla_{e_i} T|^2 + \left(\sum_{i=1}^n \langle \nabla_{e_i} T, e_i\rangle\right)^2$$

$$= -\sum_{i,j=1}^n \langle \nabla_{e_i} T, e_j\rangle^2 - \sum_{i,j=1}^n \langle \nabla_{e_j} T, e_i\rangle \langle \nabla_{e_i} T, e_j\rangle - \sum_{i=1}^n \langle \mathcal{R}_{e_i T} T, e_i\rangle$$

$$+ \operatorname{div}(\nabla_T T)^t + \left\langle (\nabla_T T)^n, \vec{H}\right\rangle + \sum_{i=1}^n |\nabla_{e_i} T|^2 + \left(T, \vec{H}\right)^2. \quad (1.3)$$

On the other hand, if $\{e_{n+1}, \ldots, e_m\}$ denotes an orthonormal set of vectors normal to N in M, then

$$\sum_{i=1}^n \langle \nabla_{e_i} T, \nabla_{e_i} T\rangle = \sum_{i,j=1}^n \langle \nabla_{e_i} T, e_j\rangle^2 + \sum_{i=1}^n \sum_{\nu=n+1}^m \langle \nabla_{e_i} T, e_\nu\rangle^2.$$

Also

$$\langle \nabla_{e_i} T, e_j\rangle = \left\langle T, \vec{II}_{ij}\right\rangle$$

$$= \langle \nabla_{e_j} T, e_i\rangle,$$

where $\vec{II}_{ij}$ denotes the second fundamental form with value in the normal bundle of N. Hence, (1.3) becomes

$$\frac{\partial^2 J}{\partial t^2}\bigg|_{t=0} = -\sum_{i,j}\left\langle T, \vec{II}_{ij}\right\rangle^2 - \sum_{i=1}^{n}\langle\mathcal{R}_{e_i}T, e_i\rangle + \mathrm{div}(\nabla_T T)^t$$

$$+ \left\langle (\nabla_T T)^n, \vec{H}\right\rangle + \sum_{i=1}^{n}\sum_{\nu=n+1}^{m}\langle\nabla_{e_i}T, e_\nu\rangle^2 + \left\langle T, \vec{H}\right\rangle^2.$$

Therefore, the second variational formula for area in terms of compactly supported normal variations is given by

$$\frac{d^2}{dt^2}A(N_t)\bigg|_{t=0} = \int_N \left\{-\sum_{i,j}\left\langle T, \vec{II}_{ij}\right\rangle^2 - \sum_{i=1}^{n}\langle\mathcal{R}_{e_i}T, e_i\rangle + \left\langle(\nabla_T T)^n, \vec{H}\right\rangle\right\}$$

$$+ \int_N \left\{\sum_{i=1}^{n}\sum_{\nu=n+1}^{m}\langle\nabla_{e_i}T, e_\nu\rangle^2 + \left\langle T, \vec{H}\right\rangle^2\right\}.$$

Definition 1.2 A minimally immersed submanifold $N \hookrightarrow M$ is said to be stable if the second variation for area with respect to all compactly supported normal variations is nonnegative. This means that the stability inequality

$$0 \leq -\int_N \sum_{i,j}\left\langle T, \vec{II}_{ij}\right\rangle^2 - \int_N \sum_{i=1}^{n}\langle\mathcal{R}_{e_i}T, e_i\rangle + \int_N \sum_{i=1}^{n}\sum_{\nu=n+1}^{m}\langle\nabla_{e_i}T, e_\nu\rangle^2$$

is valid for any compactly supported normal vector field T.

If we further restrict N to be an orientable codimension-1 minimal submanifold of an orientable manifold M, we can write any normal variation in the form $T = \psi e_m$, where ψ is a differentiable function on N and e_m is a unit normal vector field to N. Then the second variational formula can be written as

$$\frac{d^2}{dt^2}A(N_t)\bigg|_{t=0} = \int_N \left\{-\sum_{i,j}\left\langle T, \vec{II}_{ij}\right\rangle^2 - \mathcal{R}(T, T) + \sum_{i=1}^{n}\langle\nabla_{e_i}T, e_m\rangle^2\right\}$$

$$= \int_N \left\{-\psi^2 h_{ij}^2 - \psi^2 \mathcal{R}(e_m, e_m) + |\nabla\psi|^2\right\},$$

where $\vec{II}_{ij} = h_{ij} e_m$ with h_{ij} being the component of the second fundamental form and $\mathcal{R}(T, T)$ denotes the Ricci curvature of M in the direction of T. Here we have also used the fact that

1 First and second variational formulas for area

$$\langle \nabla_{e_i} T, e_m \rangle = \psi \langle \nabla_{e_i} e_m, e_m \rangle + e_i(\psi)\langle e_m, e_m \rangle$$
$$= e_i(\psi).$$

In particular, the stability inequality in this case is given by

$$\int_N |\nabla \psi|^2 \geq \int_N \psi^2 h_{ij}^2 + \int_N \psi^2 \mathcal{R}(e_m, e_m). \tag{1.4}$$

The last special case is again to assume that N is an oriented hypersurface in an oriented manifold M and we restrict the variation to be given by hypersurfaces which are a constant distant from N. The variational vector field is then given by e_m with $\nabla_{e_m} e_m \equiv 0$. This situation is particularly useful for the purpose of controlling the growth of the volume of geodesic balls of radius r. In this case, if we write $\vec{H} = H\, e_m$, the first variational formula for the area element becomes

$$\frac{\partial J}{\partial t}(x, 0) = H(x) J(x, 0), \tag{1.5}$$

and the second variational formula can be written as

$$\frac{\partial^2 J}{\partial t^2}(x, 0) = -\sum_{i,j=1}^{m-1} h_{ij}^2(x) J(x, 0)$$
$$- \mathcal{R}(e_m, e_m)(x) J(x, 0) + H^2(x) J(x, 0). \tag{1.6}$$

2
Volume comparison theorem

In this chapter, we will develop a volume comparison theorem originally proved by Bishop (see [BC]). Let $p \in M$ be a point in a complete Riemannian manifold of dimension m. In terms of polar normal coordinates at p, we can write the volume element as

$$J(\theta, r) dr \wedge d\theta,$$

where $d\theta$ is the area element of the unit $(m-1)$-sphere. The Gauss lemma asserts that the area element of submanifold $\partial B_p(r)$, which is the boundary of the geodesic ball of radius r, is given by $J(\theta, r) d\theta$. By the first and second variational formulas (1.5) and (1.6), if $x = (\theta, r)$ is not in the cut-locus of p, we have

$$J'(\theta, r) = \frac{\partial J}{\partial r}(\theta, r)$$
$$= H(\theta, r) J(\theta, r) \tag{2.1}$$

and

$$J''(\theta, r) = \frac{\partial^2 J}{\partial r^2}(\theta, r)$$
$$= -\sum_{i,j=1}^{m-1} h_{ij}^2(\theta, r) J(\theta, r) - \mathcal{R}_{rr}(\theta, r) J(\theta, r) + H^2(\theta, r) J(\theta, r), \tag{2.2}$$

where $\mathcal{R}_{rr} = \mathcal{R}(\partial/\partial r, \partial/\partial r)$, $H(\theta, r)$, and $(h_{ij}(\theta, r))$ denote the Ricci curvature in the radial direction, the mean curvature and the second fundamental form of $\partial B_p(r)$ at the point $x = (\theta, r)$ with respect to the unit normal vector $\partial/\partial r$, respectively.

2 Volume comparison theorem

Using the inequalities

$$\sum_{i,j=1}^{m-1} h_{ij}^2 \geq \sum_{i=1}^{m-1} h_{ii}^2$$

$$\geq \frac{\left(\sum_{i=1}^{m-1} h_{ii}\right)^2}{m-1}$$

$$= \frac{H^2}{m-1} \tag{2.3}$$

and (2.1), we can estimate (2.2) by

$$J'' \leq \frac{m-2}{m-1} H^2 J - \mathcal{R}_{rr} J$$

$$= \frac{m-2}{m-1} (J')^2 J^{-1} - \mathcal{R}_{rr} J. \tag{2.4}$$

Since any smooth metric is locally Euclidean, we have the initial conditions

$$J(\theta, r) \sim r^{m-1}$$

and

$$J'(\theta, r) \sim (m-1) r^{m-2}$$

as $r \to 0$. Let us point out that if M is a simply connected constant curvature space form with constant sectional curvature K, then all the above inequalities become equalities. In particular (2.4) becomes

$$J'' = \frac{m-2}{m-1} (J')^2 J^{-1} - (m-1) K J.$$

Theorem 2.1 (Bishop [BC]) *Let M be a complete Riemannian manifold of dimension m, and p be a fixed point of M. Let us assume that the Ricci curvature tensor of M at any point x is bounded below by $(m-1) K(r(p, x))$ for some function K depending only on the distance from p. If $J(\theta, r) d\theta$ is the area element of $\partial B_p(r)$ as defined above and $\bar{J}(r)$ is the solution of the ordinary differential equation*

$$\bar{J}'' = \frac{m-2}{m-1} (\bar{J}')^2 \bar{J}^{-1} - (m-1) K \bar{J}$$

with initial conditions

$$\bar{J}(r) \sim r^{m-1}$$

and

$$\bar{J}'(r) \sim (m-1)r^{m-2},$$

as $r \to 0$, then within the cut-locus of p the function $J(\theta, r)/\bar{J}(r)$ is a nonincreasing function of r. Also, if $\bar{H}(r) = \bar{J}'/\bar{J}$, then $H(\theta, r) \le \bar{H}(r)$ whenever (θ, r) is within the cut-locus of p. In particular, if K is a constant, then $\bar{J} d\theta$ corresponds to the area element of the sphere of radius r in the simply connected space form of constant curvature K.

Proof By setting $f = J^{1/(m-1)}$, (2.1) and (2.4) can be written as

$$f' = \frac{1}{m-1} H f$$

and

$$f'' \le \frac{-1}{m-1} \mathcal{R}_{rr} f$$

$$\le -K f.$$

The initial conditions become

$$f(\theta, 0) = 0$$

and

$$f'(\theta, 0) = 1.$$

Let $\bar{f} = \bar{J}^{1/(m-1)}$ be the corresponding function defined using $\bar{J}$. The function $\bar{f}$ satisfies

$$\bar{f}'' = -K \bar{f},$$
$$\bar{f}(0) = 0,$$

and

$$\bar{f}'(0) = 1.$$

Observe that when K is a constant, the function $\bar{f} > 0$ for all values of $r \in (0, \infty)$ when $K \le 0$, and for $r \in (0, \pi/\sqrt{K})$ when $K > 0$. In general, $\bar{f} > 0$

on an interval $(0, a)$ for some $a > 0$. At those values of r we can define

$$F(\theta, r) = \frac{f(\theta, r)}{\bar{f}(r)}.$$

We have

$$F' = \bar{f}^{-2}(f'\bar{f} - f\bar{f}')$$

and

$$F'' = \bar{f}^{-1} f'' - 2\bar{f}^{-2} f' \bar{f}' - \bar{f}^{-2} f \bar{f}'' + 2\bar{f}^{-3} f (\bar{f}')^2$$
$$\leq -2\bar{f}^{-1} \bar{f}' F',$$

hence

$$(\bar{f}^2 F')' = \bar{f}^2(F'' + 2\bar{f}^{-1} \bar{f}' F')$$
$$\leq 0.$$

Integrating from ϵ to r yields

$$F'(r) \leq F'(\epsilon) \bar{f}^2(\epsilon) \bar{f}^{-2}(r)$$
$$= (\bar{f}(\epsilon) f'(\epsilon) - f(\epsilon) \bar{f}'(\epsilon)) \bar{f}^{-2}(r).$$

Letting $\epsilon \to 0$, the initial conditions of f and $\bar{f}$ imply that

$$F'(r) \leq 0.$$

In particular, $\bar{f} f' - \bar{f}' f \leq 0$, implying

$$H(\theta, r) \leq \bar{H}(r).$$

Moreover, that F is a nonincreasing function of r implies that $J(\theta, r)/\bar{J}(r)$ is also a nonincreasing function of r. □

By computing the area element and the mean curvature of the constant curvature space form explicitly, we have the following corollary.

Corollary 2.2 *Under the assumption of Theorem 2.1, if K is a constant, then*

$$H \leq \begin{cases} (m-1)\sqrt{K} \cot\left(\sqrt{K}r\right) & \text{for } K > 0, \\ (m-1)r^{-1} & \text{for } K = 0, \\ (m-1)\sqrt{-K} \coth\left(\sqrt{-K}r\right) & \text{for } K < 0, \end{cases}$$

and
$$\frac{J(\theta, r)}{\bar{J}(r)}$$

is a nonincreasing function of r, where

$$\bar{J}(r) = \begin{cases} \left(\frac{1}{\sqrt{K}}\right)^{m-1} \sin^{m-1}\left(\sqrt{K}r\right) & \text{for } K > 0, \\ r^{m-1} & \text{for } K = 0, \\ \left(\frac{1}{\sqrt{-K}}\right)^{m-1} \sinh^{m-1}\left(\sqrt{-K}r\right) & \text{for } K < 0. \end{cases}$$

Let us take this opportunity to point out that this estimate implies that when $K > 0$, there must be a cut-point along any geodesic which has length $\pi/\sqrt{K}$. In particular, this proves Myers' theorem.

Corollary 2.3 (Myers) *Let M be an m-dimensional complete Riemannian manifold with Ricci curvature bounded from below by*

$$\mathcal{R}_{ij} \geq (m-1)K$$

for some constant $K > 0$. Then M must be compact with diameter d bounded from above by

$$d \leq \frac{\pi}{\sqrt{K}}.$$

Corollary 2.4 *Let M be an m-dimensional complete Riemannian manifold with Ricci curvature bounded from below by a constant $(m-1)K$. Suppose $\bar{M}$ is an m-dimensional simply connected space form with constant sectional curvature K. Let $A_p(r)$ be the area of the boundary of the geodesic ball $\partial B_p(r)$ centered at $p \in M$ of radius r and $\bar{A}(r)$ be the area of the boundary of a geodesic ball $\partial \bar{B}(r)$ of radius r in $\bar{M}$. Then for $0 \leq r_1 \leq r_2 < \infty$, we have*

$$A_p(r_1)\,\bar{A}(r_2) \geq A_p(r_2)\,\bar{A}(r_1). \tag{2.5}$$

If we let $V_p(r)$ and $\bar{V}(r)$ be the volumes of $B_p(r)$ and $\bar{B}(r)$, respectively, then for $0 \leq r_1 \leq r_2, r_3 \leq r_4 < \infty$ we have

$$\left(V_p(r_2) - V_p(r_1)\right)\left(\bar{V}(r_4) - \bar{V}(r_3)\right) \geq \left(V_p(r_4) - V_p(r_3)\right)\left(\bar{V}(r_2) - \bar{V}(r_1)\right). \tag{2.6}$$

Proof Let us define $C(r)$ to be a subset of the unit tangent sphere $S_p(M)$ at p such that for all $\theta \in C(r)$ the geodesic given by $\gamma(s) = \exp_p(s\theta)$ is

2 Volume comparison theorem

minimizing up to $s = r$. Clearly for $r_1 \leq r_2$ we have $C(r_2) \subset C(r_1)$. By Theorem 2.1, we have

$$J(\theta, r_1)\, \bar{J}(r_2) \geq J(\theta, r_2)\, \bar{J}(r_1)$$

for $\theta \in C(r_2)$. Integrating over $C(r_2)$ yields

$$\int_{C(r_2)} J(\theta, r_1)\, d\theta\, \bar{J}(r_2) \geq \int_{C(r_2)} J(\theta, r_2)\, d\theta\, \bar{J}(r_1)$$

$$= A_p(r_2)\, \bar{J}(r_1).$$

On the other hand,

$$A_p(r_1) = \int_{C(r_1)} J(\theta, r_1)\, d\theta$$

$$\geq \int_{C(r_2)} J(\theta, r_1)\, d\theta.$$

Taking this together with the fact that

$$\bar{A}(r) = \alpha_{m-1}\, \bar{J}(r)$$

with α_{m-1} being the area of the unit $(m-1)$-sphere, we conclude (2.5).

To see (2.6), we first assume that $r_1 \leq r_2 \leq r_3 \leq r_4$ and we simply integrate the inequality

$$A_p(t_1)\, \bar{A}(t_2) \geq A_p(t_2)\, \bar{A}(t_1)$$

over $r_1 \leq t_1 \leq r_2$ and $r_3 \leq t_2 \leq r_4$. For the case when $r_1 \leq r_3 \leq r_2 \leq r_4$, we write

$$\left(V_p(r_2) - V_p(r_1)\right)\left(\bar{V}(r_4) - \bar{V}(r_3)\right)$$
$$= \left(V_p(r_3) - V_p(r_1)\right)\left(\bar{V}(r_2) - \bar{V}(r_3)\right)$$
$$+ \left(V_p(r_3) - V_p(r_1)\right)\left(\bar{V}(r_4) - \bar{V}(r_2)\right)$$
$$+ \left(V_p(r_2) - V_p(r_3)\right)\left(\bar{V}(r_2) - \bar{V}(r_3)\right)$$
$$+ \left(V_p(r_2) - V_p(r_3)\right)\left(\bar{V}(r_4) - \bar{V}(r_2)\right)$$

$$\geq \left(V_p(r_2) - V_p(r_3)\right)\left(\bar{V}(r_3) - \bar{V}(r_1)\right)$$
$$+ \left(V_p(r_4) - V_p(r_2)\right)\left(\bar{V}(r_3) - \bar{V}(r_1)\right)$$
$$+ \left(V_p(r_2) - V_p(r_3)\right)\left(\bar{V}(r_2) - \bar{V}(r_3)\right)$$
$$+ \left(V_p(r_4) - V_p(r_2)\right)\left(\bar{V}(r_2) - \bar{V}(r_3)\right)$$
$$= \left(V_p(r_4) - V_p(r_3)\right)\left(\bar{V}(r_2) - \bar{V}(r_1)\right). \qquad \square$$

Let us point out that equality in (2.6) holds if and only if $C(r_1) = C(r_4)$ and $J(\theta, r) = \bar{J}(r)$ for all $0 \leq r \leq r_4$ and $\theta \in C(r_1)$. In particular, if $r_1 = 0$, then $J(\theta, r) = \bar{J}(r)$ for all $r \leq r_4$ and $\theta \in S_p(M)$. This implies that $B_p(r_4)$ is isometric to $\bar{B}(r_4)$.

Theorem 2.5 *Let M be an m-dimensional complete Riemannian manifold with nonnegative Ricci curvature. Then the volume growth of M must satisfy the following estimates:*

(1) *(Bishop) If α_{m-1} is the area of the unit $(m-1)$-sphere, then*

$$V_p(\rho) \leq \frac{\alpha_{m-1}}{m} \rho^m$$

for all $p \in M$ and $\rho \geq 0$.

(2) *(Yau [Y2]) For all $p \in M$, there exists a constant $C(m) > 0$ depending only on m, such that*

$$V_p(\rho) \geq C\, V_p(1)\, \rho$$

for all $\rho > 2$.

Proof Applying (2.6) to $r_1 = 0 = r_3$ and $r_4 = r$, we have

$$V_p(r_2)\, \bar{V}(r) \geq V_p(r)\, \bar{V}(r_2).$$

Observing that

$$\lim_{r_2 \to 0} \frac{V_p(r_2)}{\bar{V}(r_2)} = 1,$$

the upper bound follows.

To prove the lower bound, let $x \in \partial B_p(1 + \rho)$. Then (2.6) and the curvature assumption imply that

$$V_x(2 + \rho) - V_x(\rho) \leq V_x(\rho)\, \frac{(2+\rho)^m - \rho^m}{\rho^m}. \qquad (2.7)$$

However, since the distance between p and x is $r(p, x) = 1 + \rho$, we have $B_p(1) \subset (B_x(2 + \rho) \setminus B_x(\rho))$, hence

$$V_p(1) \le V_x(2 + \rho) - V_x(\rho). \tag{2.8}$$

Since $B_x(\rho) \subset B_p(1 + 2\rho)$, we have

$$V_x(\rho) \le V_p(1 + 2\rho),$$

therefore combining this with (2.7) and (2.8), we conclude that

$$V_p(1) \le V_p(1 + 2\rho) \frac{(2 + \rho)^m - \rho^m}{\rho^m}.$$

The lower bounded follows by observing that

$$\frac{(2 + \rho)^m - \rho^m}{\rho^m} = C \rho^{-1}$$

for $\frac{1}{2}\rho \to \infty$. □

We would like to remark that if we assume that for a sufficiently small $\epsilon > 0$ the Ricci curvature has a lower bound of the form

$$\mathcal{R}_{ij}(x) \ge -\epsilon(1 + r(p, x))^{-2},$$

then one can show that M must have infinite volume. On the other hand, if the Ricci curvature is bounded from below by

$$\mathcal{R}_{ij}(x) \ge -C_0(1 + r(p, x))^{-2-\delta}$$

for some constants $C_0, \delta > 0$, then the upper bound is also valid and is of the form

$$V_p(r) \le C r^m,$$

where $C(C_0, \delta, m) > 0$ is a constant depending on C_0, δ and m.

It is also a good exercise to show that if a complete manifold has Ricci curvature bounded from below by

$$\mathcal{R}_{ij} \ge \epsilon r(p, x)^{-2}$$

for some constant $\epsilon > \frac{1}{4}$ and for all $r > 1$, then M must be compact.

The next theorem shows that when the upper bound of the diameter given by Myers' theorem is achieved, then the manifold must be isometrically a sphere.

Theorem 2.6 (Cheng [Cg1]) *Let M be a complete m-dimensional Riemannian manifold with Ricci curvature bounded from below by*

$$\mathcal{R}_{ij} \geq (m-1)K$$

for some constant $K > 0$. If the diameter d of M satisfies

$$d = \frac{\pi}{\sqrt{K}},$$

then M is isometric to the standard sphere of radius $1/\sqrt{K}$.

Proof By scaling, we may assume that $K = 1$. Let p and q be a pair of points in M which realize the diameter. The volume comparison theorem implies that

$$V_p(d) \leq V_p\left(\frac{d}{2}\right) \frac{\bar{V}(d)}{\bar{V}(d/2)}.$$

The assumption that $d = \pi/\sqrt{K}$ implies that

$$\frac{\bar{V}(d)}{\bar{V}(d/2)} = 2,$$

hence

$$V_p(d) \leq 2V_p\left(\frac{d}{2}\right).$$

Similarly, we have

$$V_q(d) \leq 2V_q\left(\frac{d}{2}\right).$$

However, by the triangle inequality and the fact that $d = r(p,q)$, we have $B_p(d/2) \cap B_q(d/2) = \emptyset$. Therefore,

$$2V(M) = V_p(d) + V_q(d)$$

$$\leq 2\left(V_p\left(\frac{d}{2}\right) + V_q\left(\frac{d}{2}\right)\right)$$

$$\leq 2V(M),$$

where $V(M)$ denotes the volume of M. This implies that the inequalities in the volume comparison theorem are in fact equalities. Hence by the remark following Corollary 2.4 M must be the standard sphere. $\square$

3
Bochner–Weitzenböck formulas

Applying the Bochner–Weitzenböck formulas, sometimes referred to as the Bochner technique, is one of the most important techniques in the theory of geometric analysis. There are many formulas which can be derived for various situations. In this chapter, we will only derive the formula for differential forms so as to illustrate the flavor of this technique.

For convenience sake, we will also introduce the moving frame notation. Let $\{e_1, \ldots, e_m\}$ be a locally defined orthonormal frame field of the tangent bundle. Let us denote the dual coframe field by $\{\omega_1, \ldots, \omega_m\}$. They have the property that $\omega_i(e_j) = \delta_{ij}$. The connection 1-forms ω_{ij} are given by exterior differentiation of the ω_is, and are uniquely defined by Cartan's first structural equations

$$d\omega_i = \omega_{ij} \wedge \omega_j,$$

where

$$\omega_{ij} + \omega_{ji} = 0.$$

Cartan's second structural equations yield the curvature tensor

$$d\omega_{ij} = \omega_{ik} \wedge \omega_{kj} + \Omega_{ij},$$

with

$$\Omega_{ij} = \tfrac{1}{2}\mathcal{R}_{ijkl}\,\omega_l \wedge \omega_k.$$

Now let us consider the case that N is an n-dimensional submanifold of M. Let $\{e_1, \ldots, e_m\}$ be an adapted orthonormal frame field of M such that $\{e_1, \ldots, e_n\}$ are orthonormal to N. We will now adopt the indexing convention that $1 \leq i, j, k \leq n$ and $n + 1 \leq \nu, \mu \leq m$. The second fundamental form of N is given by

$$\omega_{\nu i} = h_{ij}^\nu \omega_j.$$

Relating the two notations, we have the formulas

$$\omega_{ij}(e_k) = \langle \nabla_{e_k} e_i, e_j \rangle,$$

$$\mathcal{R}_{ijkl} = \langle \mathcal{R}_{e_i e_j} e_l, e_k \rangle,$$

and

$$h_{ij}^\nu = \langle \vec{II}(e_i, e_j), e_\nu \rangle.$$

The sectional curvature of the 2-plane section spanned by e_i and e_j is given by $\mathcal{R}_{ijij}$ and the Ricci curvature is given by

$$\mathcal{R}_{ij} = \sum_{k=1}^m \mathcal{R}_{ikjk}.$$

Let $f \in C^\infty(M)$ be a smooth function defined on M. Its exterior derivative is given by

$$df = f_i \omega_i. \tag{3.1}$$

The second covariant derivative of f can be defined by

$$f_{ij} \omega_j = df_i + f_j \omega_{ji}. \tag{3.2}$$

Exterior differentiating (3.1), and applying the first structural equations, we have

$$0 = df_i \wedge \omega_i + f_i \, d\omega_i$$
$$= df_i \wedge \omega_i + f_i \omega_{ij} \wedge \omega_j$$
$$= (df_i + f_j \omega_{ji}) \wedge \omega_i$$
$$= f_{ij} \omega_j \wedge \omega_i.$$

This implies that $f_{ij} - f_{ji} = 0$ for all i and j. The symmetric 2-tensor given by $f_{ij} \omega_j \otimes \omega_i$ is called the *Hessian* of f. Taking the trace of the Hessian, we define the *Laplacian* of f by

$$\Delta f = f_{ii}.$$

3 Bochner–Weitzenböck formulas

The third covariant derivative of f is defined by

$$f_{ijk}\,\omega_k = df_{ij} + f_{kj}\,\omega_{ki} + f_{ik}\,\omega_{kj}.$$

Exterior differentiation of (3.2) gives

$$df_{ij} \wedge \omega_j + f_{ij}\,d\omega_j = df_j \wedge \omega_{ji} + f_j\,d\omega_{ji}.$$

However, the first and second structural equations imply that

$$\begin{aligned}
0 &= -df_{ij} \wedge \omega_j - f_{ij}\,d\omega_j + df_j \wedge \omega_{ji} + f_j\,d\omega_{ji} \\
&= -df_{ij} \wedge \omega_j - f_{ij}\,\omega_{jk} \wedge \omega_k + df_j \wedge \omega_{ji} + f_j\,\omega_{jk} \wedge \omega_{ki} + f_j\,\Omega_{ji} \\
&= -(df_{ij} + f_{ik}\,\omega_{kj}) \wedge \omega_j + (df_j + f_k\,\omega_{kj}) \wedge \omega_{ji} + f_j\,\Omega_{ji} \\
&= -(df_{ij} + f_{ik}\,\omega_{kj}) \wedge \omega_j + f_{jk}\,\omega_k \wedge \omega_{ji} + f_j\,\Omega_{ji} \\
&= -(df_{ij} + f_{ik}\,\omega_{kj} + f_{kj}\,\omega_{ki}) \wedge \omega_j + f_j\,\Omega_{ji} \\
&= -f_{ijk}\,\omega_k \wedge \omega_j + \tfrac{1}{2} f_j\,\mathcal{R}_{jikl}\,\omega_l \wedge \omega_k.
\end{aligned}$$

This yields the commutation formula

$$\begin{aligned}
f_{ijk} - f_{ikj} &= \tfrac{1}{2} f_l\,(\mathcal{R}_{lijk} - \mathcal{R}_{likj}) \\
&= f_l\,\mathcal{R}_{lijk}.
\end{aligned}$$

Contracting the indices k and i by setting $k=i$ and summing over $1 \leq i \leq m$, we have the Ricci identity

$$f_{iji} - f_{iij} = f_l\,\mathcal{R}_{lj}.$$

For $p \leq m$, we will now take the convention on the indices so that $1 \leq i, j, k, l \leq m$, $1 \leq \alpha, \beta, \gamma \leq p$, and $1 \leq a, b, c, d \leq p-1$. Let $\omega \in \Lambda^p(M)$ be an exterior p-form defined on M. Then in terms of the basis, we can write

$$\omega = a_{i_1 \ldots i_p}\,\omega_{i_p} \wedge \cdots \wedge \omega_{i_1},$$

where the summation is being performed over the multi-index $I = (i_1, \ldots, i_p)$. With this understanding, we can write

$$\omega = a_I\,\omega_I.$$

Exterior differentiation yields

$$\begin{aligned}
d\omega &= da_I \wedge \omega_I + a_I\,d\omega_I \\
&= da_I \wedge \omega_I + a_I\,(-1)^{p-\alpha}\,\omega_{i_p} \wedge \cdots \wedge d\omega_{i_\alpha} \wedge \cdots \wedge \omega_{i_1}
\end{aligned}$$

$$= da_I \wedge \omega_I + a_I \, \omega_{i_\alpha j_\alpha} \wedge \omega_{i_p} \wedge \cdots \wedge \omega_{j_\alpha} \wedge \cdots \wedge \omega_{i_1}$$

$$= (da_I + a_{i_1\ldots j_\alpha\ldots i_p} \, \omega_{j_\alpha i_\alpha}) \wedge \omega_I.$$

One defines the covariant derivatives $a_{i_1,\ldots i_p, j}$ by

$$\sum_{j=1}^m a_{i_1\ldots i_p, j} \, \omega_j = da_{i_1\ldots i_p} + \sum_{\substack{1 \le \alpha \le p \\ j_\alpha}} a_{i_1\ldots j_\alpha \ldots i_p} \, \omega_{j_\alpha i_\alpha}$$

for each multi-index $I = (i_1, \ldots, i_p)$. Similarly, for $(p-1)$-forms, we have

$$a_{i_1\ldots i_{p-1}, j} \, \omega_j = da_{i_1\ldots i_{p-1}} + a_{i_1\ldots j_\alpha \ldots i_{p-1}} \, \omega_{j_a i_a}.$$

Exterior differentiation yields

$$da_{i_1\ldots i_{p-1}, j} \wedge \omega_j + a_{i_1\ldots i_{p-1}, k} \, \omega_{kj} \wedge \omega_j$$

$$= da_{i_1\ldots j_a \ldots i_{p-1}} \wedge \omega_{j_a i_a} + a_{i_1\ldots j_a\ldots i_{p-1}} \, \omega_{j_a k} \wedge \omega_{k i_a}$$

$$+ \tfrac{1}{2} a_{i_1\ldots j_a \ldots i_{p-1}} \, \mathcal{R}_{j_a i_a k l} \, \omega_l \wedge \omega_k.$$

The left-hand side becomes

$$da_{i_1\ldots i_{p-1}, j} \wedge \omega_j + a_{i_1\ldots i_{p-1}, k} \, \omega_{kj} \wedge \omega_j$$

$$= a_{i_1\ldots i_{p-1}, jk} \, \omega_k \wedge \omega_j - a_{i_1\ldots k_a\ldots i_{p-1}, j} \, \omega_{k_a i_a} \wedge \omega_j$$

$$= a_{i_1\ldots i_{p-1}, jk} \, \omega_k \wedge \omega_j + da_{i_1\ldots k_a\ldots i_{p-1}} \wedge \omega_{k_a i_a}$$

$$+ \sum_{b \ne a} a_{i_1\ldots j_b\ldots k_a\ldots i_{p-1}} \, \omega_{j_b i_b} \wedge \omega_{k_a i_a}$$

$$+ a_{i_1\ldots j_a\ldots i_{p-1}} \, \omega_{j_a k_a} \wedge \omega_{k_a i_a}.$$

Equating this to the right-hand side gives

$$a_{i_1\ldots i_{p-1}, jk} \, \omega_k \wedge \omega_j + \sum_{b \ne a} a_{i_1\ldots j_b\ldots k_a\ldots i_{p-1}} \, \omega_{j_b i_b} \wedge \omega_{k_a i_a}$$

$$= \tfrac{1}{2} \mathcal{R}_{j_a i_a l k} \, a_{i_1\ldots j_a\ldots i_{p-1}} \, \omega_k \wedge \omega_l.$$

We now claim that the second term on the left-hand side is identically 0. Indeed, since

$$\sum_{b \neq a} a_{i_1 \ldots j_b \ldots k_a \ldots i_{p-1}} \omega_{j_b i_b} \wedge \omega_{k_a i_a}$$

$$= \sum_{b<a} a_{i_1 \ldots j_b \ldots k_a \ldots i_{p-1}} \omega_{j_b i_b} \wedge \omega_{k_a i_a} + \sum_{b>a} a_{i_1 \ldots j_b \ldots k_a \ldots i_{p-1}} \omega_{j_b i_b} \wedge \omega_{k_a i_a},$$

the claim follows by interchanging the roles of k_a and j_b in the second term. Hence

$$a_{i_1 \ldots i_{p-1}, jk}\, \omega_k \wedge \omega_j = \tfrac{1}{2} \mathcal{R}_{j_a i_a lk}\, a_{i_1 \ldots j_a \ldots i_{p-1}}\, \omega_k \wedge \omega_l,$$

implying that

$$a_{i_1 \ldots i_{p-1}, lk} - a_{i_1 \ldots i_{p-1}, kl} = \mathcal{R}_{j_a i_a lk}\, a_{i_1 \ldots j_a \ldots i_{p-1}}.$$

Similarly, for p-forms, we also have

$$a_{i_1 \ldots i_p, jk} - a_{i_1 \ldots i_p, kj} = \mathcal{R}_{l_\alpha i_\alpha jk}\, a_{i_1 \ldots l_\alpha \ldots i_p}. \tag{3.3}$$

Let us now compute the Laplacian $\Delta \omega = -d\delta\omega - \delta d\omega$ for p-forms. First we have

$$d\omega = a_{I,j}\, \omega_j \wedge \omega_I$$

$$= \sum_{i_1 < i_2 < \ldots < i_{p+1}} \sum_{\sigma(i_1, \ldots, i_{p+1})} \operatorname{sgn}(\sigma)\, a_\sigma\, \omega_{i_{p+1}} \wedge \ldots \wedge \omega_{i_1},$$

where $\sigma(i_1, \ldots, i_{p+1})$ denotes a permutation of the set $(i_1, \ldots, i_{p+1})$ and $\operatorname{sgn}(\sigma)$ is the sign of σ. Recall that if ω is a p-form then

$$\delta\omega = (-1)^{m(p+1)+1} * d * \omega.$$

The linear operator $* : \Lambda^p(M) \to \Lambda^{m-p}(M)$ is determined by

$$*(\omega_{i_p} \wedge \ldots \wedge \omega_{i_1}) = \operatorname{sgn}(\sigma(I, I^c))\, \omega_{i_m} \wedge \ldots \wedge \omega_{i_{p+1}},$$

where $\sigma(I, I^c)$ denotes a permutation by sending

$$(i_p, \ldots i_1, i_m, \ldots, i_{p+1}) \mapsto (1, \ldots, m).$$

Let us now define

$$\beta = *\omega$$

$$= a_{i_1 \ldots i_p} \operatorname{sgn}(\sigma(I, I^c))\, \omega_{i_m} \wedge \ldots \wedge \omega_{i_{p+1}}.$$

By setting $b_{k_1\ldots k_{m-p}} = \text{sgn}(\sigma(I, I^c)) a_{i_1\ldots i_p}$ with

$$K = (k_1, \ldots, k_{m-p}) = (i_{p+1}, \ldots, i_m) = I^c,$$

we can write

$$\beta = b_K \omega_K$$

and

$$d\beta = b_{K,j} \omega_j \wedge \omega_K$$
$$= (db_{k_1\ldots k_{m-p}} + b_{k_1\ldots j_\theta\ldots k_{m-p}} \omega_{j_\theta k_\theta}) \wedge \omega_K$$

for $1 \leq \theta \leq m - p$. On the other hand, we also have

$$d\beta = \text{sgn}(\sigma(I, I^c)) da_{i_1\ldots i_p} \wedge \omega_{i_m} \wedge \ldots \wedge \omega_{i_{p+1}}$$
$$+ b_{k_1\ldots j_\theta\ldots k_{m-p}} \omega_{j_\theta k_\theta} \wedge \omega_K$$
$$= \text{sgn}(\sigma(I, I^c)) a_{i_1\ldots i_p, j} \omega_j \wedge \omega_{i_m} \wedge \ldots \wedge \omega_{i_{p+1}}$$
$$- \text{sgn}(\sigma(I, I^c)) a_{i_1\ldots j_\alpha\ldots i_p} \omega_{j_\alpha i_\alpha} \wedge \omega_{i_m} \wedge \ldots \wedge \omega_{i_{p+1}}$$
$$+ b_{k_1\ldots j_\theta\ldots k_{m_p}} \omega_{j_\theta k_\theta} \wedge \omega_{i_m} \wedge \ldots \wedge \omega_{i_{p+1}}$$
$$= \text{sgn}(\sigma(I, I^c)) a_{i_1\ldots i_\alpha\ldots i_p, i_\alpha} \omega_{i_\alpha} \wedge \omega_{i_m} \wedge \ldots \wedge \omega_{i_{p+1}}$$
$$- \text{sgn}(\sigma(I, K)) a_{i_1\ldots j_\alpha\ldots i_p} \omega_{j_\alpha i_\alpha} \wedge \omega_{i_m} \wedge \ldots \wedge \omega_{i_{p+1}}$$
$$+ b_{k_1\ldots j_\theta\ldots k_{m-p}} \omega_{j_\theta k_\theta} \wedge \omega_{i_m} \wedge \ldots \wedge \omega_{i_{p+1}}.$$

However, if $j_\theta \neq i_\alpha$ for some $1 \leq \alpha \leq p$, this means that $j_\theta \in I^c = K$. In this case, we have either

$$b_{k_1\ldots j_\theta\ldots k_{m-p}} = 0 \quad \text{when} \quad j_\theta \neq k_\theta,$$

or

$$\omega_{j_\theta k_\theta} = 0 \quad \text{when} \quad j_\theta = k_\theta.$$

If $j_\theta = i_\alpha$ for some α, then we have

$$b_{k_1\ldots j_\theta\ldots k_{m-p}}$$
$$= \text{sgn}(\sigma(i_p, \ldots, \underset{\alpha\text{th slot}}{k_\theta}, \ldots, i_1, k_{m-p}, \ldots, j_\theta, \ldots, k_1)) a_{i_1\ldots \underset{\alpha\text{th slot}}{k_\theta}\ldots i_p}.$$

3 Bochner–Weitzenböck formulas

In particular,

$$b_{k_1\ldots j_\theta\ldots k_{m-p}}\,\omega_{j_\theta k_\theta} = -\mathrm{sgn}(\sigma(i_p,\ldots,i_1,i_m,\ldots,i_{m-p}))\,a_{i_1\ldots \underset{\alpha\text{th slot}}{k_\theta}\ldots i_p}\,\omega_{i_\alpha k_\theta}.$$

Using the skew symmetry $\omega_{i_\alpha k_\theta} = -\omega_{k_\theta i_\alpha}$, we conclude that

$$d\beta = \mathrm{sgn}(\sigma(I, I^c))\, a_{i_1\ldots i_\alpha\ldots i_p, i_\alpha}\,\omega_{i_\alpha}\wedge \omega_{i_m}\wedge\cdots\wedge\omega_{i_{p+1}},$$

hence

$$*d\beta = \mathrm{sgn}(\sigma(I, I^c))\,\mathrm{sgn}(\sigma(i_\alpha, i_m,\ldots,i_{p+1}, i_p,\ldots,\hat{i}_\alpha,\ldots,i_1))$$
$$\times a_{i_1\ldots i_\alpha\ldots i_p, i_\alpha}\,\omega_{i_p}\wedge\cdots\wedge\hat{\omega}_{i_\alpha}\wedge\cdots\wedge\omega_{i_1}$$
$$= (-1)^{m-\alpha+p(m-p)} a_{i_1\ldots i_p, i_\alpha}\,\omega_{i_p}\wedge\cdots\wedge\hat{\omega}_{i_\alpha}\wedge\cdots\wedge\omega_{i_1},$$

and

$$\delta\omega = \sum_{1\leq\alpha\leq p}(-1)^{\alpha+p^2+1}a_{i_1\ldots i_\alpha\ldots i_p, i_\alpha}\,\omega_{i_p}\wedge\cdots\wedge\hat{\omega}_{i_\alpha}\wedge\cdots\wedge\omega_{i_1}.$$

Computing directly gives

$$-\Delta\omega = \delta(a_{i_1\ldots i_p, j}\,\omega_j\wedge\omega_{i_p}\wedge\cdots\wedge\omega_{i_1})$$
$$+ d[(-1)^{\alpha+p^2+1} a_{i_1\ldots i_p, i_\alpha}\,\omega_{i_p}\wedge\cdots\wedge\hat{\omega}_{i_\alpha}\wedge\cdots\wedge\omega_{i_1}]$$
$$= (-1)^{\alpha+(p+1)^2+1} a_{i_1\ldots i_p, j i_\alpha}\,\omega_j\wedge\omega_{i_p}\wedge\cdots\wedge\hat{\omega}_{i_\alpha}\wedge\cdots\wedge\omega_{i_1}$$
$$+ (-1)^{(p+1)+(p+1)^2+1} a_{i_1\ldots i_p, jj}\,\omega_{i_p}\wedge\cdots\wedge\omega_{i_1}$$
$$+ (-1)^{\alpha+p^2+1} a_{i_1\ldots i_p, i_\alpha j}\,\omega_j\wedge\omega_{i_p}\wedge\cdots\wedge\hat{\omega}_{i_\alpha}\wedge\cdots\wedge\omega_{i_1}.$$

However, (3.3) implies that

$$a_{i_1\ldots i_p, i_\alpha j} = a_{i_1\ldots i_p, j i_\alpha} + \mathcal{R}_{k_\beta i_\beta i_\alpha j}\, a_{i_1\ldots k_\beta\ldots i_p},$$

hence

$$\Delta\omega = (-1)^{\alpha+p^2} \mathcal{R}_{k_\beta i_\beta i_\alpha j}\, a_{i_1\ldots k_\beta\ldots i_p}\,\omega_j\wedge\cdots\wedge\hat{\omega}_{i_\alpha}\wedge\cdots\wedge\omega_{i_1}$$
$$+ a_{i_1\ldots i_p, jj}\,\omega_{i_p}\wedge\cdots\wedge\omega_{i_1}$$
$$= -\mathcal{R}_{k_\beta i_\beta j_\alpha i_\alpha}\, a_{i_1\ldots k_\beta\ldots i_p}\,\omega_{i_p}\wedge\cdots\wedge\omega_{j_\alpha}\wedge\cdots\wedge\omega_{i_1} + a_{I, jj}\,\omega_I.$$

If we write

$$E(\omega) = \mathcal{R}_{k_\beta i_\beta j_\alpha i_\alpha}\, a_{i_1\ldots k_\beta\ldots i_p}\,\omega_{i_p}\wedge\cdots\wedge\omega_{j_\alpha}\wedge\cdots\wedge\omega_{i_1}$$

and we define the Bochner Laplacian by
$$\nabla^*\nabla \omega = a_{I,jj}\,\omega_I,$$
then
$$\Delta \omega = \nabla^*\nabla \omega - E(\omega).$$

Remark 3.1 If $\omega = f$ is a smooth function, then
$$\Delta f = \sum_{i=1}^m f_{ii}.$$

Remark 3.2 If $\omega = a_i\,\omega_i$ is a 1-form, then
$$E(\omega) = \mathcal{R}_{kiji}\,a_k\,\omega_j$$
$$= \mathcal{R}_{jk}\,a_k\,\omega_j.$$

Lemma 3.3 *If M is compact, the operator δ is the adjoint operator to d, i.e.,*
$$\int_M \langle d\omega, \eta\rangle = \int_M \langle \omega, \delta\eta\rangle$$
for all $\omega \in \Lambda^{p-1}(M)$ and $\eta \in \Lambda^p(M)$.

Proof Note that by the definition of $*$, we have
$$**\,(\omega_{i_p} \wedge \ldots \wedge \omega_{i_1}) = *(\text{sgn}(\sigma(I, I^c))\,\omega_{i_m} \wedge \ldots \wedge \omega_{i_{p+1}})$$
$$= \text{sgn}(\sigma(I^c, I))\,\text{sgn}(\sigma(I, I^c))\,\omega_{i_p} \wedge \ldots \wedge \omega_{i_1}.$$
On the other hand, it is clear that
$$\text{sgn}(\sigma(I, I^c)) = (-1)^{(m-p)p}\,\text{sgn}(\sigma(I^c, I)),$$
which yields the formula that $** = (-1)^{(m-p)p}$ on p-forms. We also observe that
$$\left(\omega_{i_p} \wedge \ldots \wedge \omega_{i_1}\right) \wedge *\left(\omega_{i_p} \wedge \ldots \wedge \omega_{i_1}\right)$$
$$= \text{sgn}(\sigma(I, I^c))\,\omega_{i_p} \wedge \ldots \wedge \omega_{i_1} \wedge \omega_{i_m} \wedge \ldots \wedge \omega_{i_{p+1}}$$
$$= \omega_1 \wedge \ldots \wedge \omega_m.$$
Also by linearity, if $\omega = a_I\,\omega_I$ and $\theta = b_I\,\omega_I$, then
$$\omega \wedge *\theta = a_I b_I\,\omega_1 \wedge \ldots \wedge \omega_m$$
$$= \langle \omega, \theta\rangle\,dV.$$

Let us now consider $\omega \in \Lambda^{p-1}(M)$ and $\eta \in \Lambda^p(M)$, then

$$\begin{aligned}\langle d\omega, \eta \rangle &= d\omega \wedge *\eta \\ &= d(\omega \wedge *\eta) + (-1)^p \omega \wedge d*\eta \\ &= d(\omega \wedge *\eta) + (-1)^p (-1)^{(m-p+1)(p-1)} \omega \wedge **d*\eta \\ &= d(\omega \wedge *\eta) + \omega \wedge *\delta\eta \\ &= d(\omega \wedge *\eta) + \langle \omega, \delta\eta \rangle.\end{aligned}$$

Integrating both sides over M the lemma follows from Stokes' theorem. $\square$

Lemma 3.4 *Let $\omega = \sum_I a_I \omega_I$ be a p-form on M. Then*

$$\Delta |\omega|^2 = 2\langle \Delta\omega, \omega \rangle + 2|\nabla\omega|^2 + 2\langle E(\omega), \omega \rangle.$$

Proof The norm of ω is given by

$$|\omega|^2 = \sum_I a_I^2.$$

Let us choose an orthonormal coframe in a neighborhood of $x \in M$ by parallel translation of an orthonormal frame $\{e_1, \ldots, e_m\}$ at the point x. Hence at x, we have $\omega_{ij} = 0$. Moreover, $\nabla_{e_i} e_j = 0$ along the geodesic tangent to e_i which implies that $\nabla_{e_i} \nabla_{e_i} e_j = 0$ at x. Direct differention yields

$$\begin{aligned}\Delta|\omega|^2 &= \left(a_I^2\right)_{jj} \\ &= 2a_I(a_I)_{jj} + 2((a_I)_j)^2.\end{aligned}$$

Note that, in general, $(a_I)_j \neq a_{I,j}$ and $(a_I)_{jj} \neq a_{I,jj}$, since the terms on the left denote differentiations of the function a_I, while the terms on the right denote covariant differentiations of the p-form. However, by the choice of our frame,

$$\begin{aligned}(a_I)_j &= da_I(e_j) \\ &= a_{I,j} - a_{i_1\ldots j_\alpha \ldots i_p} \omega_{j_\alpha i_\alpha}(e_j) \\ &= a_{I,j}\end{aligned}$$

at the point x. Similarly, we also have

$$a_{I,jj} = da_{i_1\ldots i_p, j}(e_j)$$
$$= d[da_I(e_j) + a_{i_1\ldots j_\alpha \ldots i_p} \omega_{j_\alpha i_\alpha}(e_j)](e_j)$$
$$= (a_I)_{jj} + a_{i_1\ldots j_\alpha \ldots i_p} e_j(\omega_{j_\alpha i_\alpha}(e_j))$$

at x. On the other hand,

$$e_j(\omega_{j_\alpha i_\alpha}(e_j)) = e_j \langle \nabla_{e_j} e_{j_\alpha}, e_{i_\alpha} \rangle$$
$$= \langle \nabla_{e_j} \nabla_{e_j} e_{j_\alpha}, e_{i_\alpha} \rangle + \langle \nabla_{e_j} e_{j_\alpha}, \nabla_{e_j} e_{i_\alpha} \rangle$$
$$= 0$$

at x. Therefore,

$$\Delta |\omega|^2 = 2\langle \nabla^* \nabla \omega, \omega \rangle + 2|\nabla \omega|^2$$
$$= 2\langle \Delta \omega, \omega \rangle + 2|\nabla \omega|^2 + 2\langle E(\omega), \omega \rangle$$

at the point x. Now we observe that both sides of the equation are globally defined, hence this formula is valid on M. □

Theorem 3.5 (Bochner [B]) *Let M be a compact m-dimensional Riemannian manifold without boundary. Suppose M has nonnegative Ricci curvature, $\mathcal{R}_{ij} \geq 0$, then any harmonic 1-form ω must be parallel and $\mathcal{R}(\omega, \omega) \equiv 0$. In particular, this implies that the first Betti number $b_1(M)$ of M must be at most m. If in addition there exists a point $x \in M$ such that the Ricci curvature is positive, then $b_1 = 0$.*

Proof Let ω be a harmonic 1-form. By the Bochner formula and Remark 3.2, we have

$$\Delta |\omega|^2 = 2|\nabla \omega|^2 + 2\langle E(\omega), \omega \rangle$$
$$= 2|\nabla \omega|^2 + 2\langle \mathcal{R}_{jk} a_j \omega_k, a_i \omega_i \rangle$$
$$= 2|\nabla \omega|^2 + 2\mathcal{R}_{ji} a_j a_i$$
$$= 2|\nabla \omega|^2 + 2\mathcal{R}(\omega, \omega), \tag{3.4}$$

which is nonnegative. Hence, by the maximum principle and the fact that M is compact, $|\omega|^2$ must be identically constant. Moreover, (3.4) implies that

$$|\nabla \omega|^2 = 0$$

and

$$\mathcal{R}(\omega, \omega) = 0$$

on M. Since the dimension of parallel 1-forms is at most m, we conclude that the dimension of harmonic 1-forms is at most m. The Betti number bound now follows from the Hodge decomposition theorem. If we further assume that $\mathcal{R}(x) > 0$ for some point $x \in M$, then the fact that $\mathcal{R}(\omega, \omega) \equiv 0$ implies that $\omega(x) = 0$. On the other hand, since $|\omega|^2$ is constant, this shows that $\omega \equiv 0$. Hence $b_1(M) = 0$. □

We would like to remark that if ω is a parallel 1-form on M, then by the de-Rham decomposition theorem, the universal covering $\tilde{M}$ of M must be isometrically a product of $\mathbb{R} \times N$, where $\mathbb{R}$ is given by ω. Hence, if M is a compact manifold with nonnegative Ricci curvature, then its universal covering $\tilde{M}$ must be a product of $\mathbb{R}^k \times N$ for some manifold N with nonnegative Ricci curvature and $k = b_1(M)$.

Definition 3.6 The curvature operator of a Riemannian manifold is a linear map $\mathcal{S} : \Lambda^2(M) \to \Lambda^2(M)$ given by

$$\mathcal{S}(\omega_i \wedge \omega_j) = \mathcal{R}_{jikl}\, \omega_l \wedge \omega_k.$$

Theorem 3.7 (Gallot–Meyer [GM]) *Let M be a compact manifold. If the curvature operator is nonnegative on M, then any harmonic p-form must be parallel. Hence the pth Betti number of M is at most $\binom{m}{p}$. Moreover, if there exists a point $x \in M$ such that $\mathcal{S}(x) > 0$, then $b_p(M) = 0$.*

Proof Similar to the case of 1-forms, we have the identity

$$\Delta |\omega|^2 = 2|\nabla \omega|^2 + 2\langle E(\omega), \omega \rangle,$$

where

$$\langle E(\omega), \omega \rangle = \mathcal{R}_{k_\beta i_\beta j_\alpha i_\alpha}\, a_{i_1 \ldots k_\beta \ldots i_p}\, a_{i_1 \ldots j_\alpha \ldots i_p}.$$

Let us now define the 2-form $\bar{\omega} = a_{i_1 \ldots j_\alpha \ldots i_p}\, \omega_{j_\alpha} \wedge \omega_{i_\alpha}$. By the definition of $\mathcal{S}$, we have

$$\mathcal{S}(\bar{\omega}) = \mathcal{R}_{i_\alpha j_\alpha kl}\, a_{i_1 \ldots j_\alpha \ldots i_p}\, \omega_l \wedge \omega_k.$$

Hence,

$$\langle S(\bar{\omega}), \bar{\omega} \rangle = \langle \mathcal{R}_{i_\alpha j_\alpha kl} a_{i_1...j_\alpha...i_p} \omega_l \wedge \omega_k, a_{i_1...j_\beta...i_p} \omega_{j_\beta} \wedge \omega_{i_\beta} \rangle$$

$$= \mathcal{R}_{i_\alpha j_\alpha i_\beta j_\beta} a_{i_1...j_\alpha...i_p} a_{i_1...j_\beta...i_p}$$

$$= \langle E(\omega), \omega \rangle.$$

The theorem now follows from the argument of Theorem 3.5. □

Definition 3.8 A vector field $X = a_i e_i$ is a Killing vector field if $a_{i,j} + a_{j,i} = 0$ for all $1 \leq i, j \leq m$.

Lemma 3.9 *The infinitesimal generator of a one-parameter family of isometries of M is a Killing vector field on M.*

Proof Let $\phi_t : M \to M$ be a one-parameter family of isometries parametrized by $t \in (-\epsilon, \epsilon)$. If $\{e_1, \ldots, e_m\}$ is an orthonormal frame at a point $x \in M$ given by the normal coordinates centered at x, then the fact that ϕ_t are isometries implies that

$$\langle d\phi_t(e_i), d\phi_t(e_j) \rangle = \langle e_i, e_j \rangle$$

$$= \delta_{ij}.$$

Differentiating with respect to t and evaluating at $t = 0$, we have

$$0 = \frac{d}{dt} \langle d\phi_t(e_i), d\phi_t(e_j) \rangle |_{t=0}$$

$$= \langle \nabla_T d\phi_t(e_i), d\phi_t(e_j) \rangle + \langle \nabla_T d\phi_t(e_j), d\phi_t(e_i) \rangle, \quad (3.5)$$

where T is the infinitesimal vector field given by ϕ_t.

However, since one can view $\{\partial/\partial t, e_1, \ldots, e_m\}$ as tangent vectors given by a coordinate system of $(-\epsilon, \epsilon) \times M$, we have the property that

$$[T, d\phi_t(e_i)] = 0$$

for all $1 \leq i \leq m$. Hence we can rewrite (3.5) as

$$0 = \langle \nabla_{e_i} T, e_j \rangle + \langle \nabla_{e_j} T, e_i \rangle$$

which is exactly the condition that T is a Killing vector field. □

Theorem 3.10 *Let M be a compact manifold with nonpositive Ricci curvature. Then any Killing vector field on M must be parallel. Moreover, if there exists a point $x \in M$ such that the Ricci curvature satisfies $\mathcal{R}(x) < 0$, then there are* *no nontrivial Killing vector fields. In particular, this implies that M does not have a one-parameter family of isometries.*

Proof Let $X = a_i\, e_i$ be a Killing vector field. Its dual 1-form is given by $\omega = a_i\, \omega_i$. The commutation formula yields

$$\Delta |\omega|^2 = 2a_{i,j}^2 + 2a_{i,jj}\, a_i$$
$$= 2|\nabla \omega|^2 - 2a_{j,ij}\, a_i$$
$$= 2|\nabla \omega|^2 - 2\mathcal{R}(X, X).$$

We now apply the same argument as in the proof of Theorem 3.5. □

4
Laplacian comparison theorem

Let M be a complete m-dimensional manifold. Suppose p is a fixed point in M and let us consider the distance function $r_p(x) = r(p, x)$ to p. When there is no ambiguity, the subscript will be deleted and we will simply write $r(x)$. The distance function in general is not smooth due to the presence of cut-points. However, it can be seen that it is a Lipschitz function with Lipschitz constant 1. In particular, we have

$$|\nabla r|^2 = 1$$

almost everywhere on M. Though r might not be a C^2 function, one can still estimate its Laplacian in the sense of distribution.

Theorem 4.1 *Let M be a complete m-dimensional Riemannian manifold with Ricci curvature bounded from below by*

$$\mathcal{R}_{ij} \geq (m-1)K$$

for some constant K. Then the Laplacian of the distance function satisfies

$$\Delta r(x) \leq \begin{cases} (m-1)\sqrt{K}\cot(\sqrt{K}r) & \text{for } K > 0, \\ (m-1)r^{-1} & \text{for } K = 0, \\ (m-1)\sqrt{-K}\coth(\sqrt{-K}r) & \text{for } K < 0 \end{cases}$$

in the sense of distribution.

Proof For a smooth function f, let $x \in M$ be a point such that $\nabla f(x) \neq 0$. Then locally the level surface N of f through the point x is a smooth hypersurface. Let $\{e_1, \ldots, e_{m-1}\}$ be an orthonormal frame tangent to N, and let e_m be the unit normal vector to N. The Laplacian of f at x is defined to be

$$\Delta f(x) = \sum_{i=1}^{m-1}(e_i e_i - \nabla_{e_i} e_i) f(x) + (e_m e_m - \nabla_{e_m} e_m) f(x)$$

$$= \sum_{i=1}^{m-1}(e_i e_i - (\nabla_{e_i} e_i)^t) f(x)$$

$$- \sum_{i=1}^{m-1}(\nabla_{e_i} e_i)^n f(x) + (e_m e_m - \nabla_{e_m} e_m) f(x)$$

$$= \Delta_N f(x) + \vec{H} f(x) + (e_m e_m - \nabla_{e_m} e_m) f(x)$$

$$= \vec{H} f(x) + (e_m e_m - \nabla_{e_m} e_m) f(x),$$

where Δ_N denotes the Laplacian of N with respect to the induced metric and $\vec{H}$ denotes the mean curvature vector of N. If we take $f = r$, then for a point x which is not in the cut-locus of p, we have $N = \partial B_p(r)$. Moreover, the unit normal vector $e_m = \partial/\partial r$. Hence, we have

$$\Delta r(x) = H(x),$$

where $H(x)$ is the mean curvature of $\partial B_p(r)$ with respect to $\partial/\partial r$. Therefore according to Corollary 2.2 the theorem is true for those points that are not in the cut-locus of p.

Using the same notation as in Theorem 2.1, let us denote

$$\bar{H}(r) = \begin{cases} (m-1)\sqrt{K}\cot(\sqrt{K}r) & \text{for } K > 0, \\ (m-1)r^{-1} & \text{for } K = 0, \\ (m-1)\sqrt{-K}\coth(\sqrt{-K}r) & \text{for } K < 0. \end{cases}$$

To show that Δr has the desired estimate in the sense of distribution, it suffices to show that for any nonnegative compactly supported smooth function ϕ we have

$$\int_M (\Delta \phi) r \leq \int_M \phi \, \bar{H}(r).$$

In terms of polar normal coordinates at p, we can write

$$\int_M \phi \, \bar{H}(r) = \int_0^\infty \int_{C(r)} \phi \, \bar{H}(r) \, J(\theta, r) \, d\theta \, dr.$$

On the other hand, for each $\theta \in S_p(M)$ if we let $R(\theta)$ be the maximum value of $r > 0$ such that the geodesic $\gamma(s) = \exp_p(s\theta)$ minimizes up to $s = r$, then by Fubini's theorem we can write

$$\int_0^\infty \int_{C(r)} \phi \, \bar{H}(r) \, J(\theta, r) \, d\theta \, dr = \int_{S_p(M)} \int_0^{R(\theta)} \phi \, \bar{H}(r) \, J(\theta, r) \, dr \, d\theta.$$

However, for $r < R(\theta)$, we have

$$\bar{H}(r) \, J(\theta, r) \geq H(\theta, r) \, J(\theta, r)$$
$$= \frac{\partial}{\partial r} J(\theta, r).$$

Therefore,

$$\int_M \phi \, \bar{H}(r) \geq \int_{S_p(M)} \int_0^{R(\theta)} \phi \, \frac{\partial J}{\partial r} \, dr \, d\theta$$
$$= -\int_{S_p(M)} \int_0^{R(\theta)} \frac{\partial \phi}{\partial r} J \, dr \, d\theta + \int_{S_p(M)} [\phi J]_0^{R(\theta)} \, d\theta$$
$$= -\int_M \frac{\partial \phi}{\partial r} + \int_{S_p(M)} \phi(\theta, R(\theta)) \, J(\theta, R(\theta)) \, d\theta$$
$$\geq -\int_M \langle \nabla \phi, \nabla r \rangle,$$

where we have used the facts that both ϕ and J are nonnegative and $J(\theta, 0) = 0$. On the other hand, since r is Lipschitz, we conclude that

$$-\int_M \langle \nabla \phi, \nabla r \rangle = \int_M (\Delta \phi) r,$$

which proves the theorem. □

We are now ready to prove a structural theorem for manifolds with nonnegative Ricci curvature. Let us first define the notions of a line and a ray in a Riemannian manifold.

Definition 4.2 A line is a normal geodesic $\gamma : (-\infty, \infty) \longrightarrow M$ such that any of its finite segments $\gamma|_a^b$ is a minimizing geodesic.

Definition 4.3 A ray is a half-line $\gamma_+ : [0, \infty) \longrightarrow M$, which is a normal minimizing geodesic.

Theorem 4.4 (Cheeger–Gromoll [CG2]) *Let M be a complete, m-dimensional, manifold with nonnegative Ricci curvature. If there exists a* line in M, then M is isometric to $\mathbb{R} \times N$, the product of a real line and an $(m-1)$-dimensional manifold N with nonnegative Ricci curvature.

Proof Let $\gamma_+ : [0, \infty) \longrightarrow M$ be a ray in M. One defines the Buseman function β_+ with respect to γ_+ by

$$\beta_+(x) = \lim_{t \to \infty} (t - r(\gamma_+(t), x)).$$

We observe that β_+ is a Lipschitz function with Lipschitz constant 1. Moreover, by the Laplacian comparison theorem,

$$\Delta \beta_+(x) \geq -\lim_{t \to \infty} (m-1) r(\gamma_+(t), x)^{-1}$$
$$= 0$$

in the sense of distribution. If γ is a line, then $\gamma_+(t) = \gamma(t)$ and $\gamma_-(t) = \gamma(-t)$ for $t \geq 0$ are rays. The corresponding Buseman functions β_+ and β_- are subharmonic in the sense of distribution. In particular, $\beta_- + \beta_+$ is also subharmonic on M. On the other hand, since γ is a line, the triangle inequality implies that

$$2t = r(\gamma(-t), \gamma(t))$$
$$\leq r(\gamma(-t), x) + r(\gamma(t), x).$$

Hence

$$t - r(\gamma_-(t), x) + t - r(\gamma_+(t), x) \leq 0,$$

and by taking the limit as $t \to \infty$ we have

$$\beta_-(x) + \beta_+(x) \leq 0.$$

Moreover, it is also clear that

$$\beta_-(x) + \beta_+(x) = 0$$

for all x on γ. However, by the strong maximum principle, since the subharmonic function $\beta_- + \beta_+$ has an interior maximum, it must be identically constant. In particular, both β_- and β_+ are harmonic and $\beta_- \equiv -\beta_+$. By regularity theory, β_+ is a smooth harmonic function with $|\nabla \beta_+| \leq 1$, and $|\nabla \beta_+| \equiv 1$ on the geodesic γ. For simplicity, let us write $\beta = \beta_+$. The Bochner formula gives

$$\Delta |\nabla \beta|^2 = 2\beta_{ij}^2 + 2\mathcal{R}_{ij} \beta_i \beta_j + 2\langle \nabla \beta, \nabla \Delta \beta \rangle$$
$$\geq 0. \tag{4.1}$$

Hence by the fact that $|\nabla \beta|^2$ achieves its maximum in the interior of M, the maximum principle for subharmonic functions again implies that $|\nabla \beta|^2 \equiv 1$ on M. Substituting this into (4.1) yields $\beta_{ij} \equiv 0$, and $\nabla \beta$ is a parallel vector field on M. This implies that M must split, which proves the theorem. □

Corollary 4.5 *Let M be a complete m-dimensional Riemannian manifold with nonnegative Ricci curvature. If M has at least two ends, then there exists a compact $(m-1)$-dimensional manifold N of nonnegative Ricci curvature such that*

$$M = \mathbb{R} \times N.$$

Proof The assumption that M has at least two ends implies that there exists a compact set $D \subset M$ such that $M \setminus D$ has at least two unbounded components. Hence there are two unbounded sequences of points $\{x_i\}_{i=1}^\infty$ and $\{y_i\}_{i=1}^\infty$, such that the minimal geodesics γ_i joining x_i to y_i must intersect D. By the compactness of D, if $p_i \in \gamma_i \cap D$ and $v_i = \gamma_i'(p_i)$, then by passing through a subsequence we have $p_i \to p$ and $v_i \to v$ for some $p \in D$ and $v \in T_p M$. We now claim that the geodesic $\gamma : (-\infty, \infty) \longrightarrow M$ given by the initial conditions $\gamma(0) = p$ and $\gamma'(0) = v$ is a line.

To see this, let us consider an arbitrary segment $\gamma|_{[a,b]}$ of γ. By the continuity of the initial conditions of the second order ordinary differential equation, we know that $\gamma_i|_{[a,b]} \to \gamma|_{[a,b]}$ because $(p_i, v_i) \to (p, v)$. However, by the assumption, $\gamma_i|_{[a,b]}$ are minimizing geodesics, hence $\gamma|_{[a,b]}$ is also minimizing, and γ is a line.

Theorem 4.4 now implies that $M = \mathbb{R} \times N$. The compactness of N follows from the assumption that M has at least two ends. □

Another application of the Laplacian comparison theorem is the eigenvalue comparison theorem of Cheng.

Theorem 4.6 (Cheng [Cg1]) *Let M be a compact Riemannian manifold of dimension m and ∂M be the boundary of M. Assume that the Ricci curvature of M is bounded by*

$$\mathcal{R}_{ij} \geq (m-1)K$$

for some constant K. Let us consider $\bar{M}$ to be the simply connected space form with constant curvature K. Let $\mu_1(M)$ be the first nonzero eigenvalue of the Dirichlet Laplacian on M and i be the inscribe radius of M. If $\bar{B}(i)$ is a geodesic ball in $\bar{M}$ with radius i and $\mu_1(\bar{B}(i))$ is its first Dirichlet eigenvalue, then

$$\mu_1(M) \leq \mu_1(\bar{B}(i)).$$

When $\partial M = \emptyset$, let $\lambda_1(M)$ be the first nonzero eigenvalue of the Laplacian and d be the diameter of M. If $\bar{B}(d/2)$ is a geodesic ball in $\bar{M}$ with radius $d/2$ and $\mu_1\left(\bar{B}(d/2)\right)$ is its first Dirichlet eigenvalue, then

$$\lambda_1(M) \leq \mu_1\left(\bar{B}\left(\frac{d}{2}\right)\right).$$

Proof Let us first consider the case when M has boundary. Let $B_p(i)$ be an inscribed ball in M. By the monotonicity of eigenvalues, it suffices to show that

$$\mu_1(B_p(i)) \leq \mu_1(\bar{B}(i)).$$

Let $\bar{u}$ be the first Dirichlet eigenfunction on $\bar{B}(i)$. By the uniqueness of $\bar{u}$, we may assume that $\bar{u} \geq 0$. If we let $\bar{p} \in \bar{B}(i)$ be its center and $\bar{r}$ be the distance function to $\bar{p}$, then $\bar{u}(\bar{r})$ must be a function of $\bar{r}$ alone. By the facts that $\bar{u} \geq 0$ and $\bar{u}(i) = 0$, the strong maximum principle implies that $(\partial \bar{u}/\partial \bar{r})(i) < 0$. If there is some value of $\bar{r} < i$ such that $\partial \bar{u}/\partial \bar{r} > 0$, then this would imply that $\bar{u}$ has a interior local minimum. However, this violates the strong maximum principle. Hence, $\bar{u}' = \partial \bar{u}/\partial \bar{r} \leq 0$.

Let us define a Lipschitz function on $B_p(i)$ by $u(r) = \bar{u}(r)$, where r denotes the distance function to p. Clearly, u satisfies the Dirichlet boundary condition. Computing the Laplacian of u gives

$$\Delta u = \bar{u}' \Delta r + \bar{u}'' |\nabla r|^2$$
$$\geq \bar{u}' \Delta r + \bar{u}''.$$

On the other hand, if we let $\bar{\Delta}$ be the Laplacian on $\bar{M}$, then

$$-\mu_1(\bar{B}(i))\bar{u} = \bar{\Delta}\bar{u}$$
$$= \bar{u}' \bar{\Delta}\bar{r} + \bar{u}''.$$

By Theorem 4.1 and the fact that $\bar{u}' \leq 0$, we conclude that

$$\Delta u \geq -\mu_1(\bar{B}(i))u$$

in the sense of distribution. Hence, by the Rayleigh principle for eigenvalues, we conclude that

$$\mu_1(B_p(i)) \leq \frac{\int_{B_p(i)} |\nabla u|^2}{\int_{B_p(i)} u^2}$$

$$= \frac{-\int_{B_p(i)} u \Delta u}{\int_{B_p(i)} u^2}$$

$$\leq \mu_1(\bar{B}(i)).$$

To prove the upper bound for the case when M has no boundary, we consider the disjoint balls $B_p(d/2)$ and $B_q(d/2)$ centered at a pair of points p and q which realize the diameter. By the above estimate,

$$\mu_1\left(B_p\left(\frac{d}{2}\right)\right) \leq \mu_1\left(\bar{B}\left(\frac{d}{2}\right)\right)$$

and

$$\mu_1\left(B_q\left(\frac{d}{2}\right)\right) \leq \mu_1\left(\bar{B}\left(\frac{d}{2}\right)\right).$$

We now claim that $\lambda_1(M) \leq \max\{\mu_1(M_1), \mu_1(M_2)\}$ for any disjoint pair of open subsets $M_1, M_2 \subset M$. This will establish the theorem by setting $M_1 = B_p(d/2)$ and $M_2 = B_q(d/2)$.

To prove the claim, let u_1 and u_2 be nonnegative first Dirichlet eigenfunctions on M_1 and M_2 respectively. By multiplying u_i by a constant, we may assume that $\int_{M_i} u_i = 1$ for $i = 1, 2$. Let us define a Lipschitz function on M by

$$u = \begin{cases} u_1 & \text{on } M_1, \\ -u_2 & \text{on } M_2, \\ 0 & \text{on } M \setminus (M_1 \cup M_2). \end{cases}$$

Clearly, $\int_M u = 0$, hence after applying the Rayleigh principle, we have

$$\lambda_1(M) \left(\int_{M_1} u_1^2 + \int_{M_2} u_2^2 \right) = \lambda_1(M) \int_M u^2$$

$$\leq \int_M |\nabla u|^2$$

$$= \int_{M_1} |\nabla u_1|^2 + \int_{M_2} |\nabla u_2|^2$$

$$= \mu_1(M_1) \int_{M_1} u_1^2 + \mu_1(M_2) \int_{M_2} u_2^2$$

$$\leq \max\{\mu_1(M_1), \mu_1(M_2)\} \left(\int_{M_1} u_1^2 + \int_{M_2} u_2^2 \right).$$

This establishes the claim. $\square$

5
Poincaré inequality and the first eigenvalue

In this chapter, we will obtain lower estimates for the first eigenvalue of the Laplacian on a compact manifold. For the moment, we will primarily be concerned with the case when M has no boundary. The following lower bound was proved by Lichnerowicz [Lz], while Obata [O] considered the case when the estimate is achieved.

Theorem 5.1 (Lichnerowicz and Obata) *Let M be an m-dimensional compact manifold without boundary. Suppose that the Ricci curvature of M is bounded from below by*

$$\mathcal{R}_{ij} \geq (m-1)K$$

for some constant $K > 0$, then the first nonzero eigenvalue of the Laplacian on M must satisfy

$$\lambda_1 \geq mK.$$

Moreover, equality holds if and only if M is isometric to a standard sphere of radius $1/\sqrt{K}$.

Proof Let u be a nonconstant eigenfunction satisfying

$$\Delta u = -\lambda u,$$

with $\lambda > 0$. Consider the smooth function

$$Q = |\nabla u|^2 + \frac{\lambda}{m} u^2.$$

5 Poincaré inequality and the first eigenvalue

defined on M. Computing its Laplacian

$$\Delta Q = \left(2u_j u_{ji} + \frac{2\lambda}{m} u u_i\right)_i$$

$$= 2u_{ji}^2 + 2u_j u_{jii} + \frac{2\lambda}{m} u_i^2 + \frac{2\lambda}{m} u u_{ii}$$

$$= 2u_{ji}^2 + 2\mathcal{R}_{ij} u_i u_j + 2u_j(\Delta u)_j + \frac{2\lambda}{m} |\nabla u|^2 + \frac{2\lambda}{m} u(\Delta u),$$

where we have used the Ricci identity and the convention that summation is performed on repeated indices. On the other hand,

$$\sum_{i,j=1}^m u_{ji}^2 \geq \sum_{i=1}^m u_{ii}^2$$

$$\geq \frac{\left(\sum_{i=1}^m u_{ii}\right)^2}{m}$$

$$= \frac{(\Delta u)^2}{m}$$

$$= \frac{\lambda^2 u^2}{m}.$$

Hence, by the assumption on the Ricci curvature, we have

$$\Delta Q \geq \frac{2\lambda^2 u^2}{m} + 2(m-1)K|\nabla u|^2 - 2\lambda|\nabla u|^2 + \frac{2\lambda}{m}|\nabla u|^2 - \frac{2\lambda^2 u^2}{m}$$

$$= 2(m-1)\left(K - \frac{\lambda}{m}\right)|\nabla u|^2. \tag{5.1}$$

If $\lambda \leq mK$, then Q is a subharmonic function. By the compactness of M and the maximum principle, Q must be identically constant and all the above inequalities are equalities. In particular, the right-hand side of (5.1) must be identically 0. Hence $\lambda = mK$ because u is nonconstant. Moreover,

$$|\nabla u|^2 + \frac{\lambda}{m} u^2 = \frac{\lambda}{m} |u|_\infty^2,$$

where $|u|_\infty = \sup_M |u|$. If we normalize u such that $|u|_\infty = 1$, and observe that at the maximum and minimum points of u its gradient must vanish, then we conclude that $\max u = 1 = -\min u$ and

$$\frac{|\nabla u|}{\sqrt{1 - u^2}} = \sqrt{K}.$$

Integrating this along a minimal geodesic γ joining the points where $u = 1$ and $u = -1$, we have

$$d\sqrt{K} \geq \int_\gamma \frac{|\nabla u|}{\sqrt{1-u^2}}$$

$$\geq \int_{-1}^{1} \frac{du}{\sqrt{1-u^2}}$$

$$= \pi,$$

where d denotes the diameter of M. However, Cheng's theorem (Theorem 2.6) implies that M must be the standard sphere. $\square$

We will now give a sharp lower bound for the first eigenvalue on manifolds with nonnegative Ricci curvature. The estimate of Lichnerowicz becomes trivial in this case, since the Ricci curvature does not have a positive lower bound. However, one could still estimate the first eigenvalue in terms of the diameter of M alone.

Let λ_1 be the least nontrivial eigenvalue of a compact manifold and let ϕ be the corresponding eigenfunction. By multiplying ϕ by a constant it is possible to arrange that

$$a - 1 = \inf_M \phi, \quad a + 1 = \sup_M \phi,$$

where $0 \leq a(\phi) < 1$ is the median of ϕ.

Lemma 5.2 (Li–Yau [LY1]) *Suppose M^m is a compact manifold without boundary and its Ricci curvature is nonnegative. Then the first nontrivial eigenvalue satisfies*

$$\lambda_1 \geq \frac{\pi^2}{(1+a)d^2},$$

where d is the diameter of M.

Proof Setting $\lambda = \lambda_1$ and $u = \phi - a$, the eigenvalue equation becomes

$$\Delta u = -\lambda(u+a).$$

Let $P = |\nabla u|^2 + cu^2$, where $c = \lambda(1+a)$. Suppose $x_0 \in M$ is the point where P achieves its maximum. If $|\nabla u(x_0)| \neq 0$ we may rotate the frame so that $u_1(x_0) = |\nabla u(x_0)|$. Differentiating in the e_i direction yields

$$\tfrac{1}{2} P_i = u_j u_{ji} + cuu_i.$$

5 Poincaré inequality and the first eigenvalue

At the point x_0, this yields

$$0 = u_1(u_{11} + cu)$$

and

$$u_{ji}u_{ji} \geq u_{11}^2 = c^2u^2. \tag{5.2}$$

Covariant differentiating with respect to e_i again, using the commutation formula (5.2), the definition of P, and evaluating at x_0, we have

$$0 \geq \tfrac{1}{2}\Delta P$$

$$= u_{ji}u_{ji} + u_j u_{jii} + cu_i^2 + cu\Delta u$$

$$\geq c^2 u^2 + u_j(\Delta u)_j + R_{ji}u_j u_i + cu_i^2 - c\lambda u(u+a)$$

$$\geq c^2 u^2 - \lambda u_i^2 + cu_i^2 - c\lambda u(u+a)$$

$$= (c - \lambda)\left(u_i^2 + cu^2\right) - ac\lambda u$$

$$\geq a\lambda P(x_0) - ac\lambda.$$

Hence, for all $x \in M$

$$|\nabla u(x)|^2 \leq \lambda(1+a)\left(1 - u(x)^2\right). \tag{5.3}$$

Note that (5.3) is trivially satisfied if $\nabla u(x_0) = 0$, hence the assumption $\nabla u(x_0) \neq 0$ is not necessary.

Let γ be the shortest geodesic from the minimizing point of u to the maximizing point. The length of γ is at most d. Integrating the gradient estimate (5.3) along this segment with respect to arc-length, we obtain

$$d\sqrt{\lambda(1+a)} \geq \sqrt{\lambda(1+a)}\int_\gamma ds \geq \int_\gamma \frac{|\nabla u|ds}{\sqrt{1-u^2}} \geq \int_{-1}^1 \frac{du}{\sqrt{1-u^2}} = \pi. \quad \Box$$

In view of Lemma 5.2 and known examples, Li and Yau conjectured that the sharp estimate

$$\lambda_1 \geq \frac{\pi^2}{d^2}$$

should hold for compact manifolds with nonnegative Ricci curvature. In fact, if the first eigenspace has multiplicity greater than 1, this was verified in [L3]. This conjecture was finally proved by Zhong and Yang by applying the maximum principle to a judicious choice of test function. We will now present this argument.

Lemma 5.3 *The function*

$$z(u) = \frac{2}{\pi}\left(\arcsin(u) + u\sqrt{1-u^2}\right) - u$$

defined on $[-1, 1]$ *satisfies*

$$\dot{z}u + \ddot{z}(1 - u^2) + u = 0, \tag{5.4}$$

$$\dot{z}^2 - 2z\ddot{z} + \dot{z} \geq 0, \tag{5.5}$$

$$2z - \dot{z}u + 1 \geq 0, \tag{5.6}$$

and

$$(1 - u^2) \geq 2|z|. \tag{5.7}$$

Proof Differentiating the definition of $z(u)$ yields

$$\dot{z} = \frac{4}{\pi}\sqrt{1 - u^2} - 1$$

and

$$\ddot{z} = \frac{-4u}{\pi\sqrt{1 - u^2}}.$$

Clearly (5.4) is satisfied.

To see (5.5), we have

$$\dot{z}^2 - 2z\ddot{z} + \dot{z} = \frac{4}{\pi\sqrt{1-u^2}}\left\{\frac{4}{\pi}\left(\sqrt{1-u^2} + u\arcsin(u)\right) - (1+u^2)\right\}.$$

Since the right-hand side is an even function, it suffices to check that

$$\frac{4}{\pi}\left(\sqrt{1-u^2} + u\arcsin(u)\right) - (1+u^2) \geq 0$$

on $[0, 1]$. Computing its derivative

$$\frac{d}{du}\left\{\frac{4}{\pi}\left(\sqrt{1-u^2} + u\arcsin(u)\right) - (1+u^2)\right\} = \frac{4}{\pi}\arcsin(u) - 2u,$$

which is nonpositive on [0, 1]. Hence

$$\frac{4}{\pi}\left(\sqrt{1-u^2}+u\arcsin(u)\right)-(1+u^2)$$

$$\geq \left[\frac{4}{\pi}\left(\sqrt{1-u^2}+u\arcsin(u)\right)-(1+u^2)\right]_{u=1}$$

$$= 0.$$

Inequality (5.6) follows easily because

$$2z - \dot{z}u + 1 = \frac{4}{\pi}\arcsin(u) + 1 - u,$$

which is obviously nonnegative.

To see (5.7), we will consider the cases $-1 \leq u \leq 0$ and $0 \leq u \leq 1$ separately. It is clearly that the inequality is valid at -1, 0, and 1. Let us set

$$f(u) = 1 - u^2 - \frac{4}{\pi}\left(\arcsin(u) + u\sqrt{1-u^2}\right) + 2u.$$

Then

$$\dot{f} = -2u - \frac{4}{\pi}\left(2\sqrt{1-u^2}\right) + 2,$$

$$\ddot{f} = -2 + \frac{8u}{\pi\sqrt{1-u^2}},$$

and

$$\dddot{f} = \frac{8}{\pi(1-u^2)^{\frac{3}{2}}}.$$

When $-1 \leq u \leq 0$, we have $\ddot{f} \leq 0$, hence $f(u) \geq \min\{f(-1), f(0)\} = 0$. For the case $0 \leq u \leq 1$, we have $\dddot{f} \geq 0$, hence

$$\dot{f} \leq \max\{\dot{f}(0), \dot{f}(1)\}$$

$$= \max\left\{2 - \frac{8}{\pi}, 0\right\}$$

$$= 0.$$

Therefore $f(u) \geq f(1)$ proving $(1-u^2) \geq 2z$ on $[-1, 1]$. Note that $(1-u^2) \geq 2|z|$ follows from the previous inequality because z is an odd function while $(1-u^2)$ is even. □

Lemma 5.4 *Suppose M is a compact manifold without boundary whose Ricci curvature is nonnegative. Assume that a nontrivial eigenfunction ϕ corresponding to the eigenvalue λ is normalized so that for $0 \leq a < 1$, $a + 1 = \sup \phi$ and $a - 1 = \inf_M \phi$. By setting $u = \phi - a$, its gradient must satisfy the estimate*

$$|\nabla u|^2 \leq \lambda(1 - u^2) + 2a\lambda z(u), \tag{5.8}$$

where

$$z(u) = \frac{2}{\pi}\left(\arcsin(u) + u\sqrt{1 - u^2}\right) - u. \tag{5.9}$$

Proof We will first prove an estimate similar to (5.8) for $u = \epsilon(\phi - a)$, where $0 < \epsilon < 1$. The lemma will follow by letting $\epsilon \to 1$. By the definition of u, we have

$$\Delta u = -\lambda(u + \epsilon a)$$

with $-\epsilon \leq u \leq \epsilon$. By (5.3) we may assume $a > 0$. Consider the function

$$Q = |\nabla u|^2 - c(1 - u^2) - 2a\lambda z(u),$$

and because of (5.3) and (5.7) we can choose c large enough so that $\sup_M Q = 0$. The lemma follows if $c \leq \lambda$, for a sequence of $\epsilon \to 1$, hence we may assume that $c > \lambda$.

Let the maximum point of Q be x_0. We claim that $|\nabla u(x_0)| > 0$ since otherwise $\nabla u(x_0) = 0$ and

$$0 = Q(x_0)$$
$$= -c(1 - u^2)(x_0) - 2a\lambda z(x_0)$$
$$\leq -(c - a\lambda)(1 - \epsilon^2)$$

by (5.7), which is a contradiction.

Differentiating in the e_i direction gives

$$\tfrac{1}{2} Q_i = u_j u_{ji} + cuu_i - a\lambda \dot{z} u_i. \tag{5.10}$$

At x_0, we can rotate the frame so $u_1(x_0) = |\nabla u(x_0)|$ and using $Q_i = 0$ we have

$$u_{ji} u_{ji} \geq u_{11}^2 = (cu - a\lambda \dot{z})^2. \tag{5.11}$$

5 Poincaré inequality and the first eigenvalue

Differentiating again, using the commutation formula, $Q(x_0) = 0$, (5.7), (5.10), and (5.11), we get

$$0 \geq \tfrac{1}{2}\Delta Q(x_0)$$
$$= u_{ji}u_{ji} + u_j u_{jii} + cu_i^2 + cu\Delta u - a\lambda \ddot{z} u_i^2 - a\lambda \dot{z}\Delta u$$
$$= u_{ji}u_{ji} + u_j(\Delta u)_j + R_{ji}u_j u_i + (c - a\lambda\ddot{z})u_i^2 + (cu - a\lambda\dot{z})\Delta u$$
$$\geq (cu - a\lambda\dot{z})^2 + (c - \lambda - a\lambda\ddot{z})\left[c\left(1 - u^2\right) + 2a\lambda z\right]$$
$$\quad - \lambda(cu - a\lambda\dot{z})(u + \epsilon a)$$
$$= -ac\lambda\left\{(1 - u^2)\ddot{z} + u\dot{z} + \epsilon u\right\} + a^2\lambda^2\left\{-2z\ddot{z} + \dot{z}^2 + \epsilon\dot{z}\right\}$$
$$\quad + a\lambda(c - \lambda)\{-u\dot{z} + 2z + 1\} + (c - \lambda)(c - a\lambda).$$

However by (5.4), (5.5), and (5.6), we conclude that

$$0 \geq ac\lambda(1 - \epsilon)u - a^2\lambda^2(1 - \epsilon)\dot{z} + (c - \lambda)(c - a\lambda)$$
$$\geq -ac\lambda(1 - \epsilon) - a^2\lambda^2(1 - \epsilon)\left(\frac{4}{\pi} - 1\right) + (c - \lambda)(c - a\lambda)$$
$$\geq -(c + \lambda)\lambda(1 - \epsilon) + (c - \lambda)^2.$$

This implies that

$$c \leq \lambda\left\{\frac{2 + (1 - \epsilon) + \sqrt{(1 - \epsilon)(9 - \epsilon)}}{2}\right\}.$$

Clearly, when $\epsilon \to 1$ this yields the desired estimate. $\square$

Theorem 5.5 (Zhong–Yang [ZY]) *Suppose M is a compact manifold without boundary whose Ricci curvature is nonnegative. Let $a \geq 0$ be the median of a normalized first eigenfunction with $a + 1 = \sup\phi$ and $a - 1 = \inf\phi$, and let d be the diameter. Then the first nontrivial eigenvalue satisfies*

$$d^2\lambda_1 \geq \pi^2 + \frac{6}{\pi}\left(\frac{\pi}{2} - 1\right)^4 a^2 \geq \pi^2(1 + 0.02a^2).$$

Proof Arguing with $u = \phi - a$ as before, let γ be the shortest geodesic from a minimum point of u to a maximum point, with length at most d. Integrating the gradient estimate (5.8) along this segment with respect to arc-length and using the oddness of the function, we have

$$d\lambda^{1/2} \geq \lambda^{1/2} \int_\gamma ds$$

$$\geq \int_\gamma \frac{|\nabla u| ds}{\sqrt{1 - u^2 + 2az(u)}}$$

$$\geq \int_0^1 \left\{ \frac{1}{\sqrt{1 - u^2 + 2az}} + \frac{1}{\sqrt{1 - u^2 - 2az}} \right\} du$$

$$\geq \int_0^1 \frac{1}{\sqrt{1 - u^2}} \left\{ 2 + \frac{3a^2 z^2}{1 - u^2} \right\} du$$

$$\geq \pi + 3a^2 \left(\int_0^1 \frac{z}{\sqrt{1 - u^2}} \right)^2$$

$$= \pi + \frac{3a^2}{\pi^2} \left(\frac{\pi}{2} - 1 \right)^4. \qquad \square$$

This technique also applies to manifolds with boundary. Let M^n be a compact manifold with smooth boundary whose Ricci curvature is nonnegative. Suppose that the second fundamental form of ∂M is nonnegative. Then the first nontrivial eigenvalue of the Laplacian with Neumann boundary conditions also satisfies inequality (5.8). The proof is the same as that of Lemma 5.4 except that the possibility that the maximum of the function Q may occur at the boundary must be considered. In fact, the boundary convexity assumption implies that the maximum of Q cannot occur on the boundary as indicated in the proof of Corollary 5.8.

The next theorem gives an estimate of the first eigenvalue for the general compact Riemannian manifold without boundary. The estimate depends on the lower bound of the Ricci curvature, the upper bound of the diameter, and the dimension of M alone. An estimate of this form was first conjectured by Yau in [Y3]. It was first proved by the author [L1] for manifolds with nonnegative Ricci curvature with an improvement by Li and Yau [LY1] as stated in Lemma 5.2. The general case was also proved by Li and Yau in [LY1] and will be presented in Theorem 5.7. The following lemma will be useful for the upcoming theorem and also for various estimates in the next chapter.

5 Poincaré inequality and the first eigenvalue

Lemma 5.6 *Let M be a complete m-dimensional Riemannian manifold. Suppose the Ricci curvature of M is bounded from below by*

$$\mathcal{R}_{ij} \geq -(m-1)R$$

for some constant $R \geq 0$. Let u be a function defined on M satisfying the equation

$$\Delta u = -\lambda u,$$

and if we define

$$Q = |\nabla \log(a+u)|^2,$$

then

$$\Delta Q - \frac{m}{2(m-1)} |\nabla Q|^2 Q^{-1}$$

$$+ \langle \nabla v, \nabla Q \rangle Q^{-1} \left(\frac{2(m-2)}{m-1} Q - \frac{2}{m-1} \frac{\lambda u}{a+u} \right) \geq \frac{2}{m-1} Q^2$$

$$+ \left(\frac{4}{m-1} \frac{\lambda u}{a+u} - \frac{2\lambda a}{a+u} - 2(m-1)R \right) Q + \frac{2}{m-1} \left(\frac{\lambda u}{a+u} \right)^2.$$

Proof If we set $v = \log(a+u)$, then a direct computation shows that v satisfies the equation

$$\Delta v = -|\nabla v|^2 - \lambda + \frac{\lambda a}{a+u}.$$

Using the Bochner formula, we compute

$$\Delta Q = 2v_{ij}^2 + 2\mathcal{R}_{ij} v_i v_j + 2\langle \nabla v, \nabla \Delta v \rangle$$

$$\geq 2v_{ij}^2 - 2(m-1)R\, Q - 2\langle \nabla v, \nabla Q \rangle - \frac{2\lambda a}{(a+u)} |\nabla v|^2. \quad (5.12)$$

Choosing an orthonormal frame $\{e_1, \ldots, e_m\}$ at a point so that $|\nabla v| e_1 = \nabla v$, we can write

$$|\nabla|\nabla v|^2|^2 = 4\sum_{j=1}^{m}\left(\sum_{i=1}^{m} v_i v_{ij}\right)^2$$

$$= 4v_1^2 \sum_{j=1}^{m} v_{1j}^2$$

$$= 4|\nabla v|^2 \sum_{j=1}^{m} v_{1j}^2. \tag{5.13}$$

On the other hand,

$$v_{ij}^2 \geq v_{11}^2 + 2\sum_{\alpha=2}^{m} v_{1\alpha}^2 + \sum_{\alpha=2}^{m} v_{\alpha\alpha}^2$$

$$\geq v_{11}^2 + 2\sum_{\alpha=2}^{m} v_{1\alpha}^2 + \frac{\left(\sum_{\alpha=2}^{m} v_{\alpha\alpha}\right)^2}{m-1}$$

$$= v_{11}^2 + 2\sum_{\alpha=2}^{m} v_{1\alpha}^2 + \frac{(\Delta v - v_{11})^2}{m-1}$$

$$= v_{11}^2 + 2\sum_{\alpha=2}^{m} v_{1\alpha}^2 + \frac{1}{m-1}\left(|\nabla v|^2 + \frac{\lambda u}{a+u} + v_{11}\right)^2$$

$$\geq \frac{m}{m-1}\sum_{j=1}^{m} v_{1j}^2 + \frac{1}{m-1}\left(|\nabla v|^2 + \frac{\lambda u}{a+u}\right)^2$$

$$+ \frac{2v_{11}}{m-1}\left(|\nabla v|^2 + \frac{\lambda u}{a+u}\right). \tag{5.14}$$

However, using the identity

$$2v_1 v_{11} = e_1(|\nabla v|^2)$$

$$= \langle \nabla|\nabla v|^2, \nabla v\rangle |\nabla v|^{-1},$$

we conclude that

$$2v_{11} = |\nabla v|^{-2} \langle \nabla|\nabla v|^2, \nabla v\rangle.$$

Substituting the above identity into (5.14), we obtain

$$v_{ij}^2 \geq \frac{m}{m-1}\sum_{j=1}^{m} v_{1j}^2 + \frac{1}{m-1}\left(|\nabla v|^2 + \frac{\lambda u}{a+u}\right)^2$$

$$+ \frac{1}{m-1}\langle \nabla|\nabla v|^2, \nabla v\rangle |\nabla v|^{-2}\left(|\nabla v|^2 + \frac{\lambda u}{a+u}\right).$$

Combining the above inequality with (5.12) and (5.13) yields

$$\Delta Q \geq \frac{m}{2(m-1)}|\nabla Q|^2 Q^{-1} - 2(m-1)R\, Q$$

$$+ \langle \nabla v, \nabla Q\rangle Q^{-1}\left(\frac{2\lambda u}{(m-1)(a+u)} - \frac{2(m-2)}{m-1}Q\right) + \frac{2}{m-1}Q^2$$

$$+ \frac{4}{m-1}\frac{\lambda u}{a+u}Q + \frac{2}{m-1}\left(\frac{\lambda u}{a+u}\right)^2 - \frac{2\lambda a}{a+u}Q. \tag{5.15}$$

$\square$

Theorem 5.7 (Li–Yau [LY1]) *Let M be a compact m-dimensional Riemannian manifold without boundary. Suppose that the Ricci curvature of M is bounded from below by*

$$\mathcal{R}_{ij} \geq -(m-1)R$$

for some constant $R \geq 0$, and d denotes the diameter of M. Then there exist constants $C_1(m)$, $C_2(m) > 0$ depending on m alone, such that the first nonzero eigenvalue of M satisfies

$$\lambda_1 \geq \frac{C_1}{d^2}\exp(-C_2 d\sqrt{R}).$$

Proof Let u be a nonconstant eigenfunction satisfying

$$\Delta u = -\lambda u.$$

By the fact that

$$-\lambda \int_M u = \int_M \Delta u = 0,$$

u must change sign. Hence we may normalize u to satisfy $\min u = -1$ and $\max u \leq 1$. Let us consider the function

$$v = \log(a+u)$$

for some constant $a > 1$.

If $x_0 \in M$ is a point where the function $Q = |\nabla v|^2$ achieves its maximum, then Lemma 5.6 and the maximum principle assert that

$$0 \geq \frac{2}{m-1} Q^2 + \left(\frac{4}{m-1} \frac{\lambda u}{a+u} - \frac{2\lambda a}{a+u} - 2(m-1)R \right) Q$$

$$+ \frac{2}{m-1} \left(\frac{\lambda u}{a+u} \right)^2$$

$$\geq \frac{2}{m-1} Q^2 + \left(\frac{4\lambda}{m-1} - \frac{2(m+1)}{m-1} \frac{\lambda a}{a+u} - 2(m-1)R \right) Q,$$

which implies that

$$Q(x) \leq Q(x_0)$$

$$\leq (m-1)^2 R + \frac{(m+1)a\lambda}{(a-1)}$$

for all $x \in M$. Integrating $Q^{1/2} = |\nabla \log(a+u)|$ along a minimal geodesic γ joining the points at which $u = -1$ and $u = \max u$, we have

$$\log \left(\frac{a}{a-1} \right) \leq \log \left(\frac{a + \max u}{a-1} \right)$$

$$\leq \int_\gamma |\nabla \log(a+u)|$$

$$\leq d \sqrt{ \frac{(m+1)a\lambda}{a-1} + (m-1)^2 R }$$

for all $a > 1$. Setting $t = (a-1)/a$, we have

$$(m+1)\lambda \geq t \left(\frac{1}{d^2} \left(\log \frac{1}{t} \right)^2 - (m-1)^2 R \right)$$

for all $0 < t < 1$. Maximizing the right-hand side as a function of t by setting

$$t = \exp(-1 - \sqrt{1 + (m-1)^2 R d^2}),$$

we obtain the estimate

$$\lambda \geq \frac{2}{(m+1)d^2} (1 + \sqrt{1 + (m-1)^2 R d^2}) \exp(-1 - \sqrt{1 + (m-1)^2 R d^2})$$

$$\geq \frac{C_1}{d^2} \exp(-C_2 d \sqrt{R})$$

as claimed. $\square$

5 *Poincaré inequality and the first eigenvalue*   53

We would like to point out that when M is a compact manifold with boundary, there are corresponding estimates for the first Dirichlet eigenvalue and the first nonzero Neumann eigenvalue using the maximum principle. In fact, Reilly [R] proved the Lichnerowicz–Obata result for the Dirichlet eigenvalue on manifolds whose boundary has nonnegative mean curvature with respect to the outward normal vector. In 1990, Escobar [E] established the Lichnerowicz–Obata result for the first nonzero Neumann eigenvalue on manifolds whose boundary is convex with respect to the second fundamental form. There are estimates analogous to that of Theorem 5.7 for both the Dirichlet and Neumann eigenvalues on manifolds with boundary. In general, the estimate for the Dirichlet eigenvalue [LY1] depends also on the lower bound of the mean curvature of the boundary with respect to the outward normal, and the estimate for the Neumann eigenvalue depends also on the lower bound of the second fundamental form of the boundary and the ϵ-ball condition (see [Cn]). However, when the boundary is convex, the Neumann eigenvalue has an estimate similar to manifolds without boundary.

Corollary 5.8 (Li–Yau [LY1]) *Let M be a compact m-dimensional Riemannian manifold whose boundary is convex in the sense that the second fundamental form is nonnegative with respect to the outward pointing normal vector. Suppose that the Ricci curvature of M is bounded from below by*

$$\mathcal{R}_{ij} \geq -(m-1)R$$

for some constant $R \geq 0$, and d denotes the diameter of M. Then there exist constants $C_1(m), C_2(m) > 0$ depending on m alone, such that the first nonzero Neumann eigenvalue of M satisfies

$$\lambda_1 \geq \frac{C_1}{d^2} \exp(-C_2 d \sqrt{R}).$$

Proof In view of the proof of Theorem 5.7, it suffices to show that the maximum value for the functional Q does not occur on the boundary of M. Supposing the contrary that the maximum point for Q is $x_0 \in \partial M$, let us denote the outward pointing unit normal vector by e_m, and assume that $\{e_1, \ldots, e_{m-1}\}$ are orthonormal tangent vectors to ∂M. Since Q satisfies the differential inequality (5.15), the strong maximum principle implies that

$$e_m(Q)(x_0) > 0.$$

Using the Neumann boundary condition on u, we conclude that $e_m v = 0$. Moreover, since the second covariant derivative of v is defined by

$$v_{ji} = \left(e_i e_j - \nabla_{e_i} e_j\right) v,$$

we have

$$e_m(Q) = 2\sum_{i=1}^{m}(e_i v)(e_m e_i v)$$

$$= 2\sum_{\alpha=1}^{m-1}(e_\alpha v)(v_{\alpha m} + \nabla_{e_m} e_\alpha v)$$

$$= 2\sum_{\alpha=1}^{m-1}(e_\alpha v)\left(v_{m\alpha} + \nabla_{e_m} e_\alpha v\right)$$

$$= 2\sum_{\alpha=1}^{m-1}(e_\alpha v)\left(e_\alpha e_m v - \nabla_{e_\alpha} e_m v + \nabla_{e_m} e_\alpha v\right).$$

Using $e_m v = 0$ again and the fact that the second fundamental form is defined by

$$h_{\alpha\beta} = \langle \nabla_{e_\alpha} e_m, e_\beta \rangle,$$

we have

$$e_m(Q) = -2\sum_{\alpha,\beta=1}^{m-1}(e_\alpha v)h_{\alpha\beta}(e_\beta v) + 2\sum_{\alpha,\beta=1}^{m-1}(e_\alpha v)\langle \nabla_{e_m} e_\alpha, e_\beta \rangle(e_\beta v)$$

$$\leq 2\sum_{\alpha,\beta=1}^{m-1}(e_\alpha v)\langle \nabla_{e_m} e_\alpha, e_\beta \rangle(e_\beta v).$$

On the other hand, since

$$\langle \nabla_{e_m} e_\alpha, e_\beta \rangle = -\langle \nabla_{e_m} e_\beta, e_\alpha \rangle,$$

we conclude that

$$2\sum_{\alpha,\beta=1}^{m-1}(e_\alpha v)\langle \nabla_{e_m} e_\alpha, e_\beta \rangle(e_\beta v) = -2\sum_{\alpha,\beta=1}^{m-1}(e_\alpha v)\langle \nabla_{e_m} e_\beta, e_\alpha \rangle(e_\beta v).$$

Hence

$$e_m(Q) \leq 0,$$

which is a contradiction. $\square$

5 Poincaré inequality and the first eigenvalue

When M is a complete manifold, it is often useful to have a lower bound of the first eigenvalue for the Dirichlet Laplacian on a geodesic ball. This is provided by the next theorem.

Theorem 5.9 (Li–Schoen [LS]) *Let M be a complete manifold of dimension m. Let $p \in M$ be a fixed point such that $B_p(2\rho) \cap \partial M = \emptyset$ for $2\rho \leq d$. Assume that the Ricci curvature on $B_p(2\rho)$ satisfies*

$$\mathcal{R}_{ij} \geq -(m-1)R$$

for some constant $R \geq 0$. For any $\alpha \geq 1$, there exist constants $C_1(\alpha)$, $C_2(m, \alpha) > 0$, such that for any compactly supported function f on $B_p(\rho)$

$$\int_{B_p(\rho)} |\nabla f|^\alpha \geq C_1 \rho^{-\alpha} \exp(-C_2(1 + \rho\sqrt{R})) \int_{B_p(\rho)} |f|^\alpha.$$

In particular, the first Dirichlet eigenvalue of $B_p(\rho)$ satisfies

$$\mu_1 \geq C_1 \rho^{-2} \exp(-C_2(1 + \rho\sqrt{R})).$$

Proof Let $q \in \partial B_p(2\rho)$. By the triangle inequality $B_p(\rho) \subset (B_q(3\rho) \setminus B_q(\rho))$. Theorem 4.1 implies that

$$\Delta r \leq (m-1)\sqrt{R}\coth(r\sqrt{R})$$

$$\leq (m-1)(r^{-1} + \sqrt{R})$$

for $r(x) = r(q, x)$. For $k > m - 2$, we have

$$\Delta r^{-k} = -kr^{-k-1}\Delta r + k(k+1)r^{-k-2}$$

$$\geq -k(m-1)r^{-k-1}(r^{-1} + \sqrt{R}) + k(k+1)r^{-k-2}$$

$$= kr^{-k-1}((k+2-m)r^{-1} - (m-1)\sqrt{R})$$

$$\geq kr^{-k-1}((k+2-m)(3\rho)^{-1} - (m-1)\sqrt{R})$$

on $B_p(\rho)$. Choosing $k = m - 1 + 3(m-1)\rho\sqrt{R}$ this becomes

$$\Delta r^{-k} \geq kr^{-k-1}(3\rho)^{-1}$$

$$\geq k(3\rho)^{-k-2} \qquad (5.16)$$

on $B_p(\rho)$.

Let f be a nonnegative function supported on $B_p(\rho)$. Multiplying (5.16) with f and integrating over $B_p(\rho)$ yields

$$k(3\rho)^{-k-2} \int_{B_p(\rho)} f \leq \int_{B_p(\rho)} f \Delta r^{-k}$$

$$= -\int_{B_p(\rho)} \langle \nabla f, \nabla r^{-k} \rangle$$

$$\leq k \int_{B_p(\rho)} |\nabla f| r^{-k-1}$$

$$\leq k\rho^{-k-1} \int_{B_p(\rho)} |\nabla f|,$$

implying

$$\int_{B_p(\rho)} |\nabla f| \geq C_1 \rho^{-1} \exp(-C_2(1+\rho\sqrt{R})) \int_{B_p(\rho)} f.$$

The case when $\alpha = 1$ is shown by simply applying the above inequality to $|f|$. For $\alpha > 1$, we set $f = |g|^\alpha$. Then we have

$$\alpha \left(\int_{B_p(\rho)} |\nabla g|^\alpha \right)^{1/\alpha} \left(\int_{B_p(\rho)} |g|^\alpha \right)^{(\alpha-1)/\alpha}$$

$$\geq \alpha \int_{B_p(\rho)} |g|^{\alpha-1} |\nabla g| = \int_{B_p(\rho)} |\nabla g^\alpha|$$

$$\geq C_1 \rho^{-1} \exp(-C_2(1+\rho\sqrt{R})) \int_{B_p(\rho)} |g|^\alpha,$$

which implies the desired inequality. $\square$

6
Gradient estimate and Harnack inequality

In this chapter we discuss an important estimate that is essential to the study of harmonic functions as well as many elliptic and parabolic problems. In 1975, Yau [Y1] developed a maximum principle method to prove that complete manifolds with nonnegative Ricci curvature must have a Liouville property. His argument was later localized in his paper with Cheng [CgY] and resulted in a gradient estimate for a rather general class of elliptic equations. In 1979, the maximum principle method was used by Li [L1] in proving eigenvalue estimates for compact manifolds. This method (presented in Chapter 5) was then refined and used by many authors ([LY1], [ZY], etc.) for obtaining sharp eigenvalue estimates. In 1986, Li and Yau [LY2] used a similar philosophy to prove a parabolic version of the gradient estimate for the parabolic Schrödinger equation. This method has since been used by Hamilton, Chow, Cao, and many other authors to yield estimates for various nonlinear parabolic equations. In [LW6] Li and Wang realized that Yau's gradient estimate is sharp and equality is achieved on a manifold with negative Ricci curvature. The sharpness of this was somewhat surprising since the parabolic gradient estimate is sharp on a manifold with nonnegative Ricci curvature.

In the following treatment, we will present a proof for the sharp gradient estimate for positive functions satisfying the equation

$$\Delta f = -\lambda f,$$

where $\lambda \geq 0$ is a constant. The argument will give both the local and global estimates. Various immediate consequences of the gradient estimate will also be derived.

Theorem 6.1 Let M^m be a complete Riemannian manifold of dimension m. Assume that the geodesic ball $B_p(2\rho) \cap \partial M = \emptyset$. Suppose that the Ricci curvature on $B_p(2\rho)$ is bounded from below by

$$\mathcal{R}_{ij} \geq -(m-1)R$$

for some constant $R \geq 0$. If u is a positive function defined on $B_p(2\rho) \subset M$ satisfying

$$\Delta u = -\lambda u$$

for some constant $\lambda \geq 0$, then there exists a constant C depending on m such that

$$\frac{|\nabla u|^2}{u^2}(x) \leq \frac{(4(m-1)^2 + 2\epsilon)R}{4 - 2\epsilon} + C((1 + \epsilon^{-1})\rho^{-2} + \lambda)$$

for all $x \in B_p(\rho)$ and for any $\epsilon < 2$.

Moreover, if $\partial M = \emptyset$ with $\mathcal{R}_{ij} \geq -(m-1)R$ everywhere and u is defined on M, then

$$\frac{|\nabla u|^2}{u^2}(x) \leq \frac{(m-1)^2 R}{2} - \lambda + \sqrt{\frac{(m-1)^4 R^2}{4} - (m-1)^2 \lambda R}$$

and

$$\lambda \leq \frac{(m-1)^2 R}{4}.$$

Proof If we set $v = \log u$ and $Q = |\nabla v|^2$, then Lemma 5.6 asserts that

$$\Delta Q - \frac{m}{2(m-1)} |\nabla Q|^2 Q^{-1} + \langle \nabla v, \nabla Q \rangle Q^{-1} \left(\frac{2(m-2)}{m-1} Q - \frac{2\lambda}{m-1} \right)$$

$$\geq \frac{2}{m-1} Q^2 + \left(\frac{4\lambda}{m-1} - 2(m-1)R \right) Q + \frac{2\lambda^2}{m-1}.$$

6 Gradient estimate and Harnack inequality

Let ϕ be a nonnegative cutoff function and $G = \phi Q$, then we have

$$\Delta G = (\Delta \phi) Q + 2 \langle \nabla \phi, \nabla Q \rangle + \phi \Delta Q$$

$$\geq \frac{\Delta \phi}{\phi} G + 2\phi^{-1} \langle \nabla \phi, \nabla G \rangle - 2|\nabla \phi|^2 \phi^{-2} G + \frac{m}{2(m-1)} |\nabla G|^2 G^{-1}$$

$$+ \frac{m}{2(m-1)} |\nabla \phi|^2 \phi^{-2} G$$

$$- \frac{m}{(m-1)} \phi^{-1} \langle \nabla \phi, \nabla G \rangle - \langle \nabla v, \nabla G \rangle Q^{-1} \left(\frac{2(m-2)}{m-1} Q - \frac{2\lambda}{m-1} \right)$$

$$+ \langle \nabla v, \nabla \phi \rangle \left(\frac{2(m-2)}{m-1} Q - \frac{2\lambda}{m-1} \right)$$

$$+ \frac{2}{m-1} \phi^{-1} G^2 + \left(\frac{4\lambda}{m-1} - 2(m-1)R \right) G + \frac{2\lambda^2}{m-1} \phi. \qquad (6.1)$$

Using the inequality

$$|\langle \nabla v, \nabla \phi \rangle| \leq |\nabla \phi| \phi^{-\frac{1}{2}} G^{\frac{1}{2}},$$

(6.1) can be written as

$$\Delta G \geq \frac{\Delta \phi}{\phi} G + \frac{m-2}{m-1} \phi^{-1} \langle \nabla \phi, \nabla G \rangle - \frac{3m-4}{2(m-1)} |\nabla \phi|^2 \phi^{-2} G$$

$$+ \frac{m}{2(m-1)} |\nabla G|^2 G^{-1} + \left(\frac{4\lambda}{m-1} - 2(m-1)R \right) G$$

$$- \frac{2(m-2)}{m-1} \langle \nabla v, \nabla G \rangle$$

$$- \frac{2(m-2)}{m-1} |\nabla \phi| \phi^{-3/2} G^{3/2} + \frac{2\lambda}{(m-1)Q} \langle \nabla v, \nabla G \rangle$$

$$- \frac{2\lambda}{m-1} |\nabla \phi| \phi^{-1/2} G^{1/2}$$

$$+ \frac{2}{m-1} \phi^{-1} G^2 + \frac{2\lambda^2}{m-1} \phi.$$

However, at the maximum point x_0 of G, the maximum principle asserts that

$$\Delta G(x_0) \leq 0$$

and

$$\nabla G(x_0) = 0.$$

Hence at x_0, we have

$$0 \geq (m-1)(\Delta\phi)G - \frac{3m-4}{2}|\nabla\phi|^2 \phi^{-1} G + \left(4\lambda - 2(m-1)^2 R\right)\phi G$$
$$- 2(m-2)|\nabla\phi|\phi^{-1/2} G^{3/2} - 2\lambda|\nabla\phi|\phi^{1/2} G^{1/2} + 2G^2 + 2\lambda^2 \phi^2. \tag{6.2}$$

Let us choose $\phi(x) = \phi(r(x))$ to be a function of the distance r to the fixed point p with the property that

$$\phi = 1 \quad \text{on} \quad B_p(\rho),$$

$$\phi = 0 \quad \text{on} \quad M \setminus B_p(2\rho),$$

$$-C\rho^{-1}\phi^{\frac{1}{2}} \leq \phi' \leq 0 \quad \text{on} \quad B_p(2\rho) \setminus B_p(\rho),$$

and

$$|\phi''| \leq C\rho^{-2} \quad \text{on} \quad B_p(2\rho) \setminus B_p(\rho).$$

Then the Laplacian comparison theorem asserts that

$$\Delta\phi = \phi' \Delta r + \phi''$$
$$\geq -C_1(\rho^{-1}\sqrt{R} + \rho^{-2}),$$

and also

$$|\nabla\phi|^2 \phi^{-1} \leq C_2 \rho^{-2}.$$

Hence (6.2) yields

$$0 \geq -(C_3 \rho^{-1}\sqrt{R} + C_4 \rho^{-2})G + (4\lambda\phi - 2(m-1)^2 R\phi)G$$
$$- C_5 \rho^{-1} G^{\frac{3}{2}} - C_6 \lambda \rho^{-1} G^{\frac{1}{2}} + 2G^2 + 2\lambda^2 \phi^2. \tag{6.3}$$

We observe that since x_0 is the maximum point of G and $\phi = 1$ on $B_p(\rho)$,

$$\phi(x_0)|\nabla v|^2(x_0) \geq \sup_{B_p(\rho)} |\nabla v|^2(x).$$

Using the fact that

$$\phi(x_0)|\nabla v|^2(x_0) \leq \phi(x_0) \sup_{B_p(2\rho)} |\nabla v|^2(x),$$

we conclude that

$$\sigma(\rho) \leq \phi(x_0) \leq 1,$$

6 Gradient estimate and Harnack inequality

where $\sigma(\rho)$ denotes

$$\sigma(\rho) = \frac{\sup_{B_p(\rho)} |\nabla v|^2(x)}{\sup_{B_p(2\rho)} |\nabla v|^2(x)}.$$

Applying this estimate to (6.3), we have

$$0 \geq -(C_3 \rho^{-1} \sqrt{R} + C_4 \rho^{-2} - 4\lambda\sigma(\rho) + 2(m-1)^2 R) G$$
$$- C_5 \rho^{-1} G^{3/2} - C_6 \lambda \rho^{-1} G^{1/2} + 2G^2 + 2\lambda^2 \sigma^2(\rho). \quad (6.4)$$

On the other hand, Schwarz inequality asserts that

$$-C_5 \rho^{-1} G^{3/2} \geq -\epsilon\, G^2 - \frac{C_5^2}{4} \epsilon^{-1} \rho^{-2} G,$$

$$-C_6 \lambda \rho^{-1} G^{1/2} \geq -\epsilon\, \lambda^2 - \frac{C_6^2}{4} \epsilon^{-1} \rho^{-2} G,$$

and

$$-C_3 \rho^{-1} \sqrt{R} \geq -\epsilon\, R - C_7 \epsilon^{-1} \rho^{-2}$$

for $\epsilon > 0$. Hence combining this with (6.4) we obtain

$$0 \geq -(C_8(1+\epsilon^{-1})\rho^{-2} - 4\lambda\sigma(\rho) + (2(m-1)^2 + \epsilon)R) G$$
$$+ (2-\epsilon)G^2 + 2\lambda^2 \sigma^2(\rho) - \epsilon\lambda^2.$$

In particular, if $\epsilon < 2$, then for any $x \in B_p(\rho)$ we conclude that

$$|\nabla v|^2(x) \leq G(x_0)$$
$$\leq \frac{B + \sqrt{B^2 - 4(2-\epsilon)\lambda^2(2\sigma^2(\rho) - \epsilon)}}{4 - 2\epsilon}, \quad (6.5)$$

where

$$B = C_8(1+\epsilon^{-1})\rho^{-2} - 4\lambda\sigma(\rho) + (2(m-1)^2 + \epsilon)R.$$

Since $0 \leq \sigma(\rho) \leq 1$, we conclude that

$$\frac{|\nabla u|^2}{u^2}(x) \leq \frac{(4(m-1)^2 + 2\epsilon)R}{4 - 2\epsilon} + C((1+\epsilon^{-1})\rho^{-2} + \lambda)$$

for all $x \in B_p(\rho)$ and for all $\epsilon < 2$.

If u is defined on M, then it follows that $|\nabla v|^2$ is a bounded function. In particular, if we take $\rho \to \infty$ in (6.5), then

$$\sigma(\rho) \to 1$$

and (6.5) becomes

$$|\nabla v|^2 \leq \frac{(2(m-1)^2 + \epsilon)R - 4\lambda}{4 - 2\epsilon}$$
$$+ \frac{\sqrt{((2(m-1)^2 + \epsilon)R - 4\lambda)^2 - 4(2-\epsilon)^2\lambda^2}}{4 - 2\epsilon}.$$

Letting $\epsilon \to 0$, we obtain

$$|\nabla v|^2 \leq \frac{(m-1)^2 R}{2} - \lambda + \sqrt{\frac{(m-1)^4 R^2}{4} - \lambda(m-1)^2 R}.$$

This gives a contradiction if $\lambda > (m-1)^2 R/4$, and the second assertion follows. □

The following Harnack type inequality is a direct consequence of the gradient estimate.

Corollary 6.2 *Let M^m be a complete manifold with Ricci curvature bounded from below by*

$$\mathcal{R}_{ij} \geq -(m-1)R$$

for some constant $R \geq 0$. If u is a positive function defined on the geodesic ball $B_p(2\rho) \subset M$ satisfying

$$\Delta u = -\lambda u$$

for some constant $\lambda \geq 0$, then there exists constants C_9, $C_{10} > 0$ depending on m such that

$$u(x) \leq u(y) C_9 \exp(C_{10} \rho \sqrt{R + \lambda})$$

for all $x, y \in B_p(\rho/2)$.

Proof Let γ be the shortest curve in $B_p(\rho)$ joining y to x, and clearly the length of γ is at most 2ρ. Integrating the quantity $|\nabla \log u|$ along γ yields

$$\log u(x) - \log u(y) \leq \int_\gamma |\nabla \log u|. \tag{6.6}$$

6 Gradient estimate and Harnack inequality

On the other hand, applying the gradient estimate of Theorem 6.1, we obtain

$$\int_\gamma |\nabla \log u| \leq \int_\gamma \left(\frac{(4(m-1)^2 + 2\epsilon)R}{4 - 2\epsilon} + C((1 + \epsilon^{-1})\rho^{-2} + \lambda) \right)^{1/2}$$

$$\leq \int_\gamma \left(C_{10}\sqrt{R + \lambda} + C_{12}\rho^{-1} \right)$$

$$\leq C_{10}\, \rho\, \sqrt{R + \lambda} + 2C_{12}.$$

The corollary follows by combining this inequality with (6.6). $\square$

An upper bound for the infimum of the spectrum that is a consequence of the comparison theorem of Cheng (Theorem 4.6) can also be recovered using the above estimate.

Definition 6.3 Let M^m be a complete Riemannian manifold without boundary. We denote the greatest lower bound for the L^2-spectrum of the Laplacian Δ by $\mu_1(M)$.

The notation $\mu_1(M)$ does not necessarily mean that it is an eigenvalue of Δ, but is motivated by the characterization of $\mu_1(M)$ by

$$\mu_1(M) = \lim_{i \to \infty} \mu_1(D_i)$$

for any compact exhaustion $\{D_i\}$ of M.

Corollary 6.4 (Cheng [Cg1]) *Let M^m be a complete Riemannian manifold without boundary. Suppose the Ricci curvature of M is bounded from below by*

$$\mathcal{R}_{ij} \geq -(m-1),$$

then the greatest lower bound of the L^2-spectrum of the Laplacian has an upper bound given by

$$\mu_1(M) \leq \frac{(m-1)^2}{4}.$$

To prove Cheng's theorem, we need the following lemma, which was first proved by Fischer-Colbrie and Schoen [FCS] for general operators of the form $L = \Delta + V(x)$, where $V(x)$ is a smooth potential function. We will provide the proof for the case V is a constant λ as an application of Theorem 6.1. However, it is important to point out that a Harnack inequality similar to Theorem 6.1 is valid in a more general setting of $L = \Delta + V(x)$ (see [CgY]), hence the argument provided below will yield the result of Fischer-Colbrie and Schoen.

We would also like to point out that using the Nash–Moser Harnack inequality of Chapter 19 the potential function V is only required to be in some L^p space.

Lemma 6.5 (Fischer-Colbrie–Schoen) *Let M^m be a complete Riemannian manifold, and $\mu_1(M)$ be the greatest lower bound of Δ. Then the value μ satisfies $\mu \leq \mu_1$ if and only if the operator $L = \Delta + \mu$ admits a positive solution of the equation*

$$Lu = 0$$

on M.

Proof Let us first assume that there exists a positive function u defined on M such that

$$Lu = 0.$$

For any smooth, compact, subdomain $D \subset M$, let f be the first Dirichlet eigenfunction for the operator $L = \Delta + \mu$, satisfying

$$Lf = -\mu_1(D, L) f \quad \text{on} \quad D,$$

with boundary condition

$$f = 0 \quad \text{on} \quad \partial D.$$

By multiplication by -1 if necessary, we may assume that f is positive in the interior of D. In particular,

$$-\mu_1(D, L) \int_D u f = \int_D u\, Lf - \int_D f\, Lu$$
$$= \int_{\partial D} u\, f_\nu.$$

The positivity of f on D and its boundary condition assert that $f_\nu < 0$, hence together with the positivity of u, we conclude that $\mu_1(D, L) > 0$. Since D is arbitrary, this implies that $\mu_1(M, L) \geq 0$, hence $\mu_1(M) \geq \mu$.

Conversely, let us assume that $\mu_1(M) \geq \mu$ and $\mu_1(M, L) \geq 0$. In particular, by monotonicity $\mu_1(D, L) \geq 0$ for any compact subdomain $D \subset M$. Let us consider a compact exhaustion of M by a sequence of subdomains $\{D_i\}$ with $D_i \subset D_{i+1}$ and $\cup_{i=1}^{\infty} D_i = M$. On each D_i, since $\mu_1(D_i, L) > 0$, one can find a positive solution to the Dirichlet problem

$$Lf_i = 0 \quad \text{on} \quad D_i,$$

6 Gradient estimate and Harnack inequality

with boundary condition

$$f_i = 1 \quad \text{on} \quad \partial D_i.$$

Note that f_i must be nonnegative on D_i since the subdomain defined by $\Omega = \{f_i < 0\}$ will satisfy $\mu_1(\Omega, L) > 0$, hence contradicting the fact that f_i is an eigenfunction with eigenvalue 0 on Ω. We claim that f_i must be positive in the interior of D_i. If not, then there exists $x \in D_i$ with $f_i(x) = 0$ which is a local minimum and hence a critical point of f_i. On the other hand, a regularity result (see [Cg2]) asserts that at x the function f_i is asymptotically a spherical harmonic in $\mathbb{R}^m$, contradicting the fact that f_i has a local minimum at x.

For a fixed point $p \in M$, after multiplication by a constant, we may assume that $f_i(p) = 1$. Of course, f_i still satisfies $Lf_i = 0$ but with a different boundary condition. For any fixed $\rho > 0$, Corollary 6.2 asserts that $0 \le f_i \le C$ on $B_p(2\rho)$, for some constant $C > 0$ independent of i. Hence a subsequence of $\{f_i\}$ converges uniformly on $B_p(2\rho)$. We also consider the nonnegative cutoff function ϕ supported on $B_p(2\rho)$ with the properties that

$$\phi = 1 \quad \text{on} \quad B_p(\rho),$$

$$\phi = 0 \quad \text{on} \quad M \setminus B_p(2\rho),$$

and

$$|\nabla \phi|^2 \le C\rho^{-2} \quad \text{on} \quad B_p(2\rho) \setminus B_p(\rho)$$

for some fixed constant $C > 0$ independent of ρ. For sufficiently large i, since $B_p(2\rho) \subset D_i$ we have

$$0 = \int_M \phi^2 f_i L f_i$$

$$= -\int_M \phi^2 |\nabla f_i|^2 - 2 \int_M \phi f_i \langle \nabla \phi, \nabla f_i \rangle + \mu \int_M \phi^2 f_i^2$$

$$\le -\frac{1}{2} \int_M \phi^2 |\nabla f_i|^2 + 2 \int_M |\nabla \phi|^2 f_i^2 + \mu \int_M \phi^2 f_i^2.$$

This implies

$$\int_{B_p(\rho)} |\nabla f_i|^2 \le \int_M \phi^2 |\nabla f_i|^2$$

$$\le 4 \int_M |\nabla \phi|^2 f_i^2 + 2\mu \int_M \phi^2 f_i^2$$

$$\le 4C(\rho^{-2} + \mu) V_p(2\rho),$$

hence we conclude that f_i is bounded in $H^{1,2}(B_p(\rho))$. In particular, we choose a subsequence of f_i converging uniformly on compact subsets and in $H^{1,2}$ to a nonnegative function f. In particular, f is a positive weak solution and by regularity is a strong solution to $Lf = 0$. □

Obviously, Corollary 6.4 follows by combining the second part of Theorem 6.1 and Lemma 6.5.

Corollary 6.6 (Yau [Y1], Cheng [Cg3]) *Let M^m be a complete manifold with nonnegative Ricci curvature. There exists a constant $C(m) > 0$, such that for any harmonic function defined on M if we denote*

$$i(\rho) = \inf_{x \in B_p(\rho)} u(x),$$

then

$$|\nabla u|(x) \leq C \rho^{-1} (u(x) - i(2\rho))$$

for all $x \in B_p(\rho)$. In particular, M does not admit any nonconstant harmonic function satisfying the growth estimate

$$\liminf_{x \to \infty} r^{-1}(x) u(x) \geq 0.$$

Proof Observe that $u - i(2\rho)$ is a positive harmonic function defined on $B_p(2\rho)$. Applying Theorem 6.1 to this function and setting $\epsilon = 1$, we obtain

$$|\nabla u|^2(x) \leq 2C\rho^{-2} (u(x) - i(2\rho))^2$$

for all $x \in B_p(\rho)$, proving the first part of the corollary. To prove the second part, we observe that the maximum principle asserts that $i(\rho)$ is achieved at some point $x \in \partial B_p(\rho)$. In particular, $\rho^{-1} i(\rho) = r^{-1}(x) u(x)$ and the growth assumption on u implies that

$$\lim_{\rho \to \infty} \rho^{-1} i(\rho) \geq 0.$$

Hence for any fixed $x \in M$, we apply the estimate

$$|\nabla u|(x) \leq \lim_{\rho \to \infty} C \rho^{-1} (u(x) - i(2\rho))$$

$$\leq 0$$

and conclude that u must be a constant function. □

The following example demonstrates the sharpness of Theorem 6.1 even on manifolds with negative curvature.

6 Gradient estimate and Harnack inequality

Example 6.7 Let $M^m = \mathbb{H}^m$ be the hyperbolic m-space with constant curvature -1. Using the upper half-space model, $\mathbb{H}^m$ is given by $\mathbb{R}^m_+ = \{(x_1, x_2, \ldots, x_m) \,|\, x_m > 0\}$ with metric

$$ds^2 = x_m^{-2}\left(dx_1^2 + \cdots + dx_m^2\right).$$

For $x = (x_1, \ldots, x_m)$ let us consider the function $u(x) = x_m^\alpha$. Direct computation yields

$$|\nabla u|^2 = x_m^2 \sum_{i=1}^{m} \left(\frac{\partial u}{\partial x_i}\right)^2$$
$$= \alpha^2 u^2$$

and

$$\Delta u = x_m^2 \sum_{i=1}^{m} \frac{\partial^2 u}{\partial x_i^2} - (m-2)x_m \frac{\partial u}{\partial x_m}$$
$$= \alpha(\alpha + 1 - m)u.$$

In particular, for $(m-1)/2 \leq \alpha \leq m-1$, we see that $\Delta u = -\lambda u$ with $\lambda = \alpha(m-1-\alpha)$. This implies that the gradient estimate of Theorem 6.1 is sharp since

$$\frac{(m-1)^2}{2} - \lambda + \sqrt{\frac{(m-1)^4}{4} - (m-1)^2 \lambda} = \alpha^2.$$

7
Mean value inequality

We will prove a version of the mean value inequality which is adapted to the theory of subharmonic functions on a Riemannian manifold. Let us begin by proving a theorem of Yau [Y2].

Lemma 7.1 (Yau) *Let M be a complete Riemannian manifold. Suppose $p \in M$ and $\rho_4 > 0$ are such that the geodesic ball $B_p(\rho_4)$ centered at p of radius ρ_4 satisfies $B_p(\rho_4) \cap \partial M = \emptyset$. Let f be a nonnegative subharmonic function defined on $B_p(\rho_4)$. Then for any constant $\alpha > 1$ and for any $0 \le \rho_1 < \rho_2 < \rho_3 < \rho_4$, we have the estimates*

$$\int_{B_p(\rho_3) \setminus B_p(\rho_2)} f^{\alpha-2} |\nabla f|^2 \le \frac{4}{(\alpha-1)^2} \left((\rho_2 - \rho_1)^{-2} \int_{B_p(\rho_2) \setminus B_p(\rho_1)} f^\alpha \right.$$

$$\left. + (\rho_4 - \rho_3)^{-2} \int_{B_p(\rho_4) \setminus B_p(\rho_3)} f^\alpha \right)$$

and

$$\int_{B_p(\rho_3)} f^{\alpha-2} |\nabla f|^2 \le \frac{4}{(\alpha-1)^2} (\rho_4 - \rho_3)^{-2} \int_{B_p(\rho_4) \setminus B_p(\rho_3)} f^\alpha.$$

In particular, if M has no boundary, then a nonconstant, nonnegative, L^α subharmonic function does not exist.

68

7 Mean value inequality

Proof Let $\phi(r(x))$ be a cutoff function defined by

$$\phi(r) = \begin{cases} 0 & \text{for } r \leq \rho_1, \\ \dfrac{r - \rho_1}{\rho_2 - \rho_1} & \text{for } \rho_1 \leq r \leq \rho_2, \\ 1 & \text{for } \rho_2 \leq r \leq \rho_3, \\ \dfrac{\rho_4 - r}{\rho_4 - \rho_3} & \text{for } \rho_3 \leq r \leq \rho_4, \\ 0 & \text{for } \rho_4 \leq r. \end{cases}$$

We now consider the integral

$$0 \leq \int_{B_p(\rho_4)} \phi^2 f^{\alpha-1} \Delta f$$

$$= -(\alpha - 1) \int_{B_p(\rho_4)} \phi^2 f^{\alpha-2} |\nabla f|^2 - 2 \int_{B_p(\rho_4)} \phi f^{\alpha-1} \langle \nabla \phi, \nabla f \rangle.$$

After applying the algebraic inequality

$$2 \left| \int_{B_p(\rho_4)} \phi f^{\alpha-1} \langle \nabla \phi, \nabla f \rangle \right|$$

$$\leq \frac{\alpha - 1}{2} \int_{B_p(\rho_4)} \phi^2 f^{\alpha-2} |\nabla f|^2 + \frac{2}{\alpha - 1} \int_{B_p(\rho_4)} |\nabla \phi|^2 f^\alpha,$$

we have

$$\frac{\alpha - 1}{2} \int_{B_p(\rho_3) \setminus B_p(\rho_2)} f^{\alpha-2} |\nabla f|^2 \leq \frac{\alpha - 1}{2} \int_{B_p(\rho_4)} \phi^2 f^{\alpha-2} |\nabla f|^2$$

$$\leq \frac{2}{\alpha - 1} \int_{B_p(\rho_4)} |\nabla \phi|^2 f^\alpha$$

$$\leq \frac{2}{(\rho_2 - \rho_1)^2 (\alpha - 1)} \int_{B_p(\rho_2) \setminus B_p(\rho_1)} f^\alpha$$

$$+ \frac{2}{(\rho_4 - \rho_3)^2 (\alpha - 1)} \int_{B_p(\rho_4) \setminus B_p(\rho_3)} f^\alpha.$$

The second inequality follows similarly by taking ϕ to be

$$\phi(r) = \begin{cases} 1 & \text{for } r \leq \rho_3, \\ \dfrac{\rho_4 - r}{\rho_4 - \rho_3} & \text{for } \rho_3 \leq r \leq \rho_4, \\ 0 & \text{for } \rho_4 \leq r. \end{cases}$$

If M has no boundary, we simply set $\rho_3 = \rho$, $\rho_4 = 2\rho$ and let $\rho \to \infty$. The fact that f is L^α implies that

$$\int_M f^{\alpha-2} |\nabla f|^2 = 0,$$

hence $|\nabla f| \equiv 0$, and f must be a constant function. □

Theorem 7.2 (Li–Schoen [LS]) *Let M be a complete Riemannian manifold of dimension m. Let $p \in M$ be a fixed point such that the geodesic ball $B_p(4\rho)$ centered at p of radius 4ρ satisfies $B_p(4\rho) \cap \partial M = \emptyset$. Suppose f is a nonnegative subharmonic function defined on $B_p(4\rho)$. Assume that the Ricci curvature on $B_p(4\rho)$ is bounded by $\mathcal{R}_{ij} \geq -(m-1)R$ for some constant $R \geq 0$. Then there exist constants $C_3, C_4(m) > 0$ with C_4 depending only on m such that*

$$\sup_{x \in B_p(\rho)} f^2 \leq C_3 (1 + \exp(C_4 \rho \sqrt{R})) V_p(4\rho)^{-1} \int_{B_p(4\rho)} f^2.$$

Proof Let h be a harmonic function on $B_p(2\rho)$ obtained by the solving the Dirichlet boundary problem

$$\Delta h = 0 \quad \text{on} \quad B_p(2\rho)$$

and

$$h = f \quad \text{on} \quad \partial B_p(2\rho).$$

Since f is nonnegative, the maximum principle implies that h is positive on the ball $B_p(\rho)$ and

$$f \leq h \quad \text{on} \quad B_p(\rho).$$

The Harnack inequality (Corollary 6.2) implies that

$$\sup_{B_p(\rho)} h \leq \left(\inf_{B_p(\rho)} h \right) \exp(C(1 + \rho\sqrt{R})).$$

Hence, in particular, we have

$$\sup_{B_p(\rho)} f^2 \leq \sup_{B_p(\rho)} h^2$$

$$\leq \exp(2C(1 + \rho\sqrt{R})) V_p(\rho)^{-1} \int_{B_p(\rho)} h^2. \tag{7.1}$$

7 Mean value inequality

We will now estimate the L^2-norm of h in terms of the L^2-norm of f. By the triangle inequality, we observe that

$$\int_{B_p(\rho)} h^2 \leq 2 \int_{B_p(\rho)} (h-f)^2 + 2 \int_{B_p(\rho)} f^2$$

$$\leq 2 \int_{B_p(2\rho)} (h-f)^2 + 2 \int_{B_p(4\rho)} f^2. \quad (7.2)$$

However, since the function $h - f$ vanishes on $\partial B_p(2\rho)$, the Poincaré inequality (Theorem 5.9) implies that

$$\int_{B_p(2\rho)} (h-f)^2 \leq C_1 \rho^2 \exp(C_2(1+\rho\sqrt{R})) \int_{B_p(2\rho)} |\nabla(h-f)|^2 \quad (7.3)$$

for some constants $C_1 > 0$ and $C_2(m) > 0$. Using the triangle inequality again, we have

$$\int_{B_p(2\rho)} |\nabla(h-f)|^2 \leq 2 \int_{B_p(2\rho)} |\nabla h|^2 + 2 \int_{B_p(2\rho)} |\nabla f|^2.$$

The fact that a harmonic function has the least Dirichlet integral of all functions with the same boundary data asserts that

$$\int_{B_p(2\rho)} |\nabla(h-f)|^2 \leq 4 \int_{B_p(2\rho)} |\nabla f|^2.$$

Now the argument in Lemma 7.1 implies that

$$\int_{B_p(2\rho)} |\nabla f|^2 \leq C\rho^{-2} \int_{B_p(4\rho)} f^2$$

for some constant $C > 0$. Taking this together with (7.1), (7.2), (7.3), and the volume comparison (Corollary 2.4), the theorem follows. $\square$

Let us point out that the constant in the mean value inequality depends only on the lower bound of the Ricci curvature and the radius of the ball is essential in some of the geometric applications. In fact, it is well known that one can prove another version of the mean value inequality by using an iteration method of Moser. This will be presented in Chapter 18 and the difference of these two methods will also be pointed out later.

We will now give an application of this mean value inequality to the study of the space of harmonic functions on a certain class of manifolds. This result can be viewed as a generalization of Yau's Liouville theorem. Let us first prove a lemma that is useful in estimating the dimension of a linear space.

Lemma 7.3 (Li [L2]) *Let $\mathcal{H}$ be a finite dimensional space of L^2 functions defined over a set D. If $V(D)$ denotes the volume of the set D, then there exists a function f_0 in $\mathcal{H}$ such that*

$$\dim \mathcal{H} \int_D f_0^2 \leq V(D) \sup_D f_0^2.$$

Proof Let $f_1, \ldots, f_k$ be an orthonormal basis for $\mathcal{H}$ with respect to the L^2 inner product. Let us consider the function

$$F(x) = \sum_{i=1}^{k} f_i^2(x),$$

which is well defined under an orthonormal change of basis. Clearly

$$\dim \mathcal{H} = \int_D F(x).$$

Now let us consider the subspace $\mathcal{H}_p$ of $\mathcal{H}$ consisting of functions that vanish at a point $p \in D$. The space is clearly of at most codimension 1, otherwise there are f_1 and f_2 in the complement of $\mathcal{H}_p$ which are linearly independent. This implies that both $f_1(p) \neq 0$ and $f_2(p) \neq 0$. However, clearly the linearly combination

$$f_1(p) f_2 - f_2(p) f_1$$

is a function in $\mathcal{H}_p$, which is a contradiction. This implies that by a change of orthonormal bases, there exist f_0 in the orthogonal complement of $\mathcal{H}_p$ that has unit L^2-norm, and satisfies the identity

$$F(p) = f_0^2(p).$$

Hence, in particular, if we choose $p \in D$ such that F achieves its maximum, then

$$\dim \mathcal{H} = \int_D F$$
$$\leq V(D) F(p)$$
$$= V(D) f_0^2(p)$$
$$= V(D) \sup_D f_0^2.$$

This proves the lemma. □

Theorem 7.4 (Li–Tam [LT6]) *Let M be an m-dimensional complete noncompact Riemannian manifold without boundary. Suppose that the Ricci curvature of M is nonnegative on $M \setminus B_p(1)$ for some unit geodesic ball centered at $p \in M$. Let us assume that the lower bound of the Ricci curvature on $B_p(1)$ is given by*

$$\mathcal{R}_{ij} \geq -(m-1)R$$

for some constant $R \geq 0$. If we let $\mathcal{H}'(M)$ be the space of functions spanned by the set of harmonic functions f that has the property that when restricted to each unbounded component of $M \setminus D$ it is bounded either from above or from below for some compact subset $D \subset M$, then $\mathcal{H}'(M)$ is of finite dimension. Moreover, there exists a constant $C(m, R) > 0$ depending only on m and R, such that

$$\dim \mathcal{H}'(M) \leq C(m, R).$$

Proof By the definition of $\mathcal{H}'(M)$, there exists $R_0 > 1$ such that

$$f = \sum_{i=1}^{k} v_i,$$

where each v_i is bounded from one side on each end of $M \setminus B_p(R_0)$. Let E be an end of $M \setminus B_p(R_0)$. If v is a harmonic function defined on M which is positive on E and if x is a point in E with $r(p, x) \geq 2R_0$, then by applying Theorem 6.1 to the ball $B_x(r(p, x)/2)$ and using the curvature assumption, there is a constant $C > 0$ independent of v, such that

$$|\nabla v|(x) \leq Cr^{-1}(p, x)v(x). \tag{7.4}$$

Since all the v_i are bounded on one side on E, there are constants $a_1, \ldots, a_k$ and $\epsilon_i = \pm 1$ such that the harmonic functions $u_i = a_i + \epsilon_i v_i$ are positive on E. Hence, by applying (7.4) to $u_1, \ldots, u_m$, we can estimate the gradient of f by

$$|\nabla f|(x) \leq \sum_{i=1}^{k} |\nabla v_i|(x)$$

$$= \sum_{i=1}^{k} |\nabla u_i|(x)$$

$$\leq Cr^{-1}(p, x) \sum_{i=1}^{k} u_i(x). \tag{7.5}$$

Using the fact that $|\nabla f|$ is a subharmonic function on $M \setminus B_p(1)$, the maximum principle implies that for any given $\delta > 0$

$$|\nabla f|(x) - \sup_{\partial E} |\nabla f| \leq C\delta \left(\sum_{i=1}^{m} u_i(x) + 1 \right)$$

for all $x \in E$. Letting $\delta \to 0$, we conclude that

$$\sup_{E} |\nabla f| \leq \sup_{\partial E} |\nabla f|.$$

Since E is an arbitrary end of $M \setminus B_p(R_0)$, we have

$$\sup_{M \setminus B_p(R_0)} |\nabla f| \leq \sup_{\partial B_p(R_0)} |\nabla f|. \tag{7.6}$$

In fact,

$$\sup_{M \setminus B_p(1)} |\nabla f| \leq \sup_{\partial B_p(1)} |\nabla f|$$

after applying the maximum principle to the subharmonic function $|\nabla f|$ on the set $M \setminus B_p(1)$ and (7.6). In particular, this implies that

$$\sup_{M} |\nabla f| \leq \sup_{B_p(1)} |\nabla f|. \tag{7.7}$$

Let us now consider the codimension-1 subspace $\mathcal{H}'_p(M)$ of $\mathcal{H}'(M)$ defined by

$$\mathcal{H}'_p(M) = \{ f \in \mathcal{H}'(M) \mid f(p) = 0 \}.$$

For any $f \in \mathcal{H}'_p(M)$, the fundamental theorem of calculus implies that

$$\sup_{B_p(4)} f^2 \leq 16 \sup_{B_p(4)} |\nabla f|^2.$$

Together with (7.7), we have

$$\sup_{B_p(4)} f^2 \leq 16 \sup_{B_p(1)} |\nabla f|^2.$$

7 Mean value inequality

Applying the gradient estimate (Theorem 6.1) to the function $f + \sup_{B_p(2)} |f|$ yields

$$\sup_{B_p(4)} f^2 \leq 16 \sup_{B_p(1)} |\nabla f|^2$$

$$\leq C(R+1) \sup_{B_p(1)} \left(f + \sup_{B_p(2)} |f| \right)^2$$

$$\leq C(R+1) \sup_{B_p(2)} f^2.$$

However, using this and the mean value inequality (Theorem 7.2) when applied to the nonnegative subharmonic function $|f|$, we conclude that there exist constants $C_3, C_4(m) > 0$ such that

$$V_p(4) \sup_{B_p(4)} f^2 \leq C_3 \exp(C_4 \sqrt{R}) \int_{B_p(4)} f^2. \tag{7.8}$$

On the other hand, Lemma 7.3 implies that for any finite dimensional subspace $\mathcal{H}$ of $\mathcal{H}'_p(M)$, there exists a function f_0 such that

$$\dim \mathcal{H} \int_{B_p(4)} f_0^2 \leq V_p(4) \sup_{B_p(4)} f_0^2.$$

Hence applying (7.8) to f_0 yields the estimate

$$\dim \mathcal{H} \leq C_3 \exp(C_4 \sqrt{R}).$$

Since this estimate holds for any finite dimensional subspace $\mathcal{H}$, this implies that

$$\dim \mathcal{H}'_p(M) \leq C_3 \exp(C_4 \sqrt{R}).$$

Therefore,

$$\dim \mathcal{H}'(M) \leq C_3 \exp(C_4 \sqrt{R}) + 1$$

as was to be proven. $\square$

Let us remark that if M has nonnegative Ricci curvature, then (7.7) can be written as

$$\sup_M |\nabla f| \leq |\nabla f|(p).$$

However, since $|\nabla f|$ is a subharmonic function on M, the maximum principle implies that $|\nabla f|$ must be identically constant. If f is not a constant function, we can apply the Bochner formula to $|\nabla f|$ again, and conclude that ∇f is a parallel vector field. This implies that M must split and that f is a linear growth harmonic function. In particular, f cannot be a positive harmonic function, hence we recover Yau's theorem.

8
Reilly's formula and applications

In this chapter we discuss some of the applications of the integral version of Bochner's formula derived by Reilly [R]. This formula is particularly useful when studying embedded minimal surfaces and surfaces with constant mean curvature. We first point out some standard formulas concerning submanifolds in $\mathbb{R}^{m+1}$ and $\mathbb{S}^{m+1}$.

Lemma 8.1 *Let $\{x_1, \ldots, x_{m+1}\}$ be rectangular coordinates of $\mathbb{R}^{m+1}$, and let us denote the position vector by $X = (x_1, \ldots, x_{m+1})$. If M is a submanifold of $\mathbb{R}^{m+1}$ with the induced metric and if $\vec{II}$ and $\vec{H}$ denote the second fundamental form and the mean curvature vector of M, then*

$$\mathrm{Hess}_M X = -\vec{II}$$

and

$$\Delta_M X = -\vec{H},$$

where $\mathrm{Hess}_M(X)$ and $\Delta_M(X)$ are the Hessian of X and the Laplacian of X computed on M.

Proof Let us first assume that $M^m \subset N^n$ is a submanifold of an arbitrary manifold N. Let us choose an adapted orthonormal frame with $\{e_i\}_{i=1}^m$ tangential to M and $\{e_\nu\}_{\nu=m+1}^n$ normal to M. Then the Hessian of M and the Hessian of N are related by

$$(\mathrm{Hess}_N f)_{ij} = (e_i e_j - \nabla_{e_i} e_j) f$$
$$= (\mathrm{Hess}_M f)_{ij} - \sum_{\nu=m+1}^{n} \langle \nabla_{e_i} e_j, e_\nu \rangle f_\nu$$
$$= (\mathrm{Hess}_M f)_{ij} + \sum_{\nu=m+1}^{n} \langle \vec{II}_{ij}, \nu \rangle f_\nu \qquad (8.1)$$

for any $1 \le i, j \le m$ and for any function f defined on a neighborhood of M. For $N = \mathbb{R}^{m+1}$, we apply (8.1) to the position vector $X = (x_1, \ldots, x_{m+1})$ and observe that

$$\mathrm{Hess}_{\mathbb{R}^{m+1}} X = 0,$$

hence concluding that

$$(\mathrm{Hess}_M X)_{ij} = -\langle \vec{II}_{ij}, e_{m+1} \rangle X_{m+1}$$
$$= -\langle \vec{II}_{ij}, e_{m+1} \rangle e_{m+1}$$
$$= -\vec{II}_{ij},$$

and the lemma follows. □

Corollary 8.2 *A submanifold M of $\mathbb{R}^{m+1}$ is minimal if and only if the coordinate functions are harmonic. In particular, there are no compact minimal submanifolds in $\mathbb{R}^{m+1}$ other than points.*

Lemma 8.3 *Let M be an m-dimensional submanifold of the standard unit sphere $\mathbb{S}^n$, then M is minimal if and only if all the coordinate functions of $\mathbb{S}^n \subset \mathbb{R}^{n+1}$ are eigenfunctions of M satisfying*

$$\Delta_M X = -mX.$$

Proof By Lemma 8.1, and using the fact that the position vector X is also the unit normal vector on $\mathbb{S}^n$, we have

$$(\mathrm{Hess}_{\mathbb{S}^n} X)_{ij} = -\vec{II}_{ij}$$
$$= -\delta_{ij} X.$$

Applying (8.1) to the pair $M \subset \mathbb{S}^n$ yields

$$(\mathrm{Hess}_{\mathbb{S}^n} X)_{\alpha\beta} = (\mathrm{Hess}_M X)_{\alpha\beta} + (\vec{II}_M)_{\alpha\beta}$$

8 Reilly's formula and applications

for tangent vectors e_α and e_β which are tangential to M. Taking the trace of both sides gives

$$-mX = \Delta_M X + \vec{H}_M,$$

which proves the lemma. $\square$

The following integral formula was established by Reilly [R] in his proof of Aleksandrov's theorem.

Theorem 8.4 (Reilly) *Let D be a manifold of dimension $m+1$ whose boundary is given by a smooth m-dimensional manifold M. Suppose f is a function defined on D satisfying*

$$\Delta f = g \quad \text{on} \quad D$$

and

$$f = u \quad \text{on} \quad M,$$

then

$$\frac{m}{m+1}\int_D g^2 \geq \int_M H\, f_\nu^2 + 2\int_M f_\nu \Delta_M u$$
$$+ \int_M \sum_{\alpha,\beta=1}^{m} h_{\alpha\beta} u_\alpha u_\beta + \int_D \mathcal{R}_{ij} f_i f_j,$$

where H and $h_{\alpha\beta}$ denote the mean curvature and the second fundamental form of M with respect to the outward unit normal ν, Δ_M is the Laplacian on M, and $\mathcal{R}_{ij}$ is the Ricci curvature of D. Moreover, equality holds if and only if

$$f_{ij} = \frac{g\, \delta_{ij}}{m+1}$$

on D.

Proof Let us consider the Bochner formula

$$\tfrac{1}{2}\Delta|\nabla f|^2 = f_{ij}^2 + f_i f_{ijj}$$
$$= f_{ij}^2 + f_i(\Delta f)_i + \mathcal{R}_{ij} f_i f_j$$
$$= f_{ij}^2 + \langle \nabla f, \nabla g\rangle + \mathcal{R}_{ij} f_i f_j.$$

Using the inequality

$$\sum_{i,j=1}^{m+1} f_{ij}^2 \geq \frac{\left(\sum_{i=1}^{m+1} f_{ii}\right)^2}{m+1}$$

$$= \frac{g^2}{m+1},$$

we have

$$\tfrac{1}{2}\Delta|\nabla f|^2 \geq \frac{g^2}{m+1} + \langle \nabla f, \nabla g\rangle + \mathcal{R}_{ij} f_i f_j.$$

Integrating this over D yields

$$\tfrac{1}{2}\int_D \Delta|\nabla f|^2 \geq \frac{1}{m+1}\int_D g^2 + \int_D \langle \nabla f, \nabla g\rangle + \int_D \mathcal{R}_{ij} f_i f_j. \quad (8.2)$$

On the other hand, after integration by parts, the second term on the right-hand side becomes

$$\int_D \langle \nabla f, \nabla g\rangle = -\int_D g^2 + \int_M g f_\nu,$$

where ν is the outward unit normal to M, hence (8.2) becomes

$$\tfrac{1}{2}\int_D \Delta|\nabla f|^2 \geq -\frac{m}{m+1}\int_D g^2 + \int_M g f_\nu + \int_D \mathcal{R}_{ij} f_i f_j. \quad (8.3)$$

If we pick orthonormal frame $\{e_1, \ldots, e_{m+1}\}$ near the boundary of D such that $\{e_1, \ldots, e_m\}$ are tangential to M, and $\nu = e_{m+1}$ is the outward unit normal vector, then the divergence theorem implies that

$$\tfrac{1}{2}\int_D \Delta|\nabla f|^2 = \int_M \sum_{i=1}^{m+1} (e_i f)(e_{m+1} e_i f).$$

Using the boundary data of f, and choosing $\nabla_{e_{m+1}} e_{m+1} = 0$ at a point, we conclude that

$$\sum_{i=1}^{m+1}(e_i f)(e_{m+1}e_i f) = (e_{m+1}f)(e_{m+1}e_{m+1}f) + \sum_{\alpha=1}^{m}(e_\alpha f)(e_{m+1}e_\alpha f)$$

$$= (e_{m+1}f)\left(\Delta f - \sum_{\alpha=1}^{m} f_{\alpha\alpha}\right) + \sum_{\alpha=1}^{m}(e_\alpha f)(e_{m+1}e_\alpha f)$$

$$= f_\nu(g - Hf_\nu - \Delta_M u) + \sum_{\alpha=1}^{m}(e_\alpha f)(e_{m+1}e_\alpha f), \quad (8.4)$$

8 Reilly's formula and applications

where Δ_M is the Laplacian on M and H is the mean curvature of M with respect to the unit normal ν. However,

$$e_{m+1}e_\alpha f = e_\alpha e_{m+1}f + \nabla_{e_{m+1}}e_\alpha f - \nabla_{e_\alpha}e_{m+1}f$$
$$= e_\alpha e_{m+1}f + \sum_{\beta=1}^{m}\langle \nabla_{e_{m+1}}e_\alpha, e_\beta\rangle f_\beta - \sum_{\beta=1}^{m}\langle \nabla_{e_\alpha}e_{m+1}, e_\beta\rangle f_\beta, \quad (8.5)$$

because

$$\langle \nabla_{e_{m+1}}e_\alpha, e_{m+1}\rangle = -\langle e_\alpha, \nabla_{e_{m+1}}e_{m+1}\rangle$$
$$= 0$$

and

$$\langle \nabla_{e_\alpha}e_{m+1}, e_{m+1}\rangle = \tfrac{1}{2}e_\alpha|e_{m+1}|^2$$
$$= 0.$$

Using (8.4), (8.5), and the fact that

$$\langle \nabla_{e_\alpha}e_{m+1}, e_\beta\rangle = -\langle e_{m+1}, \nabla_{e_\alpha}e_\beta\rangle$$
$$= h_{\alpha\beta},$$

we can write

$$\tfrac{1}{2}\int_D \Delta|\nabla f|^2$$
$$= \int_M g f_\nu - \int_M H f_\nu^2 - \int_M f_\nu \Delta_M u + \int_M \sum_{\alpha=1}^{m}(e_\alpha f)(e_\alpha e_{m+1}f)$$
$$+ \sum_{\alpha,\beta=1}^{m}\int_M \langle \nabla_{e_{m+1}}e_\alpha, e_\beta\rangle f_\alpha f_\beta - \int_M \sum_{\alpha,\beta=1}^{m} h_{\alpha\beta}u_\alpha u_\beta. \quad (8.6)$$

On the other hand,

$$\int_M \sum_{\alpha,\beta=1}^{m}\langle \nabla_{e_{m+1}}e_\alpha, e_\beta\rangle f_\alpha f_\beta = -\int_M \sum_{\alpha,\beta=1}^{m}\langle e_\alpha, \nabla_{e_{m+1}}e_\beta\rangle f_\alpha f_\beta$$
$$= -\int_M \sum_{\alpha,\beta=1}^{m}\langle \nabla_{e_{m+1}}e_\alpha, e_\beta\rangle f_\alpha f_\beta$$

implies that both sides are identically 0. Also integrating by parts yields

$$\int_M \sum_{\alpha=1}^m (e_\alpha f)(e_\alpha e_{m+1} f) = -\int_M f_\nu \Delta_M u.$$

Combining this with (8.3) and (8.6), we have

$$-\int_M H f_\nu^2 - 2 \int_M f_\nu \Delta_M u - \int_M \sum_{\alpha,\beta=1}^m h_{\alpha\beta} u_\alpha u_\beta$$

$$\geq -\frac{m}{m+1} \int_D g^2 + \int_D \mathcal{R}_{ij} f_i f_j,$$

which was to be proved. The equality case is clear from the Above argument. □

Theorem 8.5 (Aleksandrov–Reilly) *Any compact embedded hypersurface of constant mean curvature in $\mathbb{R}^{m+1}$ is a standard sphere.*

Proof Let M^m be a compact embedded hypersurface in $\mathbb{R}^{m+1}$ with constant mean curvature H. By compactness, we claim that $H > 0$. Indeed, if $p \in M$ is the maximum point for the function $|X|^2$ on M, then the identity

$$\Delta_M |X|^2 = 2\langle \Delta_M X, X \rangle + 2|\nabla X|^2$$

and the maximum principle implies that

$$0 \geq \langle \Delta_M X, X \rangle$$
$$= -\langle \vec{H}, X \rangle$$

at the point p. On the other hand, it is clear that X must be a multiple of the unit normal vector ν at p, hence

$$0 \geq -H |X|$$

and $H \geq 0$. Since Corollary 8.2 asserts that $H \neq 0$, this confirms the claim that $H > 0$.

After scaling, we may assume that $H = m$. The assumption that M is embedded implies that M must enclose a bounded domain D in $\mathbb{R}^{m+1}$. Let us now consider the solution f of the boundary value problem

$$\Delta f = -1 \quad \text{on} \quad D$$

and

$$f = 0 \quad \text{on} \quad M = \partial D.$$

Applying Theorem 8.4 to f, we have

$$\frac{V(D)}{m+1} \geq \int_M f_v^2. \tag{8.7}$$

Schwarz's inequality now implies that

$$A(M) \int_M f_v^2 \geq \left(\int_M f_v\right)^2$$

$$= \left(\int_D \Delta f\right)^2$$

$$= V^2(D),$$

where $A(M)$ is the area of M. Therefore, taking this together with (8.7), we obtain the inequality

$$A(M) \geq (m+1)V(D). \tag{8.8}$$

On the other hand, Lemma 8.1 asserts that $\Delta_M X = -m\,v$, hence

$$0 = \int_D \langle X, \Delta X \rangle$$

$$= -\int_D |\nabla X|^2 + \int_M \langle X, X_v \rangle$$

$$= -(m+1)V(D) - \frac{1}{m}\int_M \langle X, \Delta_M X \rangle$$

$$= -(m+1)V(D) + \frac{1}{m}\int_M |\nabla_M X|^2$$

$$= -(m+1)V(D) + A(M),$$

where we have used the fact that $|\nabla X|^2 = m+1$ and $|\nabla_M X|^2 = m$. This implies that (8.8) is in fact an equality, hence all the inequalities that were used to derive (8.8) are in fact equalities. In particular,

$$f_{ij} = -\frac{\delta_{ij}}{m+1} \quad \text{on} \quad D, \tag{8.9}$$

and f_{m+1} must be identically constant on M. Let us denote the maximum point of f in D by p. Since p is a critical point, (8.9) implies that it is an isolated maximum. On the other hand, by choosing p to be the origin of $\mathbb{R}^{m+1}$, the function

$$g = -\frac{|X|^2}{2(m+1)}$$

has a maximum at p and its Hessian is given by

$$g_{ij} = -\frac{\delta_{ij}}{m+1}.$$

In particular, this implies that $f - g$ must be a constant function on D. Since $f = 0$ on M, we conclude M must be the sphere given by the level set of $|X|^2$. In particular, M must be the standard sphere since the mean curvature of M is m. □

Theorem 8.6 (Choi–Wang [CW]) *Let M^m be a compact, embedded, oriented minimal hypersurface in a compact, oriented Riemannian manifold N^{m+1}. Suppose that the Ricci curvature of N is bounded from below by*

$$\mathcal{R}_{ij} \geq mR$$

for some constant $R > 0$. Then the first nonzero eigenvalue of M has a lower bound given by

$$\lambda_1(M) \geq \frac{mR}{2}.$$

Proof The assumption that N has positive Ricci curvature implies that its first homology group $H^1(N, \mathbb{R})$ is trivial. By an exact sequence argument, we conclude that M divides N into two connected components N_1 and N_2 with $\partial N_1 = M = \partial N_2$. Let D be one of the components to be chosen later. If u is the first nonconstant eigenfunction on M satisfying

$$\Delta_M u = -\lambda_1(M) u,$$

then let f be the solution of

$$\Delta f = 0 \quad \text{on} \quad D$$

with boundary condition

$$f = u \quad \text{on} \quad M.$$

Applying Theorem 8.4, we have

$$0 \geq -2\lambda_1(M) \int_M u f_\nu + \int_M h_{\alpha\beta} u_\alpha u_\beta + mR \int_D |\nabla f|^2.$$

On the other hand,
$$2\int_M uf_\nu = 2\int_M ff_\nu$$
$$= \int_D \Delta(f^2)$$
$$= 2\int_D |\nabla f|^2.$$

Hence, we have
$$(2\lambda_1(M) - mR)\int_D |\nabla f|^2 \geq \int_M h_{\alpha\beta} u_\alpha u_\beta. \tag{8.10}$$

Let us observe that the right-hand side is independent of the extended function f. If we choose a different component of $N \setminus M$ to perform this computation, the second fundamental form will differ by a sign, hence we may choose a component, say N_1, so that
$$\int_M h_{\alpha\beta} u_\alpha u_\beta \geq 0.$$

Hence taking this together with (8.10), we conclude that either $\lambda_1(M) \geq mR/2$, or $\nabla f = 0$ on N_1. However, the latter is impossible because f has a boundary value u which is nonconstant. This proves the estimate. $\square$

9
Isoperimetric inequalities and Sobolev inequalities

In this chapter, we will show that a class of isoperimetric inequalities that occurs in geometry is in fact equivalent to a class of Sobolev type inequalities. The relationship between these inequalities was exploited in the study of eigenvalues of the Laplacian as early as the 1920s by Faber [F] and Krahn [K]. The equivalence was first formally established by Federer and Fleming [FF] (also see [Bm]) in 1960. In 1970, Cheeger [C] observed that the same argument can apply to estimating the first eigenvalue of the Laplacian.

We will first define the isoperimetric and Sobolev constants on a manifold. Let us assume that M is a compact Riemannian manifold with or without boundary ∂M.

Definition 9.1 If $\partial M \neq \phi$, we define the Dirichlet α-isoperimetric constant of M by

$$ID_\alpha(M) = \inf_{\substack{\Omega \subset M \\ \partial \Omega \cap \partial M = \emptyset}} \frac{A(\partial \Omega)}{V(\Omega)^{\frac{1}{\alpha}}},$$

where the infimum is taken over all subdomains $\Omega \subset M$ with the property that $\partial \Omega$ is a hypersurface not intersecting ∂M.

Similarly, we define the Neumann α-isoperimetric constant of M.

Definition 9.2 The Neumann α-isoperimetric constant of M is defined by

$$IN_\alpha(M) = \inf_{\substack{\partial \Omega_1 = S = \partial \Omega_2 \\ M = \Omega_1 \cup S \cup \Omega_2}} \frac{A(S)}{\min\{V(\Omega_1), V(\Omega_2)\}^{1/\alpha}},$$

where the infimum is taken over all hypersurfaces S dividing M into two parts, denoted Ω_1 and Ω_2. Note that in this case there is no assumption on whether M has boundary or not.

9 Isoperimetric inequalities and Sobolev inequalities

Definition 9.3 If $\partial M \neq \emptyset$, we define the Dirichlet α-Sobolev constant of M by

$$SD_\alpha(M) = \inf_{\substack{f \in H_{1,1}(M) \\ f|_{\partial M}=0}} \frac{\int_M |\nabla f|}{(\int_M |f|^\alpha)^{1/\alpha}},$$

where the infimum is taken over all functions f in the first Sobolev space with Dirichlet boundary condition.

We also define the Neumann α-Sobolev constant of M.

Definition 9.4 The Neumann α-Sobolev constant of M is defined by

$$SN_\alpha(M) = \inf_{f \in H_{1,1}(M)} \frac{\int_M |\nabla f|}{(\inf_{k \in \mathbb{R}} \int_M |f - k|^\alpha)^{1/\alpha}},$$

where the first infimum is taken over all functions f in the first Sobolev space, and the second infimum is taken over all real numbers k. Again, there is no assumption on whether M has boundary or not.

Theorem 9.5 *For any $\alpha > 0$, we have $ID_\alpha(M) = SD_\alpha(M)$.*

Proof To see that $ID_\alpha(M) \leq SD_\alpha(M)$, it suffices to show that for any Lipschitz function f defined on M with boundary condition $f|_{\partial M} \equiv 0$ must satisfy

$$\int_M |\nabla f| \geq ID_\alpha(M) \left(\int_M |f|^\alpha \right)^{1/\alpha}.$$

Without loss of generality, we may assume that $f \geq 0$. Let us define $M_t = \{x \in M | f(x) > t\}$ to be the sublevel set of f. By the coarea formula,

$$\int_M |\nabla f| = \int_0^\infty A(\partial M_t) dt$$

$$\geq ID_\alpha(M) \int_0^\infty V(M_t)^{1/\alpha} dt. \tag{9.1}$$

We now claim that for any $s \geq 0$, we have the inequality

$$\left(\int_0^s V(M_t)^{1/\alpha} dt \right)^\alpha \geq \alpha \int_0^s t^{\alpha-1} V(M_t) dt.$$

This is obvious for the case $s = 0$. Differentiating both sides as functions of s, we obtain

$$\frac{d}{ds}\left(\int_0^s V(M_t)^{1/\alpha} dt \right)^\alpha = \alpha \left(\int_0^s V(M_t)^{1/\alpha} dt \right)^{\alpha-1} V(M_s)^{1/\alpha} \tag{9.2}$$

and

$$\frac{d}{ds}\left(\alpha \int_0^s t^{\alpha-1} V(M_t) dt\right) = \alpha s^{\alpha-1} V(M_s). \tag{9.3}$$

Observing that $\int_0^s V(M_t)^{1/\alpha} dt \geq s V(M_s)^{1/\alpha}$, because $M_s \subset M_t$ for $t \leq s$, we conclude that (9.2) is greater than or equal to (9.3). Integrating from 0 to s yields the inequality as claimed.

Applying this inequality to (9.1) yields

$$\int_M |\nabla f| \geq ID_\alpha(M) \left(\alpha \int_0^\infty t^{\alpha-1} V(M_t) dt\right)^{1/\alpha}.$$

However, the coarea formula implies that

$$\alpha \int_0^\infty t^{\alpha-1} V(M_t) dt = \int_0^\infty \frac{d(t^\alpha)}{dt} \int_t^\infty \int_{\partial M_s} \frac{dA_s}{|\nabla f|} ds\, dt$$

$$= \int_0^\infty t^\alpha \int_{\partial M_t} \frac{dA_t}{|\nabla f|} dt$$

$$= \int_M f^\alpha.$$

This proves $ID_\alpha(M) \leq SD_\alpha(M)$.

We will now prove that $ID_\alpha(M) \geq SD_\alpha(M)$. Let Ω be a subdomain of M with smooth boundary $\partial \Omega$ such that $\partial \Omega \cap \partial M = \phi$. We denote the ϵ-neighborhood of $\partial \Omega$ in Ω by

$$N_\epsilon = \{x \in \Omega | d(x, \partial \Omega) < \epsilon\}.$$

Note that for $\epsilon > 0$ sufficiently small, the distance function $d(\cdot, \partial \Omega)$ to $\partial \Omega$ is a smooth function on N_ϵ. Let us define the function

$$f_\epsilon(x) = \begin{cases} 0 & \text{on } M \setminus \Omega, \\ \frac{1}{\epsilon} d(x, \partial \Omega) & \text{on } N_\epsilon, \\ 1 & \text{on } \Omega \setminus N_\epsilon. \end{cases}$$

Clearly f_ϵ is a Lipschitz function defined on M with Dirichlet boundary condition. Moreover,

$$\int_M |\nabla f_\epsilon| = \int_0^\epsilon \frac{1}{\epsilon} A(\partial N_t \setminus \partial \Omega) dt.$$

9 Isoperimetric inequalities and Sobolev inequalities

On the other hand, we have

$$\int_M |\nabla f_\epsilon| \geq SD_\alpha(M) \left(\int_M |f_\epsilon|^\alpha \right)^{1/\alpha}$$

$$\geq SD_\alpha(M) V(\Omega \setminus N_\epsilon)^{\frac{1}{\alpha}}.$$

Hence

$$\frac{1}{\epsilon} \int_0^\epsilon A(\partial N_t \setminus \partial\Omega) dt \geq SD_\alpha(M) V(\Omega \setminus N_\epsilon)^{1/\alpha},$$

and letting $\epsilon \to 0$ yields

$$A(\partial\Omega) \geq SD_\alpha(M) V(\Omega)^{1/\alpha}.$$

Since Ω is arbitrary, this proves $ID_\alpha(M) \geq SD_\alpha(M)$. $\square$

Theorem 9.6 *For any $\alpha > 0$, we have*

$$\min\{1, 2^{1-1/\alpha}\} IN_\alpha(M) \leq SN_\alpha(M)$$

and

$$SN_\alpha(M) \leq \max\{1, 2^{1-1/\alpha}\} IN_\alpha(M).$$

Proof Let f be a Lipschitz function defined on M and $k \in \mathbb{R}$ be chosen such that

$$M_+ = \{x \in M | f(x) - k > 0\}$$

and

$$M_- = \{x \in M | f(x) - k < 0\}$$

satisfy the conditions that $V(M_+) \leq \frac{1}{2} V(M)$ and $V(M_-) \leq \frac{1}{2} V(M)$. To show that

$$SN_\alpha(M) \geq \min\{1, 2^{1-1/\alpha}\} IN_\alpha(M),$$

it suffices to show that

$$\int_M |\nabla u| \geq \min\{1, 2^{1-1/\alpha}\} IN_\alpha(M) \left(\int_M |u|^\alpha \right)^{1/\alpha}$$

for $u = f - k$. Note that if

$$M_t = \{x \in M | u(x) > t\},$$

then for $t > 0$, we have

$$V(M_t) \leq V(M_+) \leq \tfrac{1}{2}V(M).$$

This implies that

$$\min\{V(M_t), V(M \setminus M_t))\} = V(M_t),$$

and $A(\partial M_t) \geq IN_\alpha(M)V(M_t)^{1/\alpha}$. Therefore by the same argument as in the proof of Theorem 9.5, we have

$$\int_{M_+} |\nabla u| \geq IN_\alpha(M) \left(\int_{M_+} |u|^\alpha\right)^{1/\alpha}.$$

Similarly, we also obtain

$$\int_{M_-} |\nabla u| \geq IN_\alpha(M) \left(\int_{M_-} |u|^\alpha\right)^{1/\alpha}.$$

Hence

$$\int_M |\nabla u| \geq IN_\alpha(M) \left[\left(\int_{M_+} |u|^\alpha\right)^{1/\alpha} + \left(\int_{M_-} |u|^\alpha\right)^{1/\alpha}\right]$$

$$\geq \min\{1, 2^{1-1/\alpha}\} IN_\alpha(M) \left(\int_M |u|^\alpha\right)^{1/\alpha}.$$

This proves $SN_\alpha(M) \geq \min\{1, 2^{1-1/\alpha}\} IN_\alpha(M)$.

To prove that $\max\{1, 2^{1-1/\alpha}\} IN_\alpha(M) \geq SN_\alpha(M)$, we consider any hypersurface S dividing M into two components denoted by Ω_1 and Ω_2. Let us assume that $V(\Omega_2) \leq V(\Omega_1)$. For $\epsilon > 0$ sufficiently small, let us define

$$N_\epsilon = \{x \in \Omega_2 \mid d(x, S) < \epsilon\}$$

and the function

$$f_\epsilon(x) = \begin{cases} 1 & \text{on } \Omega_1, \\ 1 - \dfrac{1}{\epsilon}d(x, S) & \text{on } N_\epsilon, \\ 0 & \text{on } \Omega_2 - N_\epsilon. \end{cases}$$

Let $0 \leq k_\epsilon \leq 1$ be chosen such that

$$\int_M |f_\epsilon - k_\epsilon|^\alpha = \inf_{k \in \mathbb{R}} \int_M |f_\epsilon - k|^\alpha.$$

9 Isoperimetric inequalities and Sobolev inequalities

By using a similar argument to that in the proof of Theorem 9.5, we have

$$\int_M |\nabla f_\epsilon| = \int_{N_\epsilon} |\nabla f_\epsilon|$$

$$\geq SN_\alpha(M) \left(\int_M |f_\epsilon - k_\epsilon|^\alpha \right)^{1/\alpha}$$

$$\geq SN_\alpha(M) \left(\int_{\Omega_1} |f_\epsilon - k_\epsilon|^\alpha + \int_{\Omega_2 \setminus N_\epsilon} |f_\epsilon - k_\epsilon|^\alpha \right)^{1/\alpha}$$

$$\geq SN_\alpha(M) \left((1 - k_\epsilon)^\alpha V(\Omega_1) + k_\epsilon^\alpha V(\Omega_2 \setminus N_\epsilon) \right)^{1/\alpha}$$

$$\geq SN_\alpha(M) \left((1 - k_\epsilon)^\alpha + k_\epsilon^\alpha \right)^{1/\alpha} V(\Omega_2 \setminus N_\epsilon)^{1/\alpha}. \tag{9.4}$$

We now observe that $(1-k)^\alpha + k^\alpha \geq 2^{1-\alpha}$ for all $0 \leq k \leq 1$ and $\alpha > 1$, also $(1-k)^\alpha + k^\alpha \geq 1$ for all $0 \leq k \leq 1$ and $\alpha \leq 1$. Hence by taking $\epsilon \to 0$, the left-hand side of (9.4) tends to $A(S)$, while the right-hand side of (9.4) is bounded from below by $SN_\alpha(M) \min\{1, 2^{(1-\alpha)/\alpha}\} V(\Omega_2)^{1/\alpha}$. This establishes the inequality $\max\{1, 2^{1-1/\alpha}\} IN_\alpha(M) \geq SN_\alpha(M)$. $\square$

Let us point out that when the dimension of M is m and $\alpha > m/(m-1)$, then by the fact that the volume of geodesic balls of radius r behaves like

$$V(r) \sim \frac{\alpha_{m-1}}{m} r^m$$

and the area of their boundary is asymptotic to

$$A(r) \sim \alpha_{m-1} r^{m-1},$$

it is clear that $ID_\alpha(M) = 0 = IN_\alpha(M)$. Hence it is only interesting to consider those $\alpha \leq m/(m-1)$.

Corollary 9.7 (Cheeger [C]) *Let M be a compact Riemannian manifold. If $\partial M \neq \emptyset$, let $\mu_1(M)$ be the first Dirichlet eigenvalue on M and $\lambda_1(M)$ be its first nonzero Neumann eigenvalue for the Laplacian. When $\partial M = \emptyset$, we will denote the first nonzero eigenvalue of M by $\lambda_1(M)$ also. Then*

$$\mu_1(M) \geq \frac{ID_1(M)^2}{4}$$

and

$$\lambda_1(M) \geq \frac{IN_1(M)^2}{4}.$$

Proof By Theorem 9.5, to see that

$$\mu_1(M) \geq \frac{ID_1(M)^2}{4},$$

it suffices to show that any Lipschitz function f with Dirichlet boundary condition must satisfy

$$\int_M |\nabla f|^2 \geq \frac{SD_1(M)^2}{4} \int_M f^2.$$

Applying the definition of $SD_1(M)$ to the function f^2, we have

$$\int_M |\nabla f^2| \geq SD_1(M) \int_M f^2. \tag{9.5}$$

On the other hand,

$$\int_M |\nabla f^2| = 2 \int_M |f||\nabla f|$$

$$\leq 2 \left(\int_M f^2 \right)^{1/2} \left(\int_M |\nabla f|^2 \right)^{1/2}.$$

Hence, the desired inequality follows from this and (9.5).

For the Neumann eigenvalue, we simply observe that if u is the first eigenfunction satisfying

$$\Delta u = -\lambda_1(M) u,$$

then u must change sign. If we let $M_+ = \{x \in M | u(x) > 0\}$ and $M_- = \{x \in M | u(x) < 0\}$, then

$$\mu_1(M_+) = \lambda_1(M) = \mu_1(M_-).$$

Let us assume that $V(M_+) \leq V(M_-)$. In particular, this implies that $ID_1(M_+) \geq IN_1(M)$. Hence by our previous argument,

$$\lambda_1(M) = \mu_1(M_+)$$

$$\geq \frac{ID_1(M_+)^2}{4}$$

$$\geq \frac{IN_1(M)^2}{4},$$

proving the corollary. □

9 Isoperimetric inequalities and Sobolev inequalities

Corollary 9.8 *Let M be a compact Riemannian manifold with boundary. For any function $f \in H^{1,2}(M)$ and $f|_{\partial M} \equiv 0$, we have*

$$\int_M |\nabla f|^2 \geq \left(\frac{2-\alpha}{2} ID_\alpha(M)\right)^2 \left(\int_M |f|^{2\alpha/(2-\alpha)}\right)^{(2-\alpha)/\alpha}.$$

Proof By applying Theorem 9.5 and the definition of $SD_\alpha(M)$ to the function $|f|^{2/(2-\alpha)}$, we obtain

$$\int_M |\nabla |f|^{2/(2-\alpha)}| \geq ID_\alpha(M) \left(\int_M |f|^{2\alpha/(2-\alpha)}\right)^{1/\alpha}.$$

On the other hand, Schwarz's inequality implies that

$$\int_M |\nabla |f|^{2/(2-\alpha)}| = \frac{2}{2-\alpha} \int_M |f|^{\alpha/(2-\alpha)} |\nabla f|$$

$$\leq \frac{2}{2-\alpha} \left(\int_M |f|^{2\alpha/(2-\alpha)}\right)^{1/2} \left(\int_M |\nabla f|^2\right)^{1/2}.$$

This proves the corollary. □

Corollary 9.9 *Let M be a complete Riemannian manifold with or without boundary. There exist constants $C_1, C_2 > 0$ depending only on α, such that*

$$\int_M |\nabla f|^2 \geq C_1 SN_\alpha(M)^2$$

$$\times \left(\left(\int_M |f|^{2\alpha/(2-\alpha)}\right)^{(2-\alpha)/\alpha} - C_2 V(M)^{-(2-2\alpha)/\alpha} \int_M |f|^2\right)$$

for all $f \in H^{1,2}(M)$.

Proof Let us first observe that a function g satisfies

$$\int_M \mathrm{sgn}(g) |g|^{\alpha-1} = 0$$

if and only if

$$\int_M |g|^\alpha = \inf_{k \in \mathbb{R}} \int_M |g-k|^\alpha.$$

In particular,

$$SN_\alpha(M) \left(\int_M |g|^\alpha\right)^{1/\alpha} \leq \int_M |\nabla g|. \tag{9.6}$$

For any given $f \in H^{1,2}(M)$, let us choose $k \in \mathbb{R}$ such that

$$\int_M \text{sgn}(f-k)\,|f-k|^{(2\alpha-2)/(2-\alpha)} = 0. \tag{9.7}$$

Using $g = \text{sgn}(f-k)\,|f-k|^{2/(2-\alpha)}$, (9.6) implies that

$$\frac{2}{2-\alpha} \int_M |f-k|^{\alpha/(2-\alpha)}\, |\nabla f| \geq SN_\alpha(M) \left(\int_M |f-k|^{2\alpha/(2-\alpha)} \right)^{1/\alpha}.$$

Applying Schwarz's inequality as in the proof of Corollary 9.9 yields

$$\int_M |\nabla f|^2 \geq \left(\frac{2-\alpha}{2} SN_\alpha(M)\right)^2 \left(\int_M |f-k|^{2\alpha/(2-\alpha)} \right)^{(2-\alpha)/\alpha}$$

$$\geq \left(\frac{2-\alpha}{2} SN_\alpha(M)\right)^2$$

$$\times \left(2^{1-\alpha} \int_M |f|^{2\alpha/(2-\alpha)} - V(M)\,|k|^{2\alpha/(2-\alpha)} \right)^{(2-\alpha)/\alpha}$$

$$\geq \left(\frac{2-\alpha}{2} SN_\alpha(M)\right)^2 \left(2^{(2-\alpha)(1-\alpha)/\alpha} \left(\int_M |f|^{2\alpha/(2-\alpha)} \right)^{(2-\alpha)/\alpha} \right.$$

$$\left. - V(M)^{(2-\alpha)/\alpha}\,|k|^2 \right). \tag{9.8}$$

By changing the sign of f if necessary, we may assume $k \geq 0$. We can estimate k from above as follows. Let us define

$$M_+ = \{x \in M \mid f(x) - k \geq 0\}$$

and

$$M_- = \{x \in M \mid f(x) - k < 0\}.$$

The condition (9.7) implies that

$$\int_{M_+} (f-k)^{(2\alpha-2)/(2-\alpha)} = \int_{M_-} (k-f)^{(2\alpha-2)/(2-\alpha)}.$$

However, since

$$\int_{M_+} (f-k)^{(2\alpha-2)/(2-\alpha)} \leq C_3 \int_{M_+} |f|^{(2\alpha-2)/(2-\alpha)} - V(M_+)\,k^{(2\alpha-2)/(2-\alpha)}$$

and
$$\int_{M_-} (k-f)^{(2\alpha-2)/(2-\alpha)} \geq C_4 \, V(M_-) \, k^{(2\alpha-2)/(2-\alpha)} - \int_{M_-} |f|^{(2\alpha-2)/(2-\alpha)},$$

for some constants $C_3, C_4 > 0$ depending on α only, we conclude that

$$C_3 \int_{M_+} |f|^{(2\alpha-2)/(2-\alpha)} - V(M_+) \, k^{(2\alpha-2)/(2-\alpha)}$$
$$\geq C_4 \, V(M_-) \, k^{(2\alpha-2)/(2-\alpha)} - \int_{M_-} |f|^{(2\alpha-2)/(2-\alpha)}.$$

This implies that

$$C_5 \int_M |f|^{(2\alpha-2)/(2-\alpha)} \geq V(M) \, k^{(2\alpha-2)/(2-\alpha)}$$

for some constant C_5 depending only on α. Applying the Hölder inequality to the left-hand side, we obtain

$$C_6 \int_M |f|^2 \geq V(M) \, k^2.$$

Substituting the above inequality into (9.8) yields the corollary. □

10
The heat equation

In this chapter, we will discuss the existence and some basic properties of the fundamental solution of the heat equation (heat kernel). The heat equation on $M \times (0, T)$, for some constant $0 < T \leq \infty$, is given by

$$\left(\Delta - \frac{\partial}{\partial t}\right) f(x, t) = 0.$$

Typically, we consider solving it by prescribing an initial datum f_0 on M, i.e.,

$$\lim_{t \to 0} f(x, t) = f_0(x).$$

The fundamental solution for the heat equation is a kernel function $H(x, y, t)$ defined on $M \times M \times (0, \infty)$ with the property that the function $f(x, t)$ defined by

$$f(x, t) = \int_M H(x, y, t) f_0(y) \, dy$$

solves the heat equation

$$\left(\Delta - \frac{\partial}{\partial t}\right) f(x, t) = 0 \quad \text{on} \quad M \times (0, \infty) \tag{10.1}$$

with the initial condition

$$\lim_{t \to 0} f(x, t) = f_0(x). \tag{10.2}$$

When M is a compact manifold with boundary, the two natural boundary conditions are the Dirichlet and Neumann boundary conditions. For the first case, the solution is required to satisfy

$$f(x, t) = 0 \quad \text{on} \quad \partial M \times (0, \infty) \tag{10.3}$$

10 The heat equation

in addition to (10.1) and (10.2). In the case of Neumann boundary condition, the solution should satisfy

$$\frac{\partial f}{\partial \nu}(x,t) = 0 \quad \text{on} \quad \partial M \times (0,\infty), \tag{10.4}$$

where ν is the outward normal to ∂M. No boundary condition is necessary when M is a compact manifold without boundary.

Let us first consider the Dirichlet boundary condition. Elliptic theory asserts that there is a set of eigenvalues $\{\mu_1 < \mu_2 \leq \ldots \leq \mu_i \leq \ldots\}$ with corresponding eigenfunctions $\{\phi_i\}$ satisfying

$$\Delta \phi_i = -\mu_i \phi_i,$$

such that they form an orthonormal basis with respect to the L^2-norm. In particular, any function $f_0 \in L^2(M)$ can be written in the form

$$f_0(x) = \sum_{i=1}^{\infty} a_i \phi_i(x)$$

with

$$a_i = \int_M f_0 \phi_i \, dx.$$

Formally, the function given by

$$f(x,t) = \sum_{i=1}^{\infty} e^{-\mu_i t} a_i \phi_i(x)$$

will satisfy (10.1) with boundary condition (10.3) and initial condition (10.2). This is equivalent to saying that

$$f(x,t) = \int_M H(x,y,t) f_0(y) \, dy,$$

if we define the kernel by

$$H(x,y,t) = \sum_{i=1}^{\infty} e^{-\mu_i t} \phi_i(x) \phi_i(y).$$

To justify the above discussion, we first prove a theorem on $H(x,y,t)$. The key estimates follow from that in [L2] where they were derived for the more general setting of differential forms.

It is also important to point out that $H(x,y,t)$ is obviously symmetric in the x and y variables.

Theorem 10.1 *Let M be a compact manifold with boundary. The kernel function*

$$H(x, y, t) = \sum_{i=1}^{\infty} e^{-\mu_i t} \phi_i(x) \phi_i(y)$$

is well defined on $M \times M \times (0, \infty)$. It is the unique kernel, such that for any $f_0 \in L^2(M)$, the function given by

$$f(x, t) = \int_M H(x, y, t) f_0(y) \, dy$$

solves the heat equation

$$\left(\Delta - \frac{\partial}{\partial t} \right) f(x, t) = 0 \quad on \quad M \times (0, \infty)$$

and

$$f(x, t) = 0 \quad on \quad \partial M \times (0, \infty),$$

with

$$\lim_{t \to 0} f(x, t) = f_0(x).$$

Moreover, $H(x, y, t)$ is positive on $(M \setminus \partial M) \times (M \setminus \partial M) \times (0, \infty)$ and it satisfies

$$\int_M H(x, y, t) \leq 1.$$

Proof We begin by estimating the supremum norm of an eigenfunction. Let us recall that Sobolev inequality asserts that for $m \geq 3$ there exists a constant $C_{\mathcal{SD}} > 0$ depending only on M, such that

$$\int_M |\nabla f|^2 \geq C_{\mathcal{SD}} \left(\int_M |f|^{2m/(m-2)} \right)^{(m-2)/m} \tag{10.5}$$

for all $f \in H_c^{1,2}(M)$. In fact, Corollary 9.8 implies that $C_{\mathcal{SD}} \geq ((m-2)/(2m-2) \, ID_{m/(m-1)}(M))^2$. When $m = 2$, the Sobolev inequality can be taken to be

$$\int_M |\nabla f|^2 \geq C_{\mathcal{SD}} \left(\int_M |f|^p \right)^{2/p}$$

for all $f \in H_c^{1,2}(M)$ and for any fixed $2 \leq p < \infty$. For our purpose, we do not need the precise value of the Sobolev constant $C_{\mathcal{SD}}$, beyond the fact that it only depends on the manifold M.

10 The heat equation

Let us assume that $m \geq 3$ since the case when $m = 2$ can be done similarly. For any nonnegative function $u \in H_c^{1,2}(M)$ and for any constant $k \geq 2$, integrating by parts and applying the Sobolev inequality gives

$$\int_M u^{k-1} \Delta u = -(k-1) \int_M u^{k-2} |\nabla u|^2$$

$$= -\frac{4(k-1)}{k^2} \int_M |\nabla(u^{k/2})|^2$$

$$\leq -\frac{4(k-1)C_{\mathcal{SD}}}{k^2} \left(\int_M |u|^{km/(m-2)} \right)^{(m-2)/m}$$

$$\leq -\frac{2C_{\mathcal{SD}}}{k} \left(\int_M |u|^{km/(m-2)} \right)^{(m-2)/m}. \qquad (10.6)$$

In particular, if $u = |\phi_i|$, then u satisfies

$$\Delta u \geq -\mu_i u.$$

Hence (10.6) asserts that

$$\int_M |\phi_i|^k \geq \frac{2C_{\mathcal{SD}}}{k\mu_i} \left(\int_M |\phi_i|^{km/(m-2)} \right)^{(m-2)/m},$$

which can be rewritten as

$$\left(\frac{2C_{\mathcal{SD}}}{k\mu_i} \right)^{1/k} \|\phi_i\|_{k\beta} \leq \|\phi_i\|_k$$

for all $k \geq 2$, with $\beta = m/(m-2)$ and $\|\phi_i\|_k$ denoting the L^k-norm of ϕ_i. Setting $k = 2\beta^j$ for $j = 0, 1, 2, \ldots$, we have

$$\|\phi_i\|_{2\beta^{j+1}} \leq \left(\frac{\beta^j \mu_i}{C_{\mathcal{SD}}} \right)^{1/2\beta^j} \|\phi_i\|_{2\beta^j}.$$

Iterating this estimate and using $\|\phi_i\|_2 = 1$, we conclude that

$$\|\phi_i\|_{2\beta^{j+1}} \leq \prod_{\ell=0}^{j} \left(\frac{\beta^\ell \mu_i}{C_{\mathcal{SD}}} \right)^{1/2\beta^\ell}.$$

Letting $j \to \infty$ and applying the fact that

$$\lim_{p \to \infty} \|\phi_i\|_p = \|\phi_i\|_\infty,$$

we obtain

$$\|\phi_i\|_\infty \leq C_1 \mu_i^{m/4}, \qquad (10.7)$$

where $C_1 > 0$ is a constant depending on $C_{\mathcal{SD}}$.

We will now estimate μ_k using k. Let E be the k-dimensional vector space spanned by the first k eigenfunctions $\{\phi_1, \phi_2, \ldots, \phi_k\}$. We will show that for any $u \in E$, the estimate

$$\|u\|_\infty \leq C_2 \mu_k^{(m-1)/2} V(M)^{(m-1)/m} \#u\|_2 \qquad (10.8)$$

must hold for some constant $C_2 > 0$ depending only on m and $C_{\mathcal{SD}}$. This estimate and Lemma 7.3, together imply

$$\mu_k \geq C_3 k^{1/(m-1)} V(M)^{-2/m} \qquad (10.9)$$

for some constant C_3 depending only on m and $C_{\mathcal{SD}}$.

To prove (10.8), we observe that for a C^∞ function u, since $|\nabla u|^2 = |\nabla |u||^2$, the identities

$$\Delta(u^2) = 2u\,\Delta u + 2|\nabla u|^2$$

and

$$\Delta |u|^2 = 2|u|\,\Delta |u| + 2|\nabla |u||^2$$

imply that $u\,\Delta u = |u|\,\Delta |u|$. Hence applying (10.6) to $|u|$, the inequality can be written as

$$\int_M |u|^{k-2} u\,\Delta u \leq -\frac{2C_{\mathcal{SD}}}{k} \left(\int_M |u|^{km/(m-2)}\right)^{(m-2)/m}. \qquad (10.10)$$

We claim that for any $u \in E$ and $j \geq 0$, we have the estimate

$$\#u\|_{2\beta^{j+1}} \leq \prod_{\ell=0}^{j} \left(\frac{\mu_k V(M)^{2/m} \beta^\ell}{C}\right)^{1/(2\beta^\ell - 1)} \#u\|_2. \qquad (10.11)$$

To see this, for each $j \geq 0$, let us choose $h \in E$, such that,

$$\frac{\|h\|_{2\beta^{j+1}}}{\|h\|_2} = \max_{u \in E} \frac{\|u\|_{2\beta^{j+1}}}{\|u\|_2}. \qquad (10.12)$$

Setting $u = h$ and $k = 2\beta^j$ in (10.10), we have

$$\frac{C_{\mathcal{SD}}}{\beta^j}\left(\int_M |h|^{2\beta^{j+1}}\right)^{1/\beta}$$

$$\leq -\int_M |h|^{2\beta^j - 2} h \Delta h$$

$$\leq \left(\int_M |\Delta h|^{2\beta^{j+1}}\right)^{1/(2\beta^j + 1)} \left(\int_M |h|^{2\beta^j}\right)^{(2\beta^j - 1)/2\beta^j} V(M)^{(\beta - 1)/(2\beta^j + 1)}.$$

$$(10.13)$$

To estimate the first term on the right-hand side for $h \in E$, we write

$$h = \sum_{\ell=1}^{k} a_\ell \phi_\ell,$$

hence

$$\Delta h = -\sum_{\ell=1}^{k} \mu_\ell a_\ell \phi_\ell.$$

Observe that, for $p \geq 2$ and $1 \leq i \leq k$, the function

$$F(\mu_i) = \int_M |\Delta h|^p$$

is convex as a function of μ_i because

$$\frac{\partial^2 F}{\partial \mu_i^2} = \int_M \frac{\partial^2}{\partial \mu_i^2} \left|\sum_{\ell=1}^{k} \mu_\ell a_\ell \phi_\ell\right|^p$$

$$= p(p-1) \int_M \left|\sum_{\ell=1}^{k} \mu_\ell a_\ell \phi_\ell\right|^{p-2} a_i^2 \phi_i^2$$

$$\geq 0.$$

This implies that for a fixed $1 \leq i \leq k$, $F(\mu_i)$ is bounded above by either $F(0)$ or $F(\mu_k)$. In particular, we conclude that there exists a subset $\{\alpha\}$ of $\{1, 2, \ldots, k\}$ such that

$$\int_M |\Delta h|^{2\beta^{j+1}} \leq \int_M \left|\sum_\alpha \mu_k a_\alpha \phi_\alpha\right|^{2\beta^{j+1}}. \quad (10.14)$$

Since the function $\sum_\alpha a_\alpha \phi_\alpha \in E$, the extremal property of h asserts that

$$\int_M \left|\sum_\alpha a_\alpha \phi_\alpha\right|^{2\beta^{j+1}} \leq \left(\sum_\alpha a_\alpha^2\right)^{\beta^{j+1}} \int_M |h|^{2\beta^{j+1}} \left(\int_M h^2\right)^{-\beta^{j+1}}$$

$$\leq \int_M |h|^{2\beta^{j+1}}.$$

Combining this with (10.13) and (10.14), we have

$$\frac{C}{\beta^j}\left(\int_M |h|^{2\beta^{j+1}}\right)^{1/\beta} \leq \mu_k \,\|h\|_{2\beta^{j+1}} \,\|h\|_{2\beta^j}^{2\beta^j-1} V(M)^{(\beta-1)/(2\beta^j+1)},$$

hence

$$\#h\|_{2\beta^{j+1}} \leq \left(\frac{\mu_k \, V(M)^{2/m} \, \beta^j}{C}\right)^{1/(2\beta^j-1)} \#h\|_{2\beta^j}. \tag{10.15}$$

For $j=0$, the extremal property of h implies (10.11) for that case. We now assume that (10.11) is valid for $j-1$. To show that (10.11) is valid for j, we consider h satisfying (10.12). The induction hypothesis together with (10.15) yields

$$\#h\|_{2\beta^{j+1}} \leq \left(\frac{\mu_k \, V(M)^{2/m} \, \beta^j}{C}\right)^{1/(2\beta^j-1)} \#h\|_{2\beta^j}$$

$$\leq \left(\frac{\mu_k \, V(M)^{2/m} \, \beta^j}{C}\right)^{1/(2\beta^j-1)} \prod_{\ell=0}^{j-1}\left(\frac{\mu_k \, V(M)^{2/m} \, \beta^\ell}{C}\right)^{1/(2\beta^\ell-1)} \#h\|_2$$

$$= \prod_{\ell=0}^{j}\left(\frac{\mu_k \, V(M)^{2/m} \, \beta^\ell}{C}\right)^{1/(2\beta^\ell-1)} \#h\|_2.$$

Again, the extremal property of h implies (10.11) for any j, upholding the claim. Letting $j \to \infty$, we conclude the validity of inequality (10.8), hence also of (10.9).

The convergence of the infinite series

$$\sum_{i=1}^{\infty} e^{-\mu_i t} \phi_i(x) \phi_i(y)$$

10 The heat equation

can now be seen by using (10.7) to get

$$e^{-\mu_i t} \phi_i(x) \phi_i(y) \le e^{-\mu_i t} \|\phi_i\|_\infty^2$$
$$\le C_1 e^{-\mu_i t} \mu_i^{m/2}$$
$$\le C_1 C(t) e^{-\mu_i t/2},$$

where we have used the fact that

$$e^{-xt} x^{m/2} \le C(t) e^{-xt/2}$$

for some constant $C(t) > 0$ depending only on t and for all $0 \le x < \infty$. Applying (10.8), we conclude that

$$\sum_{i=1}^\infty e^{-\mu_i t} \phi_i(x) \phi_i(y) \le C_1 C(t) \sum_{i=1}^\infty e^{-C_4 i^{1/(m-1)} t},$$

which clearly converges uniformly on $M \times M \times [a, \infty)$ for any $a > 0$. Hence the kernel

$$H(x, y, t) = \sum_{i=1}^\infty e^{-\mu_i t} \phi_i(x) \phi_i(y)$$

is well defined and satisfies the Dirichlet boundary condition in both x and y for $t > 0$. Moreover, if $f_0(x) = \sum_{i=1}^\infty a_i \phi_i(x)$ in $L^2(M)$, then

$$f(x, t) = \int_M H(x, y, t) f_0(y) \, dy$$
$$= \sum_{i=1}^\infty e^{-\mu_i t} a_i \phi_i(x),$$

which obviously satisfies the initial condition

$$\lim_{t \to 0} f(x, t) = f_0(x).$$

We also note that

$$\int_M \langle \nabla \phi_i, \nabla \phi_j \rangle = \delta_{ij} \mu_i$$

implies that the finite sum

$$\int_M \left| \sum_{i=1}^k e^{-\mu_i t} \phi_i(x) \nabla \phi_i(y) \right|^2 dy$$

$$= \int_M \sum_{i,j=1}^k e^{-(\mu_i+\mu_j)t} \phi_i(x) \phi_j(x) \langle \nabla \phi_i(y), \nabla \phi_j(y) \rangle dy$$

$$= \sum_{i=1}^k e^{2\mu_i t} \mu_i \phi_i(x) \phi_i(x)$$

converges as $k \to \infty$ according to our previous argument. Hence the truncated sum $\sum_{i=1}^k e^{-\mu_i t} \phi_i(x) \phi_i(y)$ converges to $H(x, y, t)$ weakly in $H^{1,2}(M)$. In particular, $H(x, y, t)$ is a weak solution of the heat equation since each truncated sum solves the heat equation. Regularity theory implies that $H(x, y, t)$ is a smooth solution and also $f(x, t)$ solves the Dirichlet heat equation (10.1) and (10.3), with initial condition f_0.

The strong maximum principle asserts that $f(x, t) = \int_M H(x, y, t) f_0(y) dy$ is positive on $(M \setminus \partial M) \times (0, \infty)$ whenever $f_0 \geq 0$ on M. This implies that $H(x, y, t)$ is positive on $(M \setminus \partial M) \times (M \setminus \partial M) \times (0, \infty)$. If $\bar{f}(x, t)$ is another solution of the heat equation with initial condition f_0, then the maximum principle implies that the solution of the heat equation $f(x, t) - \bar{f}(x, t)$ must be identically 0 since it vanishes on $(M \times \{0\}) \cup (\partial M \times (0, \infty))$. In particular, $H(x, y, t)$ is unique. The property that $\int_M H(x, y, t) dy \leq 1$ follows from the fact that it solves the heat equation with initial condition $f_0 = 1$ and the maximum principle.

The case of $m = 2$ can be proved in a similar manner using the two-dimensional Sobolev inequality. □

Note that the same argument can be used to prove the existence of the heat kernel for compact manifolds without boundary, or for compact manifolds with Neumann boundary condition. In either of these cases, the heat kernel is given by

$$K(x, y, t) = \sum_{i=0}^\infty e^{-\lambda_i t} \psi_i(x) \psi_i(y),$$

where ψ_i is the (Neumann) eigenfunction with eigenvalue λ_i satisfying

$$\Delta \psi_i = -\lambda_i \psi_i.$$

Since $\lambda_0 = 0$ and $\psi_0 = V(M)^{-\frac{1}{2}}$, we only need to derive estimates on ψ_i and λ_i for $i \geq 1$. To see this, we replace the Sobolev inequality by that from Corollary 9.9 setting $\alpha = m/(m-1)$ when $m \geq 3$, and $\alpha = \frac{4}{3}$ when $m = 2$. It takes the form

$$\int_M |\nabla f|^2 \geq C_1 \left(\int_M |f|^{2m/(m-2)} \right)^{(m-2)/m} - C_2 \int_M |f|^2,$$

where C_1 and C_2 are constants depending only on M. Inequality (10.6) now takes the form

$$-\lambda_i \int_M u^k \leq -\frac{2C_1}{k} \left(\int_M u^{km/(m-2)} \right)^{(m-2)/m} + \frac{2C_2}{k} \int_M u^k.$$

with $u = |\psi|$, hence

$$\|\psi_i\|_{2\beta^{j+1}} \leq \left(\frac{\beta^j \lambda_i + C_2}{C_1} \right)^{1/2\beta^j} \|\psi_i\|_{2\beta^j}$$

and

$$\|\psi_i\|_\infty \leq C_3 \left(\lambda_i^{m/4} + C_4 \right).$$

We can also estimate λ_i using a similar method and the existence of $K(x, y, t)$ is established.

The positivity of $K(x, y, t)$ can be seen by applying the Hopf boundary point lemma asserting that if $(x, t) \in \partial M \times (0, \infty)$ is a maximum point for a solution to the heat equation, $f(x, t)$, then $(\partial f/\partial \nu)(x, t) > 0$. In particular, if $f_0 \geq 0$, then $f(x, t) = \int_M K(x, y, t) f_0(y) \, dy$ must have its maximum on $M \times \{0\}$ since $\partial f/\partial \nu = 0$ on $\partial M \times (0, \infty)$, hence $f(x, t) \geq 0$. Again this implies that $K(x, y, t)$ is positive on $M \times M \times (0, \infty)$. The uniqueness of K also follows. In particular, $\int_M K(x, y, t) \, dy = 1$ because the constant 1 also solves the Neumann heat equation with 1 as the initial condition. We now have the following version of Theorem 10.1 for the Neumann heat kernel.

We would also like to point out that in a [WZ], Wang and Zhou found an elegant way to obtain the estimates for eigenfunctions using gradient estimates. In fact, their arguments also apply to eigenforms and yield an estimate on the first nonzero eigenvalues for p-forms.

Theorem 10.2 *The kernel function*

$$K(x, y, t) = \sum_{i=0}^{\infty} e^{-\lambda_i t} \psi_i(x) \psi_i(y)$$

is well defined on $M \times M \times (0, \infty)$. It is the unique kernel such that for any $f_0 \in L^2(M)$ the function given by

$$f(x, t) = \int_M K(x, y, t) f_0(y) \, dy$$

solves the heat equation

$$\left(\Delta - \frac{\partial}{\partial t}\right) f(x, t) = 0 \quad \text{on} \quad M \times (0, \infty)$$

and

$$\frac{\partial f}{\partial \nu}(x, t) = 0 \quad \text{on} \quad \partial M \times (0, \infty),$$

with

$$\lim_{t \to 0} f(x, t) = f_0(x).$$

Moreover, $K(x, y, t)$ is positive on $M \times M \times (0, \infty)$ and it has the property that

$$\int_M K(x, y, t) = 1.$$

We will now establish the fact that the Dirichlet heat kernel has a minimal property among all heat kernels.

Definition 10.3 Let M be a manifold with or without boundary. We say that $H(x, y, t)$ is a heat kernel if it is positive, symmetric in the x and y variables, and satisfies the heat equation

$$\left(\Delta - \frac{\partial}{\partial t}\right) H(x, y, t) = 0$$

with the initial condition

$$\lim_{t \to 0} H(x, y, t) = \delta_x(y),$$

where $\delta_x(y)$ denotes the point mass delta function at x.

Note that if $f(x, t)$ and $g(x, t)$ are two solutions of the heat equation, then the maximum principle allows us to compare $f(x, t)$ with $g(x, t)$ by considering the difference $f(x, t) - g(x, t)$. However, if $K(x, y, t)$ and $H(x, y, t)$ are heat kernels, it is difficult to compare their values at $t = 0$. The following proposition gives an effective way of comparing them and it is referred to as the Duhamel principle.

Proposition 10.4 *Let M be a Riemannian manifold with boundary. If $H(x, y, t)$ and $K(x, y, t)$ are two heat kernels, then*

$$K(x, z, t) - H(x, z, t) = \int_0^t \int_{\partial M} \left(H(x, y, t-s) \left(\frac{\partial}{\partial \nu_y} K(y, z, s) \right) \right.$$
$$\left. - \left(\frac{\partial}{\partial \nu_y} H(x, y, t-s) \right) K(x, y, s) \right) dy\, ds$$

on $M \times M \times (0, \infty)$, where $\partial/\partial \nu_y$ denotes the outward normal derivative with respect to the y variable.

Proof Observe that since the two kernels $H(x, y, t)$ and $K(x, y, t)$ are delta functions at $t = 0$, it follows that

$$\int_0^t \frac{\partial}{\partial s} \int_M H(x, y, t-s) K(y, z, s) \, dy\, ds = K(x, z, t) - H(x, z, t).$$

On the other hand, since they both satisfy the heat equation, we have

$$\int_0^t \frac{\partial}{\partial s} \int_M H(x, y, t-s) K(y, z, s) \, dy\, ds$$
$$= -\int_0^t \int_M \left(\frac{\partial}{\partial (t-s)} H(x, y, t-s) \right) K(y, z, s) \, dy\, ds$$
$$+ \int_0^t \int_M H(x, y, t-s) \left(\frac{\partial}{\partial s} K(y, z, s) \right) dy\, ds$$
$$= -\int_0^t \int_M \Delta_y H(x, y, t-s) K(y, z, s) \, dy\, ds$$
$$+ \int_0^t \int_M H(x, y, t-s) \Delta_y K(y, z, s) \, dy\, ds$$
$$= -\int_0^t \int_{\partial M} \left(\frac{\partial}{\partial \nu_y} H(x, y, t-s) \right) K(y, z, s) \, dy\, ds$$
$$+ \int_0^t \int_{\partial M} H(x, y, t-s) \left(\frac{\partial}{\partial \nu_y} K(y, z, s) \right) dy\, ds.$$

Hence we conclude that

$$K(x,z,t) - H(x,z,t) = \int_0^t \int_{\partial M} H(x,y,t-s) \left(\frac{\partial}{\partial \nu_y} K(y,z,s)\right) dy\, ds$$
$$- \int_0^t \int_{\partial M} \left(\frac{\partial}{\partial \nu_y} H(x,y,t-s)\right) K(y,z,s)\, dy\, ds.$$

□

Corollary 10.5 *Let M be a manifold with boundary. The Dirichlet heat kernel $H(x,y,t)$ is minimal among all heat kernels. In particular, the Neumann heat kernel $K(x,y,t)$ dominates $H(x,y,t)$ with*

$$K(x,y,t) > H(x,y,t)$$

on $M \times M \times (0,\infty)$.

Proof According to the previous proposition, the boundary condition of $H(x,y,t)$ asserts that

$$K(x,z,t) - H(x,z,t) = -\int_0^t \int_{\partial M} \left(\frac{\partial}{\partial \nu_y} H(x,y,t-s)\right) K(x,y,s)\, dy\, ds.$$

Using the facts that $K(x,y,s) > 0$, $H(x,y,t-s) > 0$ with $y \in M \setminus \partial M$, and $H(x,y,t-s) = 0$ with $y \in \partial M$, we conclude that the right-hand side is positive by the Hopf boundary lemma, this proves the inequality. □

We also have the following monotonicity property for the Dirichlet heat kernel.

Corollary 10.6 *Let Ω_1 and Ω_2 be two compact subdomains of M with the property that $\Omega_1 \subset \Omega_2$. Suppose $H_1(x,y,t)$ and $H_2(x,y,t)$ are their corresponding Dirichlet heat kernels. Then*

$$H_1(x,y,t) \le H_2(x,y,t)$$

on $\Omega_1 \times \Omega_1 \times (0,\infty)$.

11
Properties and estimates of the heat kernel

Recall that on Euclidean space $\mathbb{R}^m$, the heat kernel is given by the formula

$$H(x, y, t) = (4\pi t)^{-m/2} \exp\left(-\frac{r^2(x, y)}{4t}\right).$$

Since all manifolds are locally Euclidean, this formula is in fact an approximation of any heat kernel as $t \to 0$.

Theorem 11.1 *Let M be a complete Riemannian manifold (with or without boundary). Suppose $H(x, y, t)$ is a heat kernel. Then for any $x \in M \setminus \partial M$,*

$$H(x, y, t) \sim (4\pi t)^{-\frac{m}{2}} \exp\left(-\frac{r^2(x, y)}{4t}\right)$$

as $t \to 0$ and $r(x, y) \to 0$. Also, for fixed $x, y \in M \setminus \partial M$ with $x \neq y$, we have $H(x, y, t) \to 0$ as $t \to 0$.

Proof For any integer $k > 0$, let us define the function

$$G(x, y, t) = (4\pi t)^{-m/2} \exp\left(-\frac{r^2(x, y)}{4t}\right) \sum_{i=0}^{k} u_i(x, y) t^i.$$

Taking the Laplacian with respect to the y-variable, we have

$$\Delta G = (4\pi t)^{-m/2} \exp\left(-\frac{r^2}{4t}\right)$$

$$\times \left(\left(-\frac{r\,\Delta r}{2t} - \frac{1}{2t} + \frac{r^2}{4t^2}\right)\sum_{i=0}^{k} u_i\, t^i - \frac{r}{t}\sum_{i=0}^{k}\langle \nabla r, \nabla u_i\rangle\, t^i + \sum_{i=0}^{k}(\Delta u_i)\, t^i\right),$$

109

also

$$\frac{\partial G}{\partial t} = (4\pi t)^{-\frac{m}{2}} \exp\left(-\frac{r^2}{4t}\right) \left(\left(-\frac{m}{2t} + \frac{r^2}{4t^2}\right) \sum_{i=0}^{k} u_i t^i + \sum_{i=1}^{k} i u_i t^{i-1}\right).$$

Hence,

$$\left(\Delta - \frac{\partial}{\partial t}\right) G = (4\pi t)^{-m/2} \exp\left(-\frac{r^2}{4t}\right)$$

$$\times \left(\left(-\frac{r\Delta r}{2} + \frac{(m-1)}{2}\right) \sum_{i=-1}^{k-1} u_{i+1} t^i - r \sum_{i=-1}^{k-1} \langle \nabla r, \nabla u_{i+1} \rangle t^i \right.$$

$$\left. + \sum_{i=0}^{k} (\Delta u_i) t^i - \sum_{i=0}^{k-1} (i+1) u_{i+1} t^i \right).$$

For a fixed x, let y be within the injectivity radius of x, we now choose $u_i(x, y)$ as functions of y to satisfy

$$\left(\frac{r\Delta r}{2} - \frac{m-1}{2}\right) u_0 + r \langle \nabla r, \nabla u_0 \rangle = 0, \qquad (11.1)$$

and then inductively

$$\left(\frac{r\Delta r}{2} - \frac{m-1}{2}\right) u_{i+1} + r \langle \nabla r, \nabla u_{i+1} \rangle + (i+1) u_{i+1} = \Delta u_i \qquad (11.2)$$

for $0 \leq i \leq k-1$.

To see that this can be done, we observe that $\Delta r(x, y) = J^{-1}(y) (\partial J/\partial r)(y)$, where $J(y)$ is the area element of the sphere of radius $r(x, y)$ at the point y. Equation (11.1) can be written as

$$\frac{\partial u_0}{\partial r} + \frac{1}{2J} \frac{\partial J}{\partial r} u_0 - \frac{(m-1)}{2r} u_0 = 0,$$

and $u_0 = C r^{(m-1)/2} J^{-1/2}$, with the constant C chosen to satisfy

$$u_0(x, y) = 1, \qquad (11.3)$$

is a solution of (11.1). Equation (11.2) can be written as

$$\frac{\partial u_{i+1}}{\partial r} + \frac{1}{2J} \frac{\partial J}{\partial r} u_{i+1} - \frac{m-3-2i}{2r} u_{i+1} = r^{-1} \Delta u_i,$$

11 Properties and estimates of the heat kernel

which can be solved by setting

$$u_{i+1} = r^{(m-3-2i)/2} J^{-1/2} \int_0^r s^{(2i+1-m)/2} J^{\frac{1}{2}} \Delta u_i \, ds.$$

In particular, we obtain

$$\left(\Delta - \frac{\partial}{\partial t}\right) G = (4\pi t)^{-m/2} \exp\left(-\frac{r^2}{4t}\right) \Delta u_k \, t^k.$$

Note that these computations are valid when $y \notin \partial M$ is within the injectivity radius of x. Let us choose $\phi(y)$ to be a cutoff function satisfying

$$\phi(y) = \begin{cases} 1 & \text{on} \quad B_x(\rho), \\ 0 & \text{on} \quad M \setminus B_x(2\rho), \end{cases}$$

where 2ρ is chosen to be less than the injectivity radius at x and the distance from x to ∂M. For a fixed $x \notin \partial M$, we now define $F(x, y, t) = \phi(y) G(x, y, t)$, which is supported on $B_x(2\rho)$ and it is equal to $G(x, y, t)$ on $B_x(\rho)$. In particular, on $B_x(2\rho) \setminus B_x(\rho)$,

$$\left| \left(\Delta - \frac{\partial}{\partial t}\right) F(x, y, t) \right|$$

$$= \left| \phi \, (4\pi t)^{-m/2} \exp\left(-\frac{r^2}{4t}\right) \Delta u_k \, t^k + 2\langle \nabla \phi, \nabla G \rangle + G \, \Delta \phi \right|$$

$$\leq C_2 \, t^{-m/2} \exp\left(-\frac{\rho^2}{4t}\right). \tag{11.4}$$

Note that since all manifolds are locally Euclidean and because of (11.3), $\lim_{t \to 0} G(x, y, t) = \delta_x(y)$. In particular,

$$F(x, y, t) - H(x, y, t) = \int_0^t \frac{\partial}{\partial s} \int_M H(z, y, t-s) \, F(x, z, s) \, dz \, ds$$

$$= -\int_0^t \int_M \Delta_z H(z, y, t-s) \, F(x, z, s) \, dz \, ds$$

$$+ \int_0^t \int_M H(z, y, t-s) \left(\frac{\partial}{\partial s} F(x, z, s)\right) dz \, ds$$

$$= -\int_0^t \int_M H(z, y, t-s) \left(\Delta - \frac{\partial}{\partial s}\right) F(x, z, s) \, dz \, ds.$$

Hence using the fact that $\int_M H(x, y, t)\, dy \leq 1$ and (11.4), we have

$$|H(x, y, t) - F(x, y, t)| \leq C_1 \int_0^t s^{k-m/2} \int_{B_x(\rho)} H(z, y, s)\, dz\, ds$$

$$+ C_2 \int_0^t s^{-m/2} \exp\left(-\frac{\rho^2}{4s}\right)$$

$$\times \int_{B_x(2\rho)\setminus B_x(\rho)} H(z, y, s)\, dz\, ds$$

$$\leq C_3\, t^{k+1-m/2}.$$

If $y \notin B_x(\rho)$, since

$$F(x, y, t) = \phi(y) G(x, y, t)$$

and

$$G(x, y, t) \to 0 \text{ as } t \to 0,$$

we conclude that

$$H(x, y, t) \to 0 \text{ as } t \to 0.$$

On the other hand, if $y \in B_x(\rho)$, then

$$\left| H(x, y, t) - (4\pi t)^{-m/2} \exp\left(-\frac{r^2(x, y)}{4t}\right) \sum_{i=0}^k u_i(x, y)\, t^i \right| \leq C_3\, t^{k+1-m/2}$$

and the theorem follows because of (11.3). $\square$

Another important property of the heat kernel is the semi-group property. In fact, this follows from the definition of $H(x, y, t) = \sum_{i=1}^\infty e^{-\mu_i t} \phi_i(x) \phi_i(y)$. Using the fact that ϕ_i are orthonormal with respect to L^2, we have

$$\int_M H(x, z, t) H(z, y, s)\, dz = \sum_{i,j} e^{-\mu_i t} e^{-\mu_j s}\, \phi_i(x) \phi_j(y) \int_M \phi_i(z) \phi_j(z)\, dz$$

$$= \sum_{i=1}^\infty e^{-\mu_i (t+s)} \phi_i(x) \phi_i(y)$$

$$= H(x, y, t+s). \tag{11.5}$$

11 Properties and estimates of the heat kernel

In particular,

$$H(x, x, 2t) = \int_M H^2(x, y, t)\,dy \tag{11.6}$$

and

$$H(x, y, t+s) \leq H^{1/2}(x, x, 2t)\,H^{1/2}(y, y, 2s). \tag{11.7}$$

This can be used to estimate the heat kernel from above.

Theorem 11.2 *Let M be a Riemannian manifold with boundary. The Dirichlet heat kernel is bounded from above by*

$$H(x, y, t) \leq \left(\frac{2C_{\mathcal{SD}}}{m}\right)^{-m/2} t^{-m/2},$$

for all $x, y \in M$ and $t > 0$, where $C_{\mathcal{SD}} > 0$ is the Sobolev constant from (10.5). Moreover, the kth Dirichlet eigenvalue must satisfy the estimate

$$\mu_k \geq \frac{2e\,C_{\mathcal{SD}}}{m} \left(\frac{k}{V(M)}\right)^{2/m}$$

and the kth Dirichlet eigenfunction must satisfy

$$\|\phi_k\|_\infty^2 \leq e\left(\frac{2C_{\mathcal{SD}}}{m}\right)^{-m/2} \mu_k^{m/2}\,\|\phi_k\|_2^2.$$

Proof Differentiating (11.6), we have

$$\frac{\partial}{\partial t} H(x, x, 2t) = 2\int_M H(x, y, t)\,\frac{\partial}{\partial t} H(x, y, t)\,dy$$

$$= 2\int_M H(x, y, t)\,\Delta H(x, y, t)\,dy$$

$$= -2\int_M |\nabla H(x, y, t)|^2\,dy. \tag{11.8}$$

The Sobolev inequality (10.5) asserts that

$$\int_M |\nabla H(x, y, t)|^2\,dy \geq C_{\mathcal{SD}} \left(\int_M |H(x, y, t)|^{2m/(m-2)}\,dy\right)^{(m-2)/m}, \tag{11.9}$$

where $C_{SD} > 0$ is the Sobolev constant. On the other hand, Hölder inequality implies that

$$\int_M H^2(x, y, t)\, dy \leq \left(\int_M H^{2m/(m-2)}(x, y, t)\, dy\right)^{(m-2)/(m+2)}$$
$$\times \left(\int_M H(x, y, t)\, dy\right)^{4/(m+2)}$$
$$\leq \left(\int_M H^{2m/(m-2)}(x, y, t)\, dy\right)^{(m-2)/(m+2)}.$$

Combining this inequality with (11.8) and (11.9), we have

$$\frac{\partial}{\partial t} H(x, x, 2t) \leq -2C_{SD} H^{(m+2)/m}(x, x, 2t),$$

implying

$$\frac{\partial}{\partial s} H^{-2/m}(x, x, s) \geq \frac{2C_{SD}}{m}.$$

Integrating over $\epsilon \leq s \leq t$ yields

$$H^{-2/m}(x, x, s) - H^{-2/m}(x, x, \epsilon) \geq \frac{2C_{SD}}{m}(s - \epsilon).$$

Since Theorem 11.1 asserts that

$$H(x, x, \epsilon) \sim (4\pi\epsilon)^{-m/2},$$

we conclude that

$$\lim_{\epsilon \to 0} H^{2/m}(x, x, \epsilon) = 0,$$

hence

$$H(x, x, s) \leq \left(\frac{2C_{SD}}{m} s\right)^{-m/2}.$$

Inequality (11.7) then completes the first part of the theorem.

The second part follows from the fact that

$$\sum_{i=1}^{\infty} e^{-\mu_i t} = \int_M H(x, x, t)\, dx,$$

hence combining this identity with the estimate from the first part gives

$$\sum_{i=1}^{\infty} e^{-\mu_i t} \leq \left(\frac{2C_{SD}}{m}\right)^{-m/2} t^{-m/2} V(M).$$

11 Properties and estimates of the heat kernel

Setting $t = \mu_k^{-1}$, we have

$$ke^{-1} \leq \sum_{i=1}^{k} e^{-\mu_i/\mu_k}$$

$$\leq \left(\frac{2C_{SD}}{m}\right)^{-m/2} \mu_k^{m/2} V(M),$$

yielding the lower bound for μ_k. We also have

$$e^{-\mu_k t} \phi_k^2(x) \leq H(x,x,t)$$

$$\leq \left(\frac{2C_{SD}}{m}\right)^{-m/2} t^{-m/2},$$

where ϕ_k is the kth eigenfunction with $\|\phi_k\|_2 = 1$. Again setting $t = \mu_k^{-1}$ yields the estimate

$$\|\phi_k\|_\infty^2 \leq e \left(\frac{2C_{SD}}{m}\right)^{-m/2} \mu_k^{m/2}. \qquad \square$$

Note that the estimate on μ_k is an improvement from that obtained in the proof of Theorem 10.1. We also have the corresponding theorem for the Neumann heat kernel. In this case, the Sobolev inequality required is of the form

$$\int_M |\nabla f|^2 \geq C_{SN} \left(\int_M |f|^{2m/(m-2)}\right)^{(m-2)/m} \qquad (11.10)$$

for all $f \in H^{1,2}(M)$ satisfying $\int_M f = 0$. Observe that the Sobolev inequality given in Corollary 9.9 implies the above one since the L^2-norm of f can be estimated by

$$\lambda_1(M) \int_M f^2 \leq \int_M |\nabla f|^2.$$

Theorem 11.3 *Let M be a Riemannian manifold with boundary. The Neumann heat kernel is bounded from above and below by*

$$\frac{1}{V(M)} - \left(\frac{2C_{SN}}{m}\right)^{-m/2} t^{-m/2} \leq K(x,y,t)$$

$$\leq \frac{1}{V(M)} + \left(\frac{2C_{SN}}{m}\right)^{-m/2} t^{-m/2},$$

for all $x, y \in M$ and $t > 0$, where $C_{SN} > 0$ is the Sobolev constant from (11.10). Moreover, the kth Neumann eigenvalue must satisfy the estimate

$$\lambda_k \geq \frac{2^{(m-4)/m} e C_{SN}}{m} \left(\frac{k}{V(M)}\right)^{2/m},$$

and the kth Neumann eigenfunction must satisfy

$$\|\psi_k\|_\infty^2 \leq e \left(\frac{2^{(m-4)/m} C_{SN}}{m}\right)^{-m/2} \lambda_k^{m/2} \|\psi_k\|_2^2.$$

Proof We will estimate the function

$$K_1(x, y, t) = K(x, y, t) - \frac{1}{V(M)}$$

$$= \sum_{i=1}^{\infty} e^{-\lambda_i t} \psi_i(x) \psi_i(y)$$

instead, since $\psi_0 = V^{-\frac{1}{2}}(M)$. One checks readily that K_1 also satisfies the semi-group property. Moreover,

$$\int_M K_1(x, y, t) \, dy = 0$$

and

$$\int_M |K_1(x, y, t)| \, dy \leq \int_M K(x, y, t) \, dy + 1$$

$$\leq 2.$$

The same argument as in the proof of Theorem 11.2 now works on K_1 and produces

$$K_1(x, x, s) \leq \left(\frac{2^{(m-4)/m} C_{SN}}{m} s\right)^{-m/2}.$$

Hence the above inequality together with the inequality

$$|K_1(x, y, t)| \leq K_1^{1/2}(x, x, t) K_1^{1/2}(y, y, t)$$

yields the desired estimate on K_1. The rest of the proof is identical to that of Theorem 11.2. □

To summarize, we have proved that Sobolev inequalites of the form (10.6) and (11.10) imply upper bounds of the Dirichlet and the Neumann heat kernel,

11 Properties and estimates of the heat kernel

respectively. The converse is also true as we will demonstrate in this chapter. The following theorem was proved by Varopoulos in [V].

Let us first state the Marcinkiewicz interpolation theorem as Lemma 11.5 without proof. The interested reader should refer to Appendix B of [St] for further details.

Definition 11.4 Let T be a subadditive operator on $L^p(M)$, i.e.,

$$|T(f+g)(x)| \leq |T(f)(x)| + |T(g)(x)|,$$

for all $f, g \in L^p(M)$ and for all $x \in M$. T is of weak (p, q) type if there exists a constant A, such that for $f \in L^p(M)$,

$$m\{x \mid |T(f)(x)| > \beta\} \leq \left(\frac{A \|f\|_p}{\beta}\right)^q$$

for all $\beta > 0$. If $q = \infty$, this means

$$\|T(f)\|_\infty \leq A \|f\|_p$$

for all $f \in L^p(M)$.

Lemma 11.5 *Suppose that $1 \leq p_i \leq q_i \leq \infty$, for $i = 1, 2$ is such that $p_1 < p_2$ and $q_1 \neq q_2$. Assume that T is an operator of weak (p_1, q_1) type and weak (p_2, q_2) type. Then for any $0 < \theta < 1$ with*

$$\frac{1}{p} = \frac{1-\theta}{p_1} + \frac{\theta}{p_2} \quad \text{and} \quad \frac{1}{q} = \frac{1-\theta}{q_1} + \frac{\theta}{q_2},$$

T is a bounded operator from $L^p(M)$ to $L^q(M)$.

Theorem 11.6 *Let M be a Riemannian manifold with boundary. Suppose there is a constant $A_1 > 0$, such that the Dirichlet heat kernel satisfies the estimate*

$$H(x, y, t) \leq A_1 t^{-m/2}$$

for all $x, y \in M$ and for all $t \in (0, \infty)$. Then there exists a constant $C_1 > 0$ depending only on m such that

$$\int_M |\nabla f|^2 \geq C_1 A_1^{-2/m} \left(\int_M |f|^{2m/(m-2)}\right)^{(m-2)/m}$$

for all $f \in H^{1,2}(M)$ with $f|_{\partial M} = 0$.

If there is a constant $A_2 > 0$, such that the Neumann heat kernel satisfies the estimate

$$K(x, y, t) \leq \frac{1}{V(M)} + A_2 t^{-m/2}$$

for all $x, y \in M$ and for all $t \in (0, \infty)$, then there exists a constant $C_2 > 0$ depending only on m, such that

$$\int_M |\nabla f|^2 \geq C_2 A_2^{-2/m} \left(\int_M |f|^{2m/(m-2)} \right)^{(m-2)/m}$$

for all $f \in H^{1,2}(M)$ with $\int_M f = 0$.

Proof We will first prove the case for the Dirichlet heat kernel. Let $\{\phi_i\}$ be the set of orthonormal eigenfunctions with eigenvalues $\{\mu_i\}$. For all $f \in H^{1,2}(M)$ with $f|_{\partial M} = 0$, we write

$$f = \sum_{i=1}^{\infty} a_i \phi_i$$

and hence

$$\Delta f = -\sum_{i=1}^{\infty} \mu_i a_i \phi_i.$$

For any $\alpha \in \mathbb{R}$, we define the psuedo-differential operator $(-\Delta)^\alpha$ by

$$(-\Delta)^\alpha f = \sum_{i=1}^{\infty} \mu_i^\alpha a_i \phi_i.$$

We can write

$$\int_M |\nabla f|^2 = -\int_M f \Delta f$$

$$= \sum_{i=1}^{\infty} \mu_i a_i^2$$

$$= \int_M |(-\Delta)^{1/2} f|^2,$$

hence it suffices to show that

$$\int_M |(-\Delta)^{1/2} f|^2 \geq C_1 A_1^{-2/m} \left(\int_M |f|^{2m/(m-2)} \right)^{(m-2)/m}.$$

11 Properties and estimates of the heat kernel

Let $g = (-\Delta)^{1/2} f$, hence $f = (-\Delta)^{-1/2} g$. We need to show that the operator

$$(-\Delta)^{-1/2} : L^2(M) \to L^{2m/(m-2)}(M)$$

is bounded and satisfies the bound

$$\left(\int_M |(-\Delta)^{-1/2} g|^{2m/(m-2)} \right)^{(m-2)/m} \leq C A_1^{2/m} \int_M |g|^2.$$

Expressing $(-\Delta)^{1/2} g$ in terms of the heat kernel, if $g = \sum_{i=1}^{\infty} b_i \phi_i$ then we have

$$(-\Delta)^{-1/2} g(x) = \sum_{i=1}^{\infty} \mu_i^{-1/2} b_i \phi_i(x)$$

$$= \int_M \sum_{i=1}^{\infty} \mu_i^{-1/2} \phi_i(x) \phi_i(y) g(y) \, dy$$

$$= B \int_0^{\infty} \int_M t^{-1/2} e^{-\mu_i t} \phi_i(x) \phi_i(y) g(y) \, dy$$

$$= B \int_0^{\infty} t^{-1/2} \int_M H(x, y, t) g(y) \, dy,$$

where $B^{-1} = \int_0^{\infty} s^{-1/2} e^{-s} \, ds$. For $0 < T < \infty$, let us define the functionals T_1 and T_2 by

$$T_1(g) = B \int_0^T t^{-1/2} \int_M H(x, y, t) g(y) \, dy \, dt$$

and

$$T_2(g) = B \int_T^{\infty} t^{-1/2} \int_M H(x, y, t) g(y) \, dy \, dt,$$

hence $(-\Delta)^{-1/2} = T_1 + T_2$. For $\beta > 0$, obviously the measure of the sublevel set $\{(-\Delta)^{-1/2} g \geq \beta\}$ can be estimated by

$$m\{x \mid |(-\Delta)^{-1/2} g(x)| > \beta\} \leq m\left\{x \mid |T_1 g(x)| > \frac{\beta}{2}\right\}$$
$$+ m\left\{x \mid |T_2 g(x)| > \frac{\beta}{2}\right\}.$$

On the other hand, for $1/p + 1/q = 1$, using the upper bound of H, we have

$$|T_2 g(x)| \leq B \int_T^\infty t^{-1/2} \int_M H(x, y, t) |g(y)| \, dy \, dt$$

$$\leq B \|g\|_q \int_T^\infty t^{-1/2} \left(\int_M H^p(x, y, t) \, dt \right)^{1/p} dt$$

$$\leq B \|g\|_q \int_T^\infty t^{-1/2} \sup_{y \in M} H^{(p-1)/p}(x, y, t) \left(\int_M H(x, y, t) \, dt \right)^{1/p} dt$$

$$\leq C A_1^{1/q} T^{(q-m)/2q} \|g\|_q.$$

For $1 < q < m$, by choosing T such that

$$C A_1^{1/q} T^{(q-m)/2q} \|g\|_q = \frac{\beta}{2}, \tag{11.11}$$

we have

$$m\{x \mid |(-\Delta)^{-1/2} g(x)| > \beta\} \leq m \left\{ x \mid |T_1 g(x)| > \frac{\beta}{2} \right\}$$

$$\leq \left(\frac{\beta}{2} \right)^{-q} \|T_1(g)\|_q^q. \tag{11.12}$$

To estimate $\|T_1 g\|_q$, we consider

$$|T_1 g(x)|^q \leq B^q \left(\int_0^T t^{-1/2} \int_M H(x, y, t) |g(y)| \, dy \, dt \right)^q$$

$$\leq B^q \left(\int_0^T t^{-1/2} \, dt \right)^{q/p} \int_0^T t^{-1/2} \left(\int_M H(x, y, t) |g(y)| \, dy \right)^q dt$$

$$\leq C T^{q/2p} \int_0^T t^{-1/2} \int_M H(x, y, t) |g(y)|^q \, dy \, dt,$$

where we have used the fact that $\int_M H(x, y, t) \, dy \leq 1$. This implies that

$$\int_M |T_1 g(x)|^q \, dx \leq C T^{q/2p} \int_0^T t^{-1/2} \|g\|_q^q \, dt$$

$$= C T^{q/2} \|g\|_q^q.$$

11 Properties and estimates of the heat kernel

Combining this identity with (11.11) and (11.12), we have

$$m\{x \mid |(-\Delta)^{-1/2}g(x)| > \beta\} \leq C \left(\frac{\beta}{2}\right)^{-q} T^{q/2} \|g\|_q^q$$

$$\leq C A_1^{q/(m-q)} \beta^{-mq/(m-q)} \|g\|_q^{mq/(m-q)}.$$

This establishes that the operator $(-\Delta)^{-1/2}$ is of weak $(q, mq/(m-q))$ type, for $1 < q < m$. We now choose $1 < q_1 < 2 < q_2 < m$ and $0 < \theta < 1$, such that,

$$\frac{1}{2} = \frac{1-\theta}{q_1} + \frac{\theta}{q_2},$$

implying

$$\frac{m-2}{2m} = \frac{(1-\theta)(m-q_1)}{mq_1} + \frac{\theta(m-q_2)}{mq_2},$$

and the theorem follows from Lemma 11.5.

The same argument also works for the Neumann heat kernel by defining the operator $(-\Delta)^{-1/2}$ on the subspace of $\mathcal{S} = \{g \in L^2(M) \mid \int_M g = 0\}$ by

$$(-\Delta)^{-1/2}g(x) = \sum_{i=1}^{\infty} \lambda_i^{-1/2} \psi_i(x) \int_M \psi_i(y) g(y) \, dy,$$

where $\{\psi_i\}$ is an orthonormal basis of eigenfunctions corresponding to the nonzero eigenvalues $\{\lambda_i\}$. In this case, we also have

$$(-\Delta)^{-\frac{1}{2}}g(x) = B \int_0^{\infty} \int_M t^{-\frac{1}{2}} K_1(x, y, t) g(y) \, dy \, dt.$$

Using the estimate on K_1, the theorem follows for the Neumann boundary condition. □

12
Gradient estimate and Harnack inequality for the heat equation

In this chapter, we will derive estimates for positive solutions of the heat equation. The gradient estimate given by Theorem 12.2, and the Harnack inequality that follows (Corollary 12.3), have fundamental importance in the study of parabolic equations. There are many further developments of similar types of estimates for other nonlinear partial differential equations. The estimates presented herein were proved by Li and Yau in [LY2].

Lemma 12.1 *Let M^m be a manifold whose Ricci curvature is bounded by*

$$\mathcal{R}_{ij} \geq -(m-1)R,$$

for some constant R. Suppose $g(x, t)$ is a smooth function defined on $M \times [0, \infty)$ satisfying the differential equation

$$\left(\Delta - \frac{\partial}{\partial t}\right) g(x, t) = -|\nabla g|^2(x, t).$$

Then for any given $\alpha \geq 1$, the function

$$G(x, t) = t(|\nabla g|^2(x, t) - \alpha g_t(x, t))$$

satisfies the inequality

$$\left(\Delta - \frac{\partial}{\partial t}\right) G \geq -2\langle \nabla g, \nabla G \rangle - t^{-1} G + 2\alpha^{-2} m^{-1} t^{-1} G^2$$
$$+ 4(\alpha - 1) m^{-1} \alpha^{-2} |\nabla g|^2 G$$
$$- 2^{-1} m(m-1)^2 \alpha^2 (\alpha - 1)^{-2} t R^2.$$

Proof Let $\{e_1, \ldots e_m\}$ be a set of local orthonormal frame on M. Differentiating in the direction of e_i, we have

$$G_i = t(2g_j g_{ji} - \alpha g_{ti}).$$

Covariant differentiating again, we obtain

$$\Delta G = t\left(2g_{ji}^2 + 2g_j g_{jii} - \alpha g_{tii}\right)$$

$$\geq t\left(\frac{2}{m}(\Delta g)^2 + 2\langle \nabla g, \nabla \Delta g\rangle - 2(m-1)R|\nabla g|^2 - \alpha(\Delta g)_t\right), \quad (12.1)$$

where we have used the inequality

$$\sum_{i,j} g_{ji}^2 \geq \frac{(\sum_i g_{ii})^2}{m}$$

and the commutation formula

$$g_j g_{jii} = g_j g_{iij} + \mathcal{R}_{ij} g_i g_j$$

$$\geq \langle \nabla g, \nabla \Delta g\rangle - (m-1)R|\nabla g|^2.$$

However, since

$$\Delta g = -|\nabla g|^2 + g_t$$

$$= -\alpha^{-1}t^{-1}G - \alpha^{-1}(\alpha - 1)|\nabla g|^2$$

(12.1) becomes

$$\Delta G \geq \frac{2t\alpha^{-2}}{m}(t^{-1}G + (\alpha - 1)|\nabla g|^2)^2 - 2\alpha^{-1}\langle \nabla g, \nabla G\rangle$$

$$- 2\alpha^{-1}(\alpha - 1)t\langle \nabla g, \nabla|\nabla g|^2\rangle$$

$$- 2(m-1)Rt|\nabla g|^2 + G_t - t^{-1}G + t(\alpha - 1)(|\nabla g|^2)_t$$

$$= \frac{2t\alpha^{-2}}{m}(t^{-2}G^2 + 2(\alpha - 1)t^{-1}|\nabla g|^2 G + (\alpha - 1)^2|\nabla g|^4)$$

$$- 2\alpha^{-1}\langle \nabla g, \nabla G\rangle$$

$$- 2\alpha^{-1}(\alpha - 1)\langle \nabla g, \nabla G\rangle - 2(m-1)Rt|\nabla g|^2 + G_t - t^{-1}G.$$

The lemma follows by applying the inequality

$$(\alpha - 1)^2 \alpha^{-2} |\nabla g|^4 - m(m-1) R |\nabla g|^2$$
$$\geq -4^{-1} m^2 (m-1)^2 \alpha^2 (\alpha - 1)^{-2} R^2.$$
$\square$

Theorem 12.2 *Let M^m be a complete manifold with boundary. Assume that $p \in M$ and $\rho > 0$ so that the geodesic ball $B_p(4\rho)$ does not intersect the boundary of M. Suppose the Ricci curvature of M on $B_p(4\rho)$ is bounded from below by*

$$\mathcal{R}_{ij} \geq -(m-1) R,$$

for some constant $R \geq 0$. If $f(x, t)$ is a positive solution of the equation

$$\left(\Delta - \frac{\partial}{\partial t}\right) f(x, t) = 0$$

on $M \times [0, T]$, then for any $1 < \alpha$, the function $g(x, t) = \log f(x, t)$ must satisfy the estimate

$$|\nabla g|^2 - \alpha g_t \leq \frac{m}{2} \alpha^2 t^{-1} + C_1 \alpha^2 (\alpha - 1)^{-1} (\rho^{-2} + R)$$

on $B_p(2\rho) \times (0, T]$, where C_1 is a constant depending only on m.

Proof Let $\phi(r(x))$ be a cutoff function, depending only on the distance to p, defined on $B_p(4\rho)$ with the properties

$$\phi(r(x)) = \begin{cases} 1 & \text{on} \quad B_p(2\rho), \\ 0 & \text{on} \quad M \setminus B_p(4\rho), \end{cases}$$

$$-C \rho^{-1} \leq \phi^{-\frac{1}{2}} \phi' \leq 0,$$

and

$$\phi'' \geq -C \rho^{-2},$$

where ϕ' denotes differentiation with respect to r. For

$$G = t(|\nabla g|^2 - \alpha g_t),$$

we consider the function ϕG with support on $B_p(4\rho) \times (0, T]$. Let $(x_0, t_0) \in B_p(4\rho) \times (0, T]$ be the maximum point of ϕG. We may assume that ϕG is

12 Gradient estimate and Harnack inequality for the heat equation

positive at (x_0, t_0) otherwise the theorem is trivial. Note that at (x_0, t_0) we have the properties that

$$\nabla(\phi\, G) = 0, \tag{12.2}$$

$$\frac{\partial}{\partial t}(\phi\, G) \geq 0, \tag{12.3}$$

and

$$\Delta(\phi\, G) \leq 0. \tag{12.4}$$

Also using the comparison theorem and the properties of ϕ we have

$$\Delta\phi \geq -C_2\, \rho^{-2}\left(\rho\sqrt{R} + 1\right).$$

Applying Lemma 12.1 and this estimate to the identity

$$\Delta(\phi\, G) = (\Delta\phi)\, G + 2\langle \nabla\phi, \nabla G\rangle + \phi\,(\Delta G),$$

we obtain

$$\Delta(\phi\, G) \geq -C_2\, \rho^{-2}\left(\rho\sqrt{R} + 1\right) G + 2\phi^{-1}\langle \nabla\phi, \nabla(\phi\, G)\rangle - 2\phi^{-1}|\nabla\phi|^2 G$$

$$+ \phi\left(G_t - 2\langle \nabla g, \nabla G\rangle - t^{-1} G + 2\alpha^{-2} m^{-1} t^{-1} G^2\right.$$

$$\left. + 4(\alpha - 1) m^{-1} \alpha^{-2} |\nabla g|^2 G\right)$$

$$- 2^{-1} m (m-1)^2 \alpha^2 (\alpha - 1)^{-2} t \phi R^2.$$

Evaluating at (x_0, t_0) and using (12.2), (12.3), and (12.4) yields

$$0 \geq -C_2\, \rho^{-2}\left(\rho\sqrt{R} + 1\right) G - 2\phi^{-1}|\nabla\phi|^2 G + 2G\langle \nabla\phi, \nabla g\rangle$$

$$- \phi t_0^{-1} G + 2\alpha^{-2} m^{-1} t_0^{-1} \phi G^2 + 4(\alpha - 1) m^{-1} \alpha^{-2} |\nabla g|^2 \phi G$$

$$- 2^{-1} m(m-1)^2 \alpha^2 (\alpha-1)^{-2} t_0\, \phi\, R^2.$$

Multiplying through by $\phi\, t_0$ and using the estimate for $|\nabla\phi|$, we conclude that

$$0 \geq -C_3\, \rho^{-2}\left(\rho\sqrt{R} + 1\right) t_0\, \phi\, G - \phi^2 G - C_4\, t_0\, \phi^{3/2} \rho^{-1} G\, |\nabla g|$$

$$+ 2\alpha^{-2} m^{-1} \phi^2 G^2 + 4(\alpha - 1) m^{-1} \alpha^{-2} t_0\, |\nabla g|^2 \phi^2 G$$

$$- 2^{-1} m(m-1)^2 \alpha^2 (\alpha-1)^{-2} t_0^2\, \phi^2 R^2. \tag{12.5}$$

Observe that there exists a constant $C_5 > 0$ such that
$$4m^{-1}(\alpha - 1)\alpha^{-2}\phi |\nabla g|^2 - C_4 \rho^{-1}\phi^{\frac{1}{2}} |\nabla g| \geq -C_5 \alpha^2(\alpha - 1)^{-1} \rho^{-2}. \tag{12.6}$$

Hence combining (12.5) and (12.6), we conclude that
$$0 \geq 2m^{-1}\alpha^{-2} (\phi G)^2 - \left(C_6 \rho^{-2} t_0 \left(\rho\sqrt{R} + 1 + \alpha^2(\alpha - 1)^{-1}\right) + 1\right) \phi G$$
$$- 2^{-1} m(m - 1)^2 \alpha^2 (\alpha - 1)^{-2} t_0^2 R^2,$$

for some constant $C_6 > 0$. This implies that at the maximum point (x_0, t_0),
$$\phi G \leq \frac{m\alpha^2}{2}$$
$$+ C_7 t_0 \left(\left(\rho^{-2}\alpha^2 \left(\rho\sqrt{R} + 1 + \alpha^2(\alpha - 1)^{-1}\right) + \alpha^2 (\alpha - 1)^{-1} R\right).$$

In particular, since $t_0 \leq T$, when restricted on $B_p(2\rho) \times \{T\}$ we have
$$|\nabla g|^2 - \alpha g_t$$
$$\leq \frac{m\alpha^2}{2T} + C_7 \left(\left(\rho^{-2}\alpha^2 \left(\rho\sqrt{R} + 1 + \alpha^2(\alpha - 1)^{-1}\right) + \alpha^2 (\alpha - 1)^{-1} R\right).$$

The theorem follows by letting $t = T$. $\square$

Corollary 12.3 *Under the same hypotheses as Theorem 12.2,*
$$f(x, t_1) \leq f(y, t_2) \left(\frac{t_2}{t_1}\right)^{m\alpha/2}$$
$$\times \exp\left(\frac{\alpha r^2(x, y)}{4(t_2 - t_1)} + C_1 \alpha (\alpha - 1)^{-1}(\rho^{-2} + R) (t_2 - t_1)\right)$$

for any $x, y \in B_p(\rho)$ and $0 < t_1 < t_2 \leq T$, where $r(x, y)$ is the geodesic distance between x and y.

Proof Let $\gamma : [0, 1] \to M$ be a minimizing geodesic joining y to x. Since $x, y \in B_p(\rho)$, the triangle inequality asserts that $\ell(\gamma) = r(x, y) \leq r(x) + r(y) \leq 2\rho$, where $\ell(\gamma)$ denotes the length of γ and $r(x)$ is the distance from p to x. We now claim that $\gamma \subset B_p(2\rho)$. Indeed, if this were not the case and $\gamma(s) \notin B_p(2\rho)$ for some $s \in [0, 1]$, then the minimizing property of γ would imply that $\ell(\gamma|_{[0,s]}) > \rho$ and $\ell(\gamma|_{[s,1]}) > \rho$ violating the fact that $\ell(\gamma) = r(x, y) \leq 2\rho$. We define $\eta : [0, 1] \to B_p(\rho) \times [t_1, t_2]$ by
$$\eta(s) = (\gamma(s), (1 - s)t_2 + st_1).$$

12 Gradient estimate and Harnack inequality for the heat equation

Clearly, $\eta(0) = (y, t_2)$ and $\eta(1) = (x, t_1)$. Integrating along η, we obtain

$$\log f(x, t_1) - \log f(y, t_2) = \int_0^1 \left(\frac{d}{ds} \log f\right) ds$$

$$= \int_0^1 \left(\langle \gamma', \nabla(\log f)\rangle - (t_2 - t_1)(\log f)_t\right) ds. \tag{12.7}$$

On the other hand, Theorem 12.2 implies that

$$-(\log f)_t \leq \frac{m}{2}\alpha t^{-1} + A - \alpha^{-1}|\nabla(\log f)|^2,$$

where $A = C_1 \alpha (\alpha - 1)^{-1} (\rho^{-2} + R)$. Hence (12.7) becomes

$$\log\left(\frac{f(x, t_1)}{f(y, t_2)}\right) \leq \int_0^1 \left(|\gamma'||\nabla(\log f)| - (t_2 - t_1)\alpha^{-1}|\nabla(\log f)|^2\right) ds$$

$$+ (t_2 - t_1) \int_0^1 \left(A + \frac{m}{2}\alpha t^{-1}\right) ds,$$

where $t = (1 - s)t_2 + st_1$. Using the fact that

$$-(t_2 - t_1)\alpha^{-1} z^2 + |\gamma'| z \leq \frac{\alpha |\gamma'|^2}{4(t_2 - t_1)}$$

for any z, we conclude that

$$\log\left(\frac{f(x, t_1)}{f(y, t_2)}\right) \leq \frac{\alpha r^2(x, y)}{4(t_2 - t_1)} + \frac{m\alpha}{2} \log\left(\frac{t_2}{t_1}\right) + A(t_2 - t_1),$$

where we have used $r(x, y) = \ell(\gamma) = |\gamma'|$. The corollary follows by exponentiating both sides. □

We are now ready to prove the existence of a minimal heat kernel on any complete (not necessarily compact) manifold.

Theorem 12.4 *Let M^m be a complete, noncompact manifold. There exists a positive, symmetric heat kernel $H(x, y, t)$ on M with the property that, for any $f_0 \in L^2(M)$, the function given by*

$$f(x, t) = \int_M H(x, y, t) f_0(y) dy$$

solves the heat equation

$$\left(\Delta - \frac{\partial}{\partial t}\right) f(x, t) = 0 \quad \text{on} \quad M \times (0, \infty),$$

with the initial condition

$$\lim_{t \to 0} f(x, t) = f_0(x).$$

Moreover, $H(x, y, t)$ satisfies

$$\int_M H(x, y, t)\, dy \leq 1,$$

and it is the unique minimal heat kernel on $M \times M \times (0, \infty)$; namely, if $\bar{H}(x, y, t)$ is another positive heat kernel defined on M, then

$$H(x, y, t) \leq \bar{H}(x, y, t),$$

for all x, $y \in M$ and $t \in (0, \infty)$.

Proof Let $\{\Omega_i\}$ be a compact exhaustion of M satisfying $\Omega_i \subset \Omega_j$ for $i \leq j$ and $\cup_i \Omega_i = M$. On each Ω_i, let $H_i(x, y, t)$ be the Dirichlet heat kernel given by Theorem 10.1. The maximum principle implies that $H_i(x, y, t) \leq H_j(x, y, t)$ for $i \leq j$. For a fixed point $p \in M$ and a fixed radius $\rho > 0$, there exists i_0 sufficiently large such that $B_p(4\rho) \subset \Omega_i$ for $i \geq i_0$. Corollary 12.3 asserts that there exists a constant C depending only on ρ, t, and the lower bound of the Ricci curvature on $B_p(4\rho)$, such that

$$H_i(z, x, t) \leq C\, H_i(z, y, 2t)$$

for all $z \in \Omega_i$ and x, $y \in B_p(\rho)$. Combined with the fact that $\int_{\Omega_i} H_i(z, y, 2t)\, dy \leq 1$, this yields

$$H_i(z, x, t) \leq C\, V_p^{-1}(\rho).$$

Hence the monotonically increasing sequence $\{H_i(z, x, t)\}$ converges uniformly on compact subsets of $M \times M \times (0, \infty)$ to a kernel function $H(z, x, t)$. Obviously, $H(z, x, t)$ is positive and symmetric in the space variables, and since

$$\int_{B_p(\rho)} H(x, y, t)\, dy \leq \lim_{i \to \infty} \int_{B_p(\rho)} H_i(x, y, t)\, dy$$

$$\leq 1,$$

we also obtain

$$\int_M H(x, y, t)\, dy \leq 1.$$

Moreover, the local gradient estimate of Theorem 12.2 asserts that

$$|\nabla f|^2 - \alpha f f_t \leq C f^2 \quad \text{on} \quad B_p(\rho)$$

for $f(x, t) = H_i(z, x, t)$. In particular, if ϕ is a nonnegative cutoff function supported on $B_p(\rho)$, then

$$\int_M \phi^2 |\nabla f|^2 \leq \alpha \int_M \phi^2 f f_t + C \int_M \phi^2 f^2$$

$$\leq \alpha \int_M \phi^2 f \Delta f + C \int_M \phi^2 f^2$$

$$\leq -\alpha \int_M \phi^2 |\nabla f|^2 - 2\alpha \int_M \phi f \langle \nabla \phi, \nabla f \rangle + C \int_M \phi^2 f^2.$$

Applying the Schwarz inequality

$$-2 \int_M \phi f \langle \nabla \phi, \nabla f \rangle \leq \int_M \phi^2 |\nabla f|^2 + \int_M |\nabla \phi|^2 f^2,$$

we obtain the estimate

$$\int_{B_p(\rho/2)} |\nabla f|^2 \leq C \int_M \phi^2 f^2 + \alpha \int_M |\nabla \phi|^2 f^2$$

$$\leq C_1.$$

by choosing $\phi = 1$ on $B_p(\rho/2)$. Hence the sequence $H_i(z, x, t)$ converges in $H^{1,2}$-norm over any compact subset of M. In particular, $H(z, x, t)$ solves the heat equation.

The semi-group property of $H_i(z, x, t)$ asserts that

$$\int_{\Omega_j} H_i(z, y, t) H_i(x, y, s) \, dy \leq \int_{\Omega_i} H_i(z, y, t) H_i(x, y, s) \, dy$$

$$= H_i(z, x, t + s),$$

for $i \geq j$. Hence letting $i \to \infty$, we conclude that

$$\int_{\Omega_j} H(z, y, t) H(x, y, s) \, dy \leq H(z, x, t + s).$$

In particular, by setting $x = z$ and $t = s$, this implies that $H(z, y, t)$ is in L^2 with respect to the y variable, hence also the z variable because of symmetry. Letting $j \to \infty$, we also conclude that

$$\int_M H(z, y, t) H(x, y, s) \, dy \leq H(z, x, t + s).$$

On the other hand, since $\{H_i\}$ is monotonically increasing, we have

$$\int_{\Omega_i} H(z, y, t) H(x, y, s) \, dy \geq \int_{\Omega_i} H_i(z, y, t) H_i(x, y, s) \, dy$$
$$= H_i(z, x, t + s),$$

hence

$$\int_M H(z, y, t) H(x, y, s) \, dy \geq H(z, x, t + s).$$

Therefore H also satisfies the semi-group property

$$\int_M H(z, y, t) H(x, y, s) \, dy = H(z, x, t + s).$$

Obviously, if f_0 is a continuous function with compact support, then

$$f(x, t) = \int_M H(x, y, t) f_0(y) \, dy$$

will satisfy the heat equation with initial data f_0, since this is the case for the function

$$f_i(x, t) = \int_M H_i(x, y, t) f_0(y) \, dy$$

when i is sufficiently large. For f_0 just continuous and in $L^2(M)$, we see that this is also the case by an approximation argument since $H \in L^2(M)$. □

In the above proof, the estimates on $H_i(x, y, t)$ are interior estimates and the upper bound does not depend on the boundary condition. In particular, the same estimates are valid for Neumann heat kernels $K_i(x, y, t)$ on Ω_i. Though the sequence $\{K_i(x, y, t)\}$ is not necessarily monotonic, the kernels are uniformly bounded on compact subsets, and one can extract a convergenct subsequence replacing the original sequence $K_i(x, y, t)$ converging to a kernel $K(x, y, t)$. However, Corollary 10.5 asserts that $H_i(x, y, t) < K_i(x, y, t)$ and by taking the limit, we conclude that

$$H(x, y, t) \leq K(x, y, t). \tag{12.8}$$

An interesting question to ask is if $H(x, y, t) = K(x, y, t)$. Recall that $\int_M H(x, y, t) \, dy \leq 1$, and one can also show that $\int_M K(x, y, t) \, dy \leq 1$. Hence if

$$\int_M H(x, y, t) \, dy = 1, \tag{12.9}$$

then this identity combined with (12.8) implies that $H(x, y, t) = K(x, y, t)$. We will address the validity of (12.9) in a later chapter.

The estimates in Theorem 12.2 and Corollary 12.3 take on much simpler forms when the function is defined globally on a complete manifold with nonnegative Ricci curvature. In fact, the estimates are sharp on $\mathbb{R}^m$.

Corollary 12.5 *Let M^m be a complete manifold with nonnegative Ricci curvature. If $f(x, t)$ is a positive solution of the heat equation*

$$\left(\Delta - \frac{\partial}{\partial t}\right) f(x, t) = 0$$

on $M \times [0, \infty)$, then

$$\frac{|\nabla f|^2}{f^2} - \frac{f_t}{f} \leq \frac{m}{2t}$$

on $M \times [0, \infty)$. Moreover,

$$f(x, t_1) \leq f(y, t_2) \left(\frac{t_2}{t_1}\right)^{m/2} \exp\left(\frac{r^2(x, y)}{4(t_2 - t_1)}\right)$$

for all $x, y \in M$ and $0 < t_1 < t_2 < \infty$.

Proof Apply Theorem 12.2 by first taking $\rho \to \infty$, then $\alpha \to 1$. □

The gradient estimate is also valid for manifolds with boundary when the boundary is assumed to satisfy some convexity hypothesis. In the case of the Dirichlet boundary condition, one needs to assume that the boundary has nonnegative mean curvature with respect to the outward normal. For the Neumann boundary condition, one needs to assume that the boundary is convex in the sense that it has nonnegative second fundamental form with respect to the outward normal. We will state and prove the Neumann boundary condition case since this will be used in Chapter 14.

Theorem 12.6 *Let M be a complete manifold with convex boundary ∂M. Suppose the Ricci curvature of M is bounded from below by*

$$\mathcal{R}_{ij} \geq -(m-1)R$$

for some constant $R \geq 0$. If $f(x, t)$ is a positive solution of the equation

$$\left(\Delta - \frac{\partial}{\partial t}\right) f(x, t) = 0$$

on $M \times [0, T]$ with Neumann boundary condition

$$f_\nu(x, t) = 0 \quad \text{on} \quad \partial M \times (0, T],$$

then for any $1 < \alpha$, the function $g(x, t) = \log f(x, t)$ satisfies the estimate

$$|\nabla g|^2 - \alpha g_t \leq \frac{m}{2} \alpha^2 t^{-1} + C_1 \alpha^2 (\alpha - 1)^{-1} R$$

on $M \times (0, T]$, where C_1 is a constant depending only on m. Moreover,

$$f(x, t_1) \leq f(y, t_2) \left(\frac{t_2}{t_1}\right)^{m\alpha/2} \exp\left(\frac{\alpha \ell^2(\gamma)}{4(t_2 - t_1)} + C_1 \alpha (\alpha - 1)^{-1} R (t_2 - t_1)\right)$$

for any $x, y \in M$ and $0 < t_1 < t_2 \leq T$, where $\ell(\gamma)$ is the length of the shortest curve $\gamma \subset M$ joining x to y.

Proof The proof for the gradient estimate is again based on applying the maximum principle to the function

$$G(x, t) = t(|\nabla g|^2(x, t) - \alpha g_t(x, t)).$$

In this case, since M is compact, we will not need to use a cutoff function. In particular, if (x_0, t_0) is the maximum point for G on $M \times [0, T_0]$ with $T_0 \leq T$ and if $t_0 = 0$, then

$$G(x, t) \leq 0$$

for all $(x, t) \in M \times [0, T_0]$. Hence we may assume that $t_0 > 0$. If $x_0 \in \partial M$, then the Hopf boundary point lemma asserts that

$$G_\nu(x_0, t_0) > 0, \qquad (12.10)$$

where ν is the outward normal to ∂M. However, since

$$t(2g_{i\nu} g_i - \alpha g_{t\nu})$$
$$= t(2 f^{-2} f_{i\nu} f_i - 2 f^{-3} |\nabla f|^2 f_\nu - \alpha f^{-1} f_{t\nu} + \alpha f^{-2} f_t f_\nu),$$

the boundary condition on f implies that

$$G_\nu(x_0, t_0) = 2t_0 f^{-2}(x_0, t_0) f_{i\nu}(x_0, t_0) f_i(x_0, t_0).$$

On the other hand, the computation in the proof of Corollary 5.8 implies that

$$2 f_{i\nu} f_i = -2h_{\beta\gamma} f_\beta f_\gamma$$
$$\leq 0,$$

where $h_{\beta\gamma}$ is the second fundamental form of ∂M. This gives a contradiction to (12.10), hence $x_0 \notin \partial M$.

12 Gradient estimate and Harnack inequality for the heat equation

We can now apply the maximum principle to G at the point (x_0, t_0). At this point, Lemma 12.1 implies that

$$0 \geq -G + 2\alpha^{-2} m^{-1} G^2 - 2^{-1} m(m-1)^2 \alpha^2 (\alpha - 1)^{-2} t_0^2 R^2,$$

hence we conclude that

$$G(x, t) \leq G(x_0, t_0)$$
$$\leq \frac{m}{2}\alpha^2 + C_1 \alpha^2 (\alpha - 1)^{-1} t_0 R.$$

In particular,

$$|\nabla g|^2(x, T_0) - \alpha g_t(x, T_0) \leq \frac{m}{2}\alpha^2 T_0^{-1} + C_1 \alpha^2 (\alpha - 1)^{-1} R,$$

and the first part of the theorem follows.

The Harnack inequality follows from the same argument as in the proof of Corollary 12.3. □

13
Upper and lower bounds for the heat kernel

In this chapter, we will derive estimates for positive solutions of the heat equation. In particular, upper and lower bounds for the fundamental solution of the heat equation will be established for manifolds with Ricci curvature bounded from below. Much of the following argument was developed by Li and Yau in [LY2]. However, part of the proof has since been simplified and generalized. In particular, the integral estimate of the heat kernel given in Lemma 13.3 is due to Davies [D].

Lemma 13.1 *Let M be a compact domain with boundary. Suppose $f(x,t)$ is a nonnegative subsolution for the heat equation on $M \times (0, T)$ with the Dirichlet boundary condition*

$$f(x,t) = 0 \quad on \quad \partial M \times (0, T).$$

Assume that $g(x, t)$ is a function defined on $M \times (0, T)$ satisfying the differential inequality

$$|\nabla g|^2 + g_t \leq 0,$$

and $\mu_1(M) > 0$ denotes the first eigenvalue of the Dirichlet Laplacian on M. Then the function

$$F(t) = \exp(2\mu_1(M)\,t) \int_M \exp(2g(x,t))\, f^2(x,t)\, dx$$

is a nonincreasing function of $t \in (0, T)$.

13 Upper and lower bounds for the heat kernel

Proof Using the variational characterization of $\mu_1(M)$, we have that

$$\mu_1(M) \int_M \exp(2g) \, f^2 \leq \int_M |\nabla(\exp(g) \, f)|^2$$

$$= \int_M |\nabla \exp(g)|^2 \, f^2$$

$$+ 2 \int_M \exp(g) \, f \, \langle \nabla \exp(g), \nabla f \rangle$$

$$+ \int_M \exp(2g) \, |\nabla f|^2$$

$$= \int_M |\nabla \exp(g)|^2 \, f^2$$

$$+ \tfrac{1}{2} \int_M \langle \nabla \exp(2g), \nabla(f^2) \rangle + \int_M \exp(2g) \, |\nabla f|^2.$$

Integration by parts yields

$$\int_M \langle \nabla \exp(2g), \nabla(f^2) \rangle = - \int_M \exp(2g) \, \Delta(f^2)$$

$$= -2 \int_M \exp(2g) \, f \Delta f - 2 \int_M \exp(2g) \, |\nabla f|^2,$$

hence

$$\mu_1(M) \int_M \exp(2g) \, f^2 \leq \int_M |\nabla \exp(g)|^2 \, f^2 - \int_M \exp(2g) \, f \, \Delta f$$

$$\leq \int_M \exp(2g) \, f^2 \, |\nabla g|^2 - \int_M \exp(2g) \, f \, f_t. \quad (13.1)$$

On the other hand,

$$\frac{\partial}{\partial t} \left(\int_M \exp(2g) \, f^2 \right) = 2 \int_M \exp(2g) \, f^2 \, g_t + 2 \int_M \exp(2g) \, f \, f_t$$

$$\leq -2 \int_M \exp(2g) \, f^2 \, |\nabla g|^2 + 2 \int_M \exp(2g) \, f \, f_t,$$

hence combining this inequality with (13.1), we have

$$2\mu_1(M) \int_M \exp(2g) \, f^2 \leq -\frac{\partial}{\partial t} \left(\int_M \exp(2g) \, f^2 \right).$$

This implies that

$$\frac{\partial}{\partial t}\left(\exp(2\mu_1(M)t)\int_M \exp(2g)\,f^2\right) \leq 0$$

and the lemma follows. □

Corollary 13.2 *Let M be a complete Riemannian manifold and $f(x,t) = \int_M H(x,y,t)u(y)\,dy$ be a solution of the heat equation defined on $M \times [0,\infty)$. Suppose $\mu_1(M) \geq 0$ denotes the greatest lower bound of the L^2-spectrum of the Laplacian on M. Assume that $g(x,t)$ is a function defined on $M \times (0,\infty)$ satisfying*

$$|\nabla g|^2 + g_t \leq 0,$$

then the function

$$F(x,t) = \exp(2\mu_1(M)t)\int_M \exp(2g(x,t))\,f^2(x,t)\,dx$$

is nonincreasing in $t \in (0,\infty)$.

Proof Let $\{\Omega_i\}$ be a compact exhaustion of M such that

$$\Omega_i \subset \Omega_{i+1}$$

and

$$\cup_i \Omega_i = M.$$

Define $f_i(x,t) = \int_{\Omega_i} H_i(x,y,t)u(y)\,dy$, where $H_i(x,y,t)$ is the Dirichlet heat kernel defined on Ω_i. According to Lemma 13.1, the function

$$F_i(x,t) = \exp(2\mu_1(\Omega_i)t)\int_{\Omega_i} \exp(2g(x,t))\,f_i^2(x,t)\,dx$$

is a nonincreasing function in t. The corollary follows by letting $i \to \infty$ and observing that $\mu_1(\Omega_i) \to \mu_1(M)$ and $f_i \to f$. □

Lemma 13.3 (Davies) *Let M be a complete manifold. Suppose B_1 and B_2 are bounded subsets of M. Then*

$$\int_{B_1}\int_{B_2} H(x,y,t)\,dy\,dx \leq \exp(-\mu_1(M)t)V^{1/2}(B_1)V^{1/2}(B_2)$$

$$\times \exp\left(\frac{-d^2(B_1,B_2)}{4t}\right),$$

13 Upper and lower bounds for the heat kernel

where $V(B_i)$ denotes the volume of the set B_i for $i = 1, 2$ and $d(B_1, B_2)$ denotes the distance between the sets B_1 and B_2.

If M is a manifold with boundary, and $H(x, y, t)$ is the Dirichlet heat kernel on M, then the same estimate still holds provided that $d(B_1, B_2)$ is now interpreted as the distance between the two sets of curves in M. In particular, $d(B_1, B_2)$ is given by the infimum of the length of all curves contained in M joining B_1 to B_2.

Proof For $i = 1, 2$, let

$$f_i(x, t) = \int_{B_i} H(x, y, t) \, dy$$

and

$$g_i(x, t) = \frac{d^2(x, B_i)}{4(t + \epsilon)},$$

where $d(x, B_i)$ is the distance between x and the set B_i. A direct computation shows that

$$|\nabla g_i|^2 = \left| \frac{d \nabla d}{2(t + \epsilon)} \right|^2$$

$$\leq \left| \frac{d^2(x, B_i)}{4(t + \epsilon)^2} \right|$$

and

$$\frac{\partial g_i}{\partial t} = \frac{-d^2(x, B_i)}{4(t + \epsilon)^2}.$$

Applying Corollary 13.2 to f_i and g_i, we conclude that

$$\int_M \exp(2g_i(x, t)) \, f_i^2(x, t) \, dx \leq \exp(-2\mu_1(M) t)$$

$$\times \int_M \exp(2g_i(x, 0)) \, f_i^2(x, 0) \, dx.$$

However, since

$$f_i(x, 0) = \int_{B_i} H(x, y, 0) \, dy$$

$$= \chi(B_i)$$

is the characteristic function of B_i, we conclude that

$$\int_M \exp(2g_i(x,0)) f_i^2(x,0) = V(B_i)$$

and

$$\int_M \exp(2g_i(x,t)) f_i^2(x,t) \, dx \leq V(B_i) \exp(-2\mu_1(M)t) \qquad (13.2)$$

for $i = 1, 2$. On the other hand, the triangle inequality asserts that

$$d^2(B_1, B_2) \leq (d(x, B_1) + d(x, B_2))^2$$
$$\leq 2d^2(x, B_1) + 2d^2(x, B_2),$$

hence implying that

$$\exp\left(\frac{d^2(B_1, B_2)}{8(t+\epsilon)}\right) \leq \exp(g_1 + g_2).$$

Combining the above estimate with (13.2), we have

$$\int_M \exp\left(\frac{d^2(B_1, B_2)}{8(t+\epsilon)}\right) f_1 f_2 \leq \int_M \exp(g_1) \exp(g_2) f_1 f_2$$

$$\leq \left(\int_M \exp(2g_1) f_1^2\right)^{1/2} \left(\int_M \exp(2g_2) f_2^2\right)^{1/2}$$

$$\leq V^{1/2}(B_1) V^{1/2}(B_2) \exp(-2\mu_1(M)t). \qquad (13.3)$$

Using the semi-group property, the left-hand side can be written as

$$\int_M \exp\left(\frac{d^2(B_1, B_2)}{8(t+\epsilon)}\right) f_1 f_2$$

$$= \exp\left(\frac{d^2(B_1, B_2)}{8(t+\epsilon)}\right) \int_M \int_{B_1} \int_{B_2} H(x, z, t) H(x, y, t) \, dz \, dy \, dx$$

$$= \exp\left(\frac{d^2(B_1, B_2)}{8(t+\epsilon)}\right) \int_{B_1} \int_{B_2} H(y, z, 2t) \, dz \, dy.$$

Lemma 13.3 follows by combing this with (13.3) and letting $\epsilon \to 0$. □

We are now ready to give an upper bound on the fundamental solution of the heat equation.

13 Upper and lower bounds for the heat kernel

Theorem 13.4 (Li–Yau) *Let M^m be a complete manifold and $H(x, y, t)$ denote the minimal symmetric heat kernel defined on $M \times M \times (0, \infty)$ with the properties that*

$$\left(\Delta_y - \frac{\partial}{\partial t}\right) H(x, y, t) = 0$$

and

$$\lim_{t \to 0} H(x, y, t) = \delta_x(y).$$

For any $p \in M$, $\epsilon > 0$, and $\rho > 0$, if the Ricci curvature of M on $B_p(4\rho)$ is bounded from below by

$$\mathcal{R}_{ij} \geq -(m-1)R,$$

then there exist constants $C_1, C_2 > 0$ depending only on m and ϵ, such that

$$H(x, y, t) \leq C_1 \exp(-\mu_1(M)t) \, V_x^{-1/2}\left(\sqrt{t}\right)$$

$$\times V_y^{-1/2}\left(\sqrt{t}\right) \exp\left(-\frac{r^2(x, y)}{4(1+2\epsilon)t} + C_2 \sqrt{(\rho^{-2}+R)t}\right)$$

for any $x, y \in B_p(\rho)$ and $t \leq \rho^2/4$. In particular, if the Ricci curvature of M is bounded by

$$\mathcal{R}_{ij} \geq -(m-1)R,$$

then

$$H(x, y, t) \leq C_1 \exp(-\mu_1(M)t) \, V_x^{-1/2}\left(\sqrt{t}\right)$$

$$\times V_y^{-1/2}\left(\sqrt{t}\right) \exp\left(-\frac{r^2(x, y)}{4(1+2\epsilon)t} + C_2 \sqrt{Rt}\right)$$

for all $x, y \in M$ and $t \in (0, \infty)$.

Proof For a fixed $y \in B_p(\rho)$ and $\delta > 0$, applying Corollary 12.3 to the positive solution $f(x, t) = H(x, y, t)$ by taking $t_1 = t$ and $t_2 = (1+\delta)t$, we have

$$H(x, y, t) \leq (1+\delta)^{m\alpha/2} \exp\left(\frac{\alpha \, r^2(x, x')}{4\delta t} + A\delta t\right) H(x', y, (1+\delta)t),$$

where $A = C_1 \alpha(\alpha - 1)^{-1}(\rho^{-2} + R)$. Integrating over $x' \in B_x(\sqrt{t})$, this gives

$$H(x, y, t) \le (1+\delta)^{m\alpha/2} \exp\left(\frac{\alpha}{4\delta} + A\delta t\right) V_x^{-1}(\sqrt{t})$$

$$\times \int_{B_x(\sqrt{t})} H(x', y, (1+\delta)t)\,dx'. \qquad (13.4)$$

Applying Corollary 12.3 and the same argument to the positive solution

$$f(y, s) = \int_{B_x(\sqrt{t})} H(x', y, s)\,dx'$$

by taking $t_1 = (1+\delta)t$ and $t_2 = (1+2\delta)t$, we obtain

$$\int_{B_x(\sqrt{t})} H(x', y, (1+\delta)t)\,dx'$$

$$\le \left(\frac{1+2\delta}{1+\delta}\right)^{m\alpha/2} \exp\left(\frac{\alpha}{4\delta} + A\delta t\right) V_y^{-1}(\sqrt{t})$$

$$\times \int_{B_y(\sqrt{t})} \int_{B_x(\sqrt{t})} H(x', y', (1+2\delta)t)\,dx'\,dy'.$$

Hence combining the upper bound with (13.4), we have

$$H(x, y, t) \le (1+2\delta)^{m\alpha/2} \exp\left(\frac{\alpha}{2\delta} + 2A\delta t\right)$$

$$\times V_x^{-1}(\sqrt{t}) V_y^{-1}(\sqrt{t})$$

$$\times \int_{B_y(\sqrt{t})} \int_{B_x(\sqrt{t})} H(x', y', (1+2\delta)t)\,dx'\,dy'. \qquad (13.5)$$

On the other hand, Lemma 13.3 implies that

$$\int_{B_y(\sqrt{t})} \int_{B_x(\sqrt{t})} H(x', y', (1+2\delta)t)\,dx'\,dy'$$

$$\le \exp(-\mu_1(1+2\delta)t) V_x^{1/2}(\sqrt{t}) V_y^{1/2}(\sqrt{t})$$

$$\times \exp\left(\frac{-d^2(B_x(\sqrt{t}), B_y(\sqrt{t}))}{4(1+2\delta)t}\right). \qquad (13.6)$$

13 Upper and lower bounds for the heat kernel

Observing that

$$d\left(B_x\left(\sqrt{t}\right), B_y\left(\sqrt{t}\right)\right) = \begin{cases} 0 & \text{if } r(x,y) \leq 2\sqrt{t}, \\ r(x,y) - 2\sqrt{t} & \text{if } r(x,y) > 2\sqrt{t}, \end{cases}$$

hence we obtain the estimate

$$\frac{-d^2\left(B_x\left(\sqrt{t}\right), B_y\left(\sqrt{t}\right)\right)}{4(1+2\delta)t} = 0$$

$$\leq 1 - \frac{r^2(x,y)}{4(1+4\delta)t},$$

when $r(x,y) \leq 2\sqrt{t}$, and

$$\frac{-d^2\left(B_x\left(\sqrt{t}\right), B_y\left(\sqrt{t}\right)\right)}{4(1+2\delta)t} \leq \frac{-\left(r(x,y) - 2\sqrt{t}\right)^2}{4(1+2\delta)t}$$

$$\leq \frac{-r^2(x,y)}{4(1+4\delta)t} + \frac{1}{2\delta}$$

when $r(x,y) > 2\sqrt{t}$. In any case, there exists a constant $C_3 > 0$, such that when combining the above estimate with (13.5) and (13.6) we obtain

$$H(x,y,t) \leq C_3 (1+2\delta)^{m\alpha/2} \exp\left(\frac{(\alpha+1)}{2\delta} + (2\delta At - \mu_1(1+2\delta)t\right)$$

$$\times V_x^{-1/2}(\sqrt{t}) V_y^{-1/2}(\sqrt{t}) \exp\left(\frac{-r^2(x,y)}{4(1+4\delta)t}\right). \tag{13.7}$$

When $At > 1$, we choose $2\delta = \epsilon/\sqrt{AT}$, and (13.7) can be estimated by

$$H(x,y,t) \leq C_3 \left(1 + \frac{\epsilon}{\sqrt{AT}}\right)^{m\alpha/2} \exp\left(\left(\frac{(\alpha+1)}{\epsilon} + \epsilon\right)\sqrt{At}\right) \exp(-\mu_1 t)$$

$$\times V_x^{-1/2}\left(\sqrt{t}\right) V_y^{-1/2}\left(\sqrt{t}\right) \exp\left(\frac{-r^2(x,y)}{4\left(1+2\epsilon/\sqrt{AT}\right)t}\right)$$

$$\leq C_3 (1+\epsilon)^{m\alpha/2} \exp\left(\left(\frac{(\alpha+1)}{\epsilon} + \epsilon\right)\sqrt{At}\right) \exp(-\mu_1 t)$$

$$\times V_x^{-1/2}\left(\sqrt{t}\right) V_y^{-1/2}\left(\sqrt{t}\right) \exp\left(\frac{-r^2(x,y)}{4(1+2\epsilon)t}\right).$$

When $At \leq 1$, we choose $2\delta = \epsilon$ and conclude from (13.7) and $At \leq \sqrt{At}$ that

$$H(x, y, t) \leq C_3 (1+\epsilon)^{m\alpha/2} \exp\left(\frac{(\alpha+1)}{\epsilon} + \epsilon\sqrt{At}\right)$$

$$\times \exp(-\mu_1 t) \, V_x^{-\frac{1}{2}}\left(\sqrt{t}\right) V_y^{-\frac{1}{2}}\left(\sqrt{t}\right) \exp\left(\frac{-r^2(x,y)}{4(1+2\epsilon)t}\right).$$

In either case, the theorem follows by choosing $\alpha = 2$. □

Due to the fact that both Corollary 12.3 and Lemma 13.3 are valid on manifolds with boundary after suitable localization, a corresponding upper bound for the Dirichlet heat kernel can also be derived following the above proof.

Corollary 13.5 *Let M^m be a manifold with boundary and $H(x, y, t)$ be its Dirichlet heat kernel. Let us assume that $B_p(4\rho) \cap \partial M = \emptyset$ and the Ricci curvature of M on $B_p(4\rho)$ is bounded from below by*

$$\mathcal{R}_{ij} \geq -(m-1)R$$

for some constant $R \geq 0$. For any $\epsilon > 0$, there exist constants $C_1 > 0$ depending only on m and ϵ, and C_2 depending only on m, such that

$$H(x, y, t) \leq C_1 \exp(-\mu_1(M)t) \, V_x^{-1/2}\left(\sqrt{t}\right) V_y^{-1/2}\left(\sqrt{t}\right)$$

$$\times \exp\left(-\frac{r^2(x,y)}{4(1+2\epsilon)t} + C_2\sqrt{(\rho^{-2}+R)t}\right)$$

for any $x, y \in B_p(\rho)$ and $t \leq \rho^2/4$.

We will now derive a lower bound for the heat kernel. The following comparison type theorems were first proved by Cheeger and Yau [CY] and adapted to the present form in [LY2].

Lemma 13.6 *Let M^m be a complete manifold with boundary. Suppose $B_p(\rho)$ is the geodesic ball centered at some fixed point $p \in M$ with radius $0 < \rho \leq d(p, \partial M)$, at most the distance from p to ∂M. Assume that the Ricci curvature of M is bounded from below by*

$$\mathcal{R}_{ij} \geq (m-1)K \quad on \quad B_p(\rho),$$

for some constant K. Suppose $\bar{M}$ is a simply connected space form with constant sectional curvature K and $B_{\bar{p}}(\rho)$ is a geodesic ball centered at some fixed point $\bar{p} \in \bar{M}$ of radius ρ. Let $H(x, y, t)$ and $\bar{H}(\bar{x}, \bar{y}, t)$ be the Dirichlet

13 Upper and lower bounds for the heat kernel

heat kernels on M and $B_{\bar{p}}(\rho)$, respectively. Then for any nonnegative function $f_0(r)$ defined on $[0, \rho]$ satisfying the properties $f_0'(0) = 0$, $f_0(\rho) = 0$, and $f_0'(r) \leq 0$, the inequality

$$\int_M H(z, y, t)\, f_0(r(z))\, dz \geq \int_{B_{\bar{p}}(\rho)} \bar{H}(\bar{z}, \bar{y}, t)\, f_0(\bar{r}(\bar{z}))\, d\bar{z}$$

is valid for all $y \in B_p(\rho)$ with $\bar{y} \in B_{\bar{p}}(\rho)$ such that $r(y) = \bar{r}(\bar{y})$, where $r(y)$ and $\bar{r}(\bar{y})$ denote the distance to $p \in M$ and the distance to $\bar{p} \in \bar{M}$, respectively. In particular, the inequalities

$$\int_{B_p(\rho_1)} H(z, y, t)\, dz \geq \int_{B_{\bar{p}}(\rho_1)} \bar{H}(\bar{z}, \bar{y}, t)\, d\bar{z}$$

for all $\rho_1 \leq \rho$, and

$$H(p, y, t) \geq \bar{H}(\bar{p}, \bar{y}, t)$$

are valid for $r(y) = \bar{r}(\bar{y})$ and for $t \in (0, \infty)$.

Proof We will prove the inequality for $\rho < d(p, \partial M)$ and the general case follows by taking the limit as $\rho \to d(p, \partial M)$. For $\rho < d(p, \partial M)$, let $f_0(x) = f_0(r(x))$ be a nonnegative function of the distance r to the center point p satisfying the stated properties. Let

$$f(x, t) = \int_M H(z, x, t)\, f_0(z)\, dz$$

be the solution of the Dirichlet heat equation with f_0 as the initial datum on M. We also consider the solution of the heat equation on $B_{\bar{p}}(\rho)$ given by

$$\bar{f}(\bar{x}, t) = \int_{B_{\bar{p}}(\rho)} \bar{H}(\bar{z}, \bar{x}, t)\, f_0(\bar{r}(\bar{z}))\, d\bar{z},$$

where $\bar{r}(\bar{z})$ is the distance function on $\bar{M}$ to the point $\bar{p}$. By the uniqueness of the heat equation and the fact that f_0 is rotationally symmetric, $\bar{f}$ must also be rotationally symmetric. Hence, we can write $\bar{f}(\bar{y}, t) = \bar{f}(\bar{r}(\bar{y}), t)$. In particular,

$$\bar{f}'(0, t) = \frac{\partial \bar{f}}{\partial \bar{r}}(0, t) = 0,$$

and also by the Dirichlet boundary condition

$$\bar{f}'(\rho, t) = \frac{\partial \bar{f}}{\partial \bar{r}}(\rho, t) < 0. \tag{13.8}$$

We claim that
$$\bar{f}'(\bar{r}, t) \le 0 \tag{13.9}$$
for all $0 \le \bar{r} \le \rho$ and for all $t \ge 0$. Indeed, since $\bar{f}$ is a function of $\bar{r}$ and t alone, the heat equation on $\bar{M}$ can be written as
$$\begin{aligned} 0 &= \left(\bar{\Delta} - \frac{\partial}{\partial t}\right) \bar{f}(\bar{r}, t) \\ &= \bar{f}' \bar{\Delta}\bar{r} + \bar{f}'' - \bar{f}_t, \end{aligned} \tag{13.10}$$
with
$$\bar{\Delta}\bar{r} = \begin{cases} (m-1)\sqrt{K} \cot\left(\sqrt{K}r\right) & \text{for } K > 0, \\ (m-1)r^{-1} & \text{for } K = 0, \\ (m-1)\sqrt{-K} \coth\left(\sqrt{-K}r\right) & \text{for } K < 0. \end{cases}$$
Differentiating this with respect to $\bar{r}$ again yields
$$0 = u' \bar{\Delta}\bar{r} + u (\bar{\Delta}\bar{r})' + u'' - u_t$$
with $u = \bar{f}'$ and
$$(\bar{\Delta}\bar{r})' = \begin{cases} -(m-1)K \csc^2\left(\sqrt{K}r\right) & \text{for } K > 0, \\ -(m-1)r^{-2} & \text{for } K = 0, \\ (m-1)K \operatorname{cosech}^2\left(\sqrt{-K}r\right) & \text{for } K < 0. \end{cases}$$

Note that $(\bar{\Delta}\bar{r})' \le 0$ for any choices of K. Hence, the boundary conditions $u(0, t) = 0$ and $u(\rho, t) < 0$ and the initial condition $u(\bar{r}, 0) \le 0$, together with the maximum principle imply that $u(\bar{r}, t) \le 0$.

We now consider the transplanted function defined by $\bar{f}(y, t) = \bar{f}(r(y), t)$. The Laplacian comparison theorem, (13.9), and (13.10) assert that
$$\begin{aligned} \left(\Delta - \frac{\partial}{\partial t}\right) \bar{f}(r(y), t) &= \bar{f}' \Delta r + \bar{f}'' - \bar{f}_t \\ &\ge \bar{f}' \bar{\Delta}\bar{r} + \bar{f}'' - \bar{f}_t \\ &= 0 \end{aligned}$$
in the sense of distribution on M, with the initial condition
$$\bar{f}(y, 0) = f_0(y).$$
The difference $F = f - \bar{f}$ is therefore a supersolution of the heat equation of $B_p(\rho) \times (0, \infty)$ which vanishes on $B_p(\rho) \times \{0\}$. Moreover, the Dirichlet

boundary condition yields

$$F(y,t) = f(y,t) \geq 0 \quad \text{on} \quad \partial B_p(\rho) \times (0, \infty).$$

The parabolic maximum principle implies that $F \geq 0$, hence

$$f(y,t) \geq \bar{f}(r(y),t).$$

The second and the third inequalities follow by approximating the characteristic function of $B_p(\rho_1)$ and the point mass function at p by a sequence of smooth functions satisfying the properties required for f_0. □

Lemma 13.7 *Let M^m be a complete manifold with boundary. Suppose there exists a point $p \in M$ such that M is geodesically star-shaped with respect to p, and ρ_0 is the maximum distance from p to any point in ∂M. Assume that the Ricci curvature of M is bounded from below by*

$$\mathcal{R}_{ij} \geq (m-1)K$$

for some constant K. Suppose $\bar{M}$ is a simply connected space form with constant sectional curvature K and $B_{\bar{p}}(\rho_0)$ is a geodesic ball centered at some fixed point $\bar{p} \in \bar{M}$ of radius ρ_0. Let $K(x,y,t)$ and $\bar{K}(\bar{x},\bar{y},t)$ be the Neumann heat kernels on M and $B_{\bar{p}}(\rho_0)$, respectively. Suppose $\rho_1 \leq d(p, \partial M)$, then for any nonnegative function $f_0(r)$ defined on $[0, \rho_1]$ satisfying properties that $f_0'(0) = 0$, $f_0(\rho_1) = 0$, and $f_0'(r) \leq 0$, the inequality

$$\int_M K(z,y,t) f_0(r(z)) \, dz \geq \int_{B_{\bar{p}}(\rho_0)} \bar{K}(\bar{z},\bar{y},t) f_0(\bar{r}(\bar{z})) d\bar{z}$$

is valid for all $y \in M$ with $\bar{y} \in B_{\bar{p}}(\rho_0)$ such that $r(y) = \bar{r}(\bar{y})$, where $r(y)$ and $\bar{r}(\bar{y})$ denote the distance to $p \in M$ and the distance to $\bar{p} \in \bar{M}$, respectively. In particular, the inequalities

$$\int_{B_p(\rho_1)} K(z,y,t) \, dz \geq \int_{B_{\bar{p}}(\rho_1)} \bar{K}(\bar{z},\bar{y},t) \, d\bar{z}$$

for all $\rho_1 \leq d(p, \partial M)$, and

$$K(p,y,t) \geq \bar{K}(\bar{p},\bar{y},t)$$

are valid for all $r(y) = \bar{r}(\bar{y})$ and for all $t \in (0, \infty)$.

Proof The proof for the Neumann boundary condition follows using the same argument as the previous lemma. By solving the heat equation with Neumann boundary condition on $B_{\bar{p}}(\rho_0)$, we consider the rotationally symmetric solution

$$\bar{f}(\bar{y}, t) = \int_{B_{\bar{p}}(\rho_0)} K(\bar{z}, \bar{y}, t) \, f_0(\bar{z}) \, d\bar{z}.$$

The Neumann boundary condition asserts that

$$\bar{f}'(\rho_0, t) = 0$$

instead of (13.8). As in the previous proof, the maximum principle implies that $u(\bar{r}, t) = \bar{f}'(\bar{r}, t) \leq 0$. Again we consider the difference

$$F(y, t) = f(y, t) - \bar{f}(r(y), t) \quad \text{on} \quad M \times (0, \infty),$$

with

$$f(y, t) = \int_M K(z, y, t) \, f_0(z) \, dz.$$

We now claim that the normal derivative $\partial F/\partial \nu$ with respect to the outward normal must be nonnegative on ∂M. Clearly, by the boundary condition of F, it suffices to check that

$$\frac{\partial \bar{f}}{\partial \nu} \leq 0.$$

Indeed, using the fact that $\bar{f}$ is rotationally symmetric, for any $y \in \partial M$ we have

$$\frac{\partial \bar{f}}{\partial \nu}(y, t) = \bar{f}'(r(y), t) \left\langle \frac{\partial}{\partial r}, \nu \right\rangle (y).$$

The assertion follows from the fact that $\bar{f}' \leq 0$, and the fact that $\langle \partial/\partial r, \nu \rangle \geq 0$ on ∂M because M is geodesically star-shaped with respect to p. Hence

$$\frac{\partial F}{\partial \nu}(y, t) \geq 0 \quad \text{on} \quad \partial M \times (0, \infty),$$

implying that the minimum of F cannot occur on $\partial M \times (0, \infty)$, because of the Hopf boundary lemma. Hence, the minimum of F must be achieved on $M \times \{0\}$, which has value 0. This implies that $f \geq \bar{f}$ and the rest of the lemma follows. □

Theorem 13.8 *Let M^m be a complete manifold with boundary. Suppose $p \in M$ and $\rho > 0$ are such that $B_p(4\rho) \cap \partial M = \emptyset$. Assume that the Ricci curvature is bounded from below by*

$$\mathcal{R}_{ij} \geq -(m-1)R \quad \text{on} \quad B_p(4\rho),$$

13 Upper and lower bounds for the heat kernel

for some constant $R \geq 0$. Suppose $H(x, y, t)$ is Dirichlet heat kernel on $B_p(4\rho)$ and $\bar{H}(\bar{x}, \bar{y}, t)$ is the Dirichlet heat kernel on $B_{\bar{x}}(3\rho)$ in the simply connected, constant $-R$ curvature space form. Then for $\epsilon > 0$, there exists constant $C_4 > 0$ depending on m, such that

$$H(x, y, t) \geq (1-\epsilon)^m \, \bar{V}_{\bar{x}}\left(\sqrt{t}\right) V_x^{-1/2}\left(\sqrt{t}\right) V_y^{-1/2}\left(\sqrt{t}\right)$$
$$\times \exp\left(-2\epsilon^{-1} - \epsilon\, C_4 (1+Rt)\right) \bar{H}(\bar{x}, \bar{y}, (1-\epsilon)t)$$

and

$$H(x, y, t) \geq (1-\epsilon)^m \, \bar{V}_{\bar{x}}\left(\sqrt{t}\right) V_x^{-1}\left(\sqrt{t}\right)$$
$$\times \exp\left(-2\epsilon^{-1} - \epsilon\, C_4 (1+Rt)\right) \bar{H}(\bar{x}, \bar{y}, (1-\epsilon)t)$$

for $x, y \in B_p(\rho)$ and $0 < t \leq \rho^2$.

Proof Let $x, y \in B_p(\rho)$. The comparison theorem of Lemma 13.6 asserts that if $\bar{H}(\bar{z}, \bar{y}, t)$ is the Dirichlet heat kernel on $B_{\bar{x}}(3\rho) \subset \bar{M}$, then

$$\int_{B_{\bar{x}}(\rho_1)} \bar{H}(\bar{z}, \bar{y}, (1-\delta)t)\, d\bar{z} \leq \int_{B_x(\rho_1)} H(z, y, (1-\delta)t)\, dz \qquad (13.11)$$

for $\rho_1 \leq 3\rho$. However, Corollary 12.3 implies that

$$H(z, y, (1-\delta)t) \leq (1-\delta)^{-\alpha m/2} \exp\left(\frac{\alpha \rho_1^2}{4\delta t} + A\delta t\right) H(x, y, t)$$

for $z \in B_x(\rho_1)$, with $A = C_1 \alpha (\alpha-1)^{-1}(\rho^{-2} + R)$. Integrating over $B_x(\rho_1)$, we conclude that

$$\int_{B_x(\rho_1)} H(z, y, (1-\delta)t)\, dz \leq (1-\delta)^{-\alpha m/2}$$
$$\times \exp\left(\frac{\alpha \rho_1^2}{4\delta t} + A\delta t\right) H(x, y, t)\, V_x(\rho_1). \qquad (13.12)$$

Applying the same argument to $\bar{H}(\bar{x}, \bar{y}, t)$, we have

$$\bar{H}(\bar{x}, \bar{y}, (1-2\delta)t)\, \bar{V}_{\bar{x}}(\rho_1) \leq \left(\frac{1-\delta}{1-2\delta}\right)^{\alpha m/2} \exp\left(\frac{\alpha \rho_1^2}{4\delta t} + A\delta t\right)$$
$$\times \int_{B_{\bar{x}}(\rho_1)} \bar{H}(\bar{z}, \bar{y}, (1-\delta)t)\, d\bar{z}.$$

Combining this estimate with (13.11) and (13.12), and setting $\rho_1 = \sqrt{t}$, we obtain

$$H(x, y, t) \geq \bar{V}_{\bar{x}}\left(\sqrt{t}\right) V_x^{-1}\left(\sqrt{t}\right) (1 - 2\delta)^{\alpha m/2} \exp\left(-\frac{\alpha}{2\delta} - 2A\delta t\right)$$
$$\times \bar{H}(\bar{x}, \bar{y}, (1 - 2\delta)t).$$

Setting $\alpha = 2$, $\epsilon = 2\delta$, and using $t \leq \rho^2$, we conclude that

$$H(x, y, t) \geq (1 - \epsilon)^m \bar{V}_{\bar{x}}\left(\sqrt{t}\right) V_x^{-1}(\sqrt{t}) \exp\left(-\frac{2}{\epsilon} - \epsilon C_4 (1 + Rt)\right)$$
$$\times \bar{H}(\bar{x}, \bar{y}, (1 - \epsilon)t)$$

for some constant $C_4 > 0$ depending only on m. If $y \in B_p(\rho)$ also, then we can symmetrize the estimate and obtain

$$H(x, y, t) \geq (1 - \epsilon)^m \bar{V}_{\bar{x}}\left(\sqrt{t}\right) V_x^{-1/2}\left(\sqrt{t}\right) V_y^{-1/2}\left(\sqrt{t}\right)$$
$$\times \exp\left(-\frac{2}{\epsilon} - \epsilon C_4 (1 + Rt)\right) \bar{H}(\bar{x}, \bar{y}, (1 - \epsilon)t). \quad \square$$

Corollary 13.9 *Let M^m be a complete, noncompact manifold without boundary. Suppose M has nonnegative Ricci curvature. For any $\epsilon > 0$, there exists a constant $C_5 > 0$ depending on m and ϵ, such that the minimal heat kernel $H(x, y, t)$ satisfies the lower bounds*

$$H(x, y, t) \geq C_5 V_x^{-1/2}\left(\sqrt{t}\right) V_y^{-1/2}\left(\sqrt{t}\right) \exp\left(-\frac{r^2(x, y)}{4(1 - \epsilon)t}\right)$$

and

$$H(x, y, t) \geq C_5 V_x^{-1}\left(\sqrt{t}\right) \exp\left(-\frac{r^2(x, y)}{4(1 - \epsilon)t}\right)$$

for all $x, y \in M$ and $t \in (0, \infty)$.

Proof According to Theorem 13.8, by setting $\rho = \infty$, it suffices to estimate the heat kernel $\bar{H}(\bar{x}, \bar{y}, t)$ on $\bar{M}$ from below. Note that if $R = 0$, then

$$\bar{H}(\bar{x}, \bar{y}, (1 - \epsilon)t) \bar{V}_{\bar{x}}\left(\sqrt{t}\right) = C \exp\left(-\frac{\bar{r}^2(\bar{x}, \bar{y})}{4(1 - \epsilon)t}\right)$$

and our estimate becomes

$$H(x, y, t) \geq C_4 V_x^{-1/2}\left(\sqrt{t}\right) V_y^{-1/2}\left(\sqrt{t}\right) \exp\left(-\frac{r^2(x, y)}{4(1 - \epsilon)t}\right)$$

since $\bar{r}(\bar{x}, \bar{y}) = r(x, y)$. $\quad \square$

14
Sobolev inequality, Poincaré inequality and parabolic mean value inequality

In this chapter, we will use the heat kernel to estimate the Sobolev constant and the first nonzero Neumann eigenvalue for geodesic balls. These estimates are useful in the studies of various partial differential equations. In particular, they will provide control over the constants involved in the De Giorgi–Nash–Moser theory in terms of the lower bound of the Ricci curvature and the diameter of the underlying geodesic ball. The Sobolev inequality was proved by Saloff-Coste [SC] adapting Varopoulos' argument (Theorem 11.6) [V] to this situation.

We will also use the heat kernel estimate to prove a parabolic mean value inequality established by Li and Tam in [LT5]. A corollary to this parabolic version will be a generalization to the mean value inequality given in Chapter 7.

We will first establish a lemma on the decay rate of the heat kernel.

Lemma 14.1 *Let M be a manifold with boundary. If $\mu_1(M)$ is the first Dirichlet eigenvalue of M, then the Dirichlet heat kernel must satisfy*

$$H(x, x, t) \leq \exp(-\mu_1(M)(t - t_0)) H(x, x, t_0)$$

for all $x \in M$ and $t \geq t_0 > 0$. If $\lambda_1(M)$ is the first nonzero Neumann eigenvalue of M, then the Neumann heat kernel must satisfy

$$K(x, x, t) - V^{-1}(M) \leq \exp(-\lambda_1(M)(t - t_0))(K(x, x, t_0) - V^{-1}(M))$$

for all $x \in M$ and $t \geq t_0 > 0$.

Proof The semi-group property of the heat kernel implies that

$$\frac{\partial}{\partial t} H(x, x, 2t) = \frac{\partial}{\partial t} \int_M H^2(x, y, t) \, dy$$

$$= 2 \int_M H(x, y, t) \, \Delta H(x, y, t) \, dy$$

$$= -2 \int_M |\nabla H|^2(x, y, t) \, dy$$

$$\leq -2\mu_1(M) \int_M H^2(x, y, t) \, dy$$

$$= -2\mu_1(M) H(x, x, 2t).$$

Rewriting the inequality in the form

$$\frac{\partial}{\partial t} (\log H(x, x, t)) \leq -\mu_1(M)$$

and integrating from t_0 to t yields that desired estimate.

The estimate on the Neumann heat kernel follows in exactly the same manner by considering

$$K_1(x, x, t) = K(x, x, t) - V^{-1}(M)$$

that has the property

$$\int_M K_1(x, y, t) \, dy = 0,$$

and using the variational principle for $\lambda_1(M)$. □

Theorem 14.2 *Let M^m be a complete manifold and $B_p(\rho)$ be a geodesic ball centered at p with radius ρ. Assume that M has Ricci curvature bounded from below by*

$$\mathcal{R}_{ij} \geq -(m-1)R$$

for some constant $R \geq 0$. Let $H(x, y, t)$ be the Dirichlet heat kernel defined on $B_p(\rho)$. Then there exist constants $C_{15}, C_{16} > 0$ depending only on m such that for all $x \in B_p(\rho)$ and $t > 0$ we have

$$H(x, x, t) \leq C_{15} t^{-m/2} V_p^{-1}(\rho) \rho^m \exp\left(C_{16} \rho \sqrt{R}\right),$$

where $V_p(\rho)$ is the volume of $B_p(\rho)$.

14 Inequalities of Sobolev, Poincaré and parabolic mean value

Proof Let $H_M(x, y, t)$ be the heat kernel on the complete manifold M. Recall that Corollary 13.5 implies that

$$H_M(x, x, t) \leq C_1 V_x^{-1}\left(\sqrt{t}\right) \exp\left(C_2\sqrt{R}t\right)$$

for some constant $C_1, C_2 > 0$. By the monotonicity property of the heat kernel, we conclude that for $x \in B_p(\rho)$ and for $t \leq \rho^2$

$$H(x, x, t) \leq C_1 \bar{V}^{-1}\left(\sqrt{t}\right)\left(\bar{V}\left(\sqrt{t}\right) V_x^{-1}\left(\sqrt{t}\right)\right) \exp(C_2\rho\sqrt{R}),$$

where $\bar{V}\left(\sqrt{t}\right)$ is the volume of the ball of radius $\sqrt{t}$ in the simply connected space form of constant curvature $-R$. By the volume comparison theorem, since $t \leq \rho^2$, we have

$$\bar{V}\left(\sqrt{t}\right) V_x^{-1}\left(\sqrt{t}\right) \leq \bar{V}(\rho) V_x^{-1}(\rho),$$

hence

$$H(x, x, t) \leq C_1 \bar{V}^{-1}\left(\sqrt{t}\right)(\bar{V}(\rho) V_x^{-1}(\rho)) \exp\left(C_2\rho\sqrt{R}\right)$$

$$\leq C_1 t^{-m/2}(\bar{V}(\rho) V_x^{-1}(\rho)) \exp\left(C_2\rho\sqrt{R}\right). \tag{14.1}$$

On the other hand, for $t \geq \rho^2$ we can use the estimate given by Lemma 14.1 and conclude that

$$H(x, x, t) \leq H(x, x, \rho^2) \exp(-\mu_1(t - \rho^2))$$

$$\leq C_1 V_x^{-1}(\rho) \exp\left(C_2\rho\sqrt{R} - \mu_1(t - \rho^2)\right). \tag{14.2}$$

Let us now consider the function

$$f(t) = t^{m/2} \exp(-\mu_1(t - \rho^2)).$$

It has a local maximum at the point $t = m/2\mu_1$ with maximum value

$$f\left(\frac{m}{2\mu_1}\right) = \left(\frac{m}{2\mu_1}\right)^{m/2} \exp\left(-\frac{m}{2} + \mu_1\rho^2\right).$$

Moreover, $f(t)$ is a decreasing function for $t > m/2\mu_1$. Hence if $m/2\mu_1 \leq \rho^2$, then

$$f(t) \leq f(\rho^2) = \rho^m \tag{14.3}$$

for $t \geq \rho^2$. On the other hand, if $m/2\mu_1 \geq \rho^2$, then

$$f(t) \leq f\left(\frac{m}{2\mu_1}\right)$$

$$= \left(\frac{m}{2\mu_1}\right)^{m/2} \exp\left(-\frac{m}{2} + \mu_1 \rho^2\right)$$

$$\leq \left(\frac{m}{2\mu_1}\right)^{m/2}. \tag{14.4}$$

However, by the estimate of Li–Schoen (Theorem 5.9), $\mu_1 \geq C_3 \rho^{-2} \exp\left(-C_4 \rho \sqrt{R}\right)$. Applying this to (14.4) and taking it together with (14.3) yields

$$f(t) \leq C_5 \rho^m \exp\left(C_6 \rho \sqrt{R}\right)$$

for all $t \geq \rho^2$. Therefore, substituting this into (14.2), we have

$$H(x, x, t) \leq C_7 t^{-m/2} V_x^{-1}(\rho) \rho^m \exp\left(C_8 \rho \sqrt{R}\right)$$

for $t \geq \rho^2$.

On the other hand, using $\bar{V}(\rho) \leq C_9 \rho^m \exp\left(C_{10} \rho \sqrt{R}\right)$ for some constants $C_9, C_{10} > 0$ depending only on m, the estimate (14.1) becomes

$$H(x, x, t) \leq C_{11} t^{-\frac{m}{2}} V_x^{-1}(\rho) \rho^m \exp(C_{12} \rho \sqrt{R}) \tag{14.5}$$

for all $t \leq \rho^2$, hence it is valid for all $t > 0$. The Bishop volume comparison theorem and the fact that $x \in B_p(\rho)$ imply that

$$V_x(\rho) \geq V_x(2\rho) \frac{\bar{V}(\rho)}{\bar{V}(2\rho)}$$

$$\geq C_{13} V_p(\rho) \exp\left(-C_{14} \rho \sqrt{R}\right).$$

The theorem follows by combining this with (14.5). $\square$

Combining Theorem 14.2 and Theorem 11.6, we obtain the following Sobolev inequality on a geodesic ball.

Theorem 14.3 *Let M^m be a complete manifold, of dimension $m \geq 3$, with Ricci curvature bounded from below by*

$$\mathcal{R}_{ij} \geq -(m-1)R$$

14 Inequalities of Sobolev, Poincaré and parabolic mean value

for some constant $R \geq 0$. There exist constants $C_{17}, C_{18} > 0$ depending only on m such that for any function $f \in H_c^{1,2}(B_p(\rho))$ with compact support in some geodesic ball $B_p(\rho)$, f must satisfy the Sobolev inequality

$$\int_{B_p(\rho)} |\nabla f|^2$$

$$\geq C_{17} \rho^{-2} \exp\left(-C_{18} \rho \sqrt{R}\right) (V_p(\rho))^{2/m} \left(\int_{B_p(\rho)} |f|^{2m/(m-2)}\right)^{(m-2)/m}.$$

Remark 1: The Sobolev inequality can be written in the form

$$\fint_{B_p(\rho)} |\nabla f|^2 \geq C_{17} \rho^{-2} \exp\left(-C_{18} \rho \sqrt{R}\right) \left(\fint_{B_p(\rho)} |f|^{2m/(m-2)}\right)^{(m-2)/m},$$

where $\fint_{B_p(\rho)} g$ denotes $V_p^{-1}(\rho) \int_{B_p(\rho)} g$. This has the advantage that the constant $C_{17} \rho^{-2} \exp\left(-C_{18} \rho \sqrt{R}\right)$ is now independent on the volume of the ball.

Remark 2: When M has nonnegative Ricci curvature, the Sobolev inequality now takes the form

$$\int_{B_p(\rho)} |\nabla f|^2 \geq C_{17} \rho^{-2} (V_p(\rho))^{2/m} \left(\int_{B_p(\rho)} |f|^{2m/(m-2)}\right)^{(m-2)/m}.$$

In particular, when M has maximal volume growth, i.e., $V_p(\rho) \geq C \rho^m$, the Sobolev constant is uniformly bounded independent of ρ, as in the case of $\mathbb{R}^m$. Conversely, we show in Chapter 16 that if a complete manifold has a uniformly bounded Sobolev constant of the form

$$\int_{B_p(\rho)} |\nabla f|^2 \geq C \left(\int_{B_p(\rho)} |f|^{2m/(m-2)}\right)^{(m-2)/m},$$

then the manifold must have volume growth bounded by

$$V_p(\rho) \geq C \rho^m.$$

Theorem 14.4 *Let M^m be a compact manifold without boundary of dimension $m \geq 3$. Suppose the Ricci curvature of M is bounded below by*

$$\mathcal{R}_{ij} \geq -(m-1)R$$

for some constant $R \geq 0$, and let $d(M)$ be diameter of M. Then there exist constants $C_1, C_2 > 0$ depending only on m, such that

$$\int_M |\nabla f|^2 + d^{-2}(M) \int_M f^2 \geq C_1 \exp\left(-C_2 d(M)\sqrt{R}\right) V^{2/m}(M)$$

$$\times \left(\int_M |f|^{2m/(m-2)}\right)^{(m-2)/m}$$

for all $f \in H^{1,2}(M)$. Moreover, there are also constants $C_3, C_4 > 0$ depending only on m, such that the inequality

$$\int_M |\nabla f|^2 \geq C_1 \exp\left(-C_2 d(M)\sqrt{R}\right) V^{2/m}(M) \left(\int_M |f|^{2m/(m-2)}\right)^{(m-2)/m}$$

is also valid for $f \in H^{1,2}(M)$ satisfying the extra condition $\int_M f = 0$.

Proof Let $\{x_i\}_{i=1}^k$ be a maximal set of points in M with the properties that $r(x_i, x_j) \geq d/4$ for all $i \neq j$, where $d = d(M)$. Clearly this implies that

$$B_{x_i}\left(\frac{d}{8}\right) \cap B_{x_j}\left(\frac{d}{8}\right) = \emptyset \tag{14.6}$$

for $i \neq j$, hence

$$\sum_{i=1}^k V_{x_i}\left(\frac{d}{8}\right) \leq V(M). \tag{14.7}$$

On the other hand, if $x \in M \setminus \bigcup_{i=1}^k B_{x_i}\left(\frac{d}{4}\right)$, then $r(x, x_i) \geq d/4$, contradicting the maximal property of the set $\{x_i\}$, hence we have

$$M = \bigcup_{i=1}^k B_{x_i}\left(\frac{d}{4}\right) \tag{14.8}$$

and

$$V(M) \leq \sum_{i=1}^k V_{x_i}\left(\frac{d}{4}\right). \tag{14.9}$$

However, the volume comparison theorem asserts that

14 Inequalities of Sobolev, Poincaré and parabolic mean value

$$V_{x_i}\left(\frac{d}{4}\right) \leq V_{x_i}\left(\frac{d}{8}\right) \bar{V}\left(\frac{d}{4}\right) \bar{V}^{-1}\left(\frac{d}{8}\right),$$

where $\bar{V}(r)$ denotes the volume of the geodesic ball of radius r in the simply connected space form with constant $-R$ curvature. Combining the volume bound with (14.7), we conclude that

$$kV(M) \leq \sum_{i=1}^{k} V_{x_i}(d)$$

$$\leq \bar{V}(d) \bar{V}^{-1}\left(\frac{d}{8}\right) \sum_{i=1}^{k} V_{x_i}\left(\frac{d}{8}\right)$$

$$\leq \bar{V}(d) \bar{V}^{-1}\left(\frac{d}{8}\right) V(M),$$

implying the bound

$$k \leq \bar{V}\left(\frac{d}{4}\right) \bar{V}^{-1}\left(\frac{d}{8}\right)$$

$$\leq C \exp\left(Cd\sqrt{R}\right).$$

To prove the Sobolev inequality, we may consider only nonnegative functions $f \in H^{1,2}(M)$. Let ϕ be a nonnegative cutoff function with the properties that

$$\phi = \begin{cases} 1 & \text{on} \quad B_{x_i}\left(\frac{d}{4}\right), \\ 0 & \text{on} \quad M \setminus B_{x_i}\left(\frac{d}{2}\right), \end{cases}$$

and

$$|\nabla \phi|^2 \leq Cd^{-2}.$$

Applying Theorem 14.3 to the function ϕf yields

$$\int_{B_{x_i}\left(\frac{d}{2}\right)} |\nabla(\phi f)|^2 \geq C_{17} d^{-2} \exp\left(-C_{18} d\sqrt{R}\right) V_{x_i}^{2/m}\left(\frac{d}{2}\right)$$

$$\times \left(\int_{B_{x_i}\left(\frac{d}{2}\right)} |\phi f|^{2m/(m-2)}\right)^{(m-2)/m}. \tag{14.10}$$

Using

$$2\int_M |\nabla f|^2 + 2Cd^{-2}\int_M f^2 \geq 2\int_{B_{x_i}\left(\frac{d}{2}\right)} \phi^2 |\nabla f|^2 + 2\int_{B_{x_i}\left(\frac{d}{2}\right)} |\nabla \phi|^2 f^2$$

$$\geq \int_{B_{x_i}\left(\frac{d}{2}\right)} |\nabla(\phi f)|^2$$

and

$$\int_{B_{x_i}(d/2)} |\phi f|^{2m/(m-2)} \geq \int_{B_{x_i}\left(\frac{d}{4}\right)} |f|^{2m/(m-2)}$$

(14.10) becomes

$$\int_M |\nabla f|^2 + Cd^{-2}\int_M f^2 \geq C_{19}d^{-2} \exp\left(-C_{18}d\sqrt{R}\right) V_{x_i}^{2/m}\left(\frac{d}{2}\right)$$

$$\times \left(\int_{B_{x_i}\left(\frac{d}{4}\right)} |f|^{2m/(m-2)}\right)^{(m-2)/m}.$$

Note that the volume comparison theorem allows us to estimate

$$V_{x_i}\left(\frac{d}{2}\right) \geq C \exp\left(-Cd\sqrt{R}\right) V_{x_i}(d)$$

$$= C \exp\left(-Cd\sqrt{R}\right) V(M).$$

Summing over $i = 1, \ldots, k$ and using the inequality

$$\sum_{i=1}^k \left(\int_{B_{x_i}\left(\frac{d}{4}\right)} |f|^{2m/(m-2)}\right)^{(m-2)/m} \geq \left(\sum_{i=1}^k \int_{B_{x_i}\left(\frac{d}{4}\right)} |f|^{2m/(m-2)}\right)^{(m-2)/m}$$

$$= \left(\int_M |f|^{2m/(m-2)}\right)^{(m-2)/m}$$

because of (14.8), we conclude that

$$k\int_M |\nabla f|^2 + kCd^{-2}\int_M f^2 \geq C_{19}d^{-2} \exp\left(-C_{20}d\sqrt{R}\right) V^{2/m}(M)$$

$$\times \left(\int_M |f|^{2m/(m-2)}\right)^{(m-2)/m}.$$

The first part of the theorem follows by applying the upper bound for k.

14 Inequalities of Sobolev, Poincaré and parabolic mean value

The second part follows from the Poincaré inequality

$$\int_M |\nabla f|^2 \geq \lambda_1(M) \int_M f^2$$

for $f \in H^{1,2}(M)$ satisfying $\int_M f = 0$, and the lower bound of $\lambda_1(M)$ provided by Theorem 5.7. $\square$

We will next present an estimate of the first nonzero Neumann eigenvalue for star-shaped domains. As a consequence, it gives a lower bound of the Neumann eigenvalue for geodesic balls. This was first proved by Buser [Bu] using a geometric argument via the isoperimetric inequality. The analytic argument which we present below was due to Chen and Li [CnL].

Theorem 14.5 *Let M be an m-dimensional manifold with boundary ∂M. Assume that M is geodesically star-shaped with respect to a point $p \in M$. Suppose that the Ricci curvature of M is bounded from below by*

$$\mathcal{R}_{ij} \geq -(m-1)R$$

for some constant $R \geq 0$. Let ρ be the radius of the largest geodesic ball centered at p contained in M, and ρ_0 be the radius of the smallest geodesic ball centered at p containing M. Then there exists a constant $C_1 > 0$ depending only on m, such that the first nonzero Neumann eigenvalue $\lambda_1(M)$ has a lower bound given by

$$\lambda_1(M) \geq \frac{\rho^m}{\rho_0^{m+2}} \exp\left(-C_1\left(1 + \rho_0\sqrt{R}\right)\right).$$

Proof By the variational principle, it suffices to show that there exists a constant $C_1 > 0$ such that

$$\int_M |\nabla f|^2 \geq \frac{\rho^m}{\rho_0^{m+2}} \exp\left(-C_1\left(1 + \rho_0\sqrt{R}\right)\right) \inf_{a \in \mathbb{R}} \int_M (f-a)^2$$

for any smooth function f_0 defined on M. Let $K(x, y, t)$ be the Neumann heat kernel defined on M. The function

$$f(x, t) = \int_M K(x, y, t) f_0(y) \, dy$$

solves the heat equation

$$\left(\Delta - \frac{\partial}{\partial t}\right) f(x, t) = 0$$

on M, with the Neumann boundary condition

$$\frac{\partial f}{\partial \nu} = 0$$

on ∂M, and the initial condition $f(x, 0) = f_0(x)$. Let us now consider the function

$$g(x, t) = \int_M K(x, y, t) (f_0(y) - f(x, t))^2 \, dy.$$

Clearly,

$$\int_M g(x, t) \, dx = \int_M \int_M K(x, y, t) f_0^2(y) \, dy \, dx - \int_M f^2(x, t) \, dx$$

$$= \int_M f_0^2(y) \, dy - \int_M f^2(x, t) \, dx$$

$$= -\int_0^t \frac{\partial}{\partial s} \int_M f^2(x, s) \, dx$$

$$= -2 \int_0^t \int_M f(x, s) \, \Delta f(x, s) \, dx$$

$$= 2 \int_0^t \int_M |\nabla f|^2(x, s) \, dx.$$

However, if we consider

$$\frac{\partial}{\partial t} \int_M |\nabla f|^2(x, t) \, dx = 2 \int_M \langle \nabla f, \nabla f_t \rangle (x, t) \, dx$$

$$= -2 \int_M \langle \Delta f, f_t \rangle (x, t) \, dx$$

$$= -2 \int_M f_t^2(x, t) \, dx,$$

we conclude that

$$\int_M |\nabla f|^2(x, t) \, dx \leq \int_M |\nabla f_0|^2(x) \, dx.$$

Hence, we have the estimate

$$\int_M g(x, t) \, dx \leq 2t \int_M |\nabla f_0|^2(x) \, dx. \qquad (14.11)$$

14 Inequalities of Sobolev, Poincaré and parabolic mean value

On the other hand, since g is also nonnegative, we have

$$\int_M g(x,t)\,dx \geq \int_{B_p(\rho/2)} g(x,t)\,dx$$

$$= \int_{B_p(\rho/2)} \int_M K(x,y,t)(f_0(y) - \bar{f}(x,t))^2\,dy\,dx$$

$$\geq \int_{B_p(\rho/2)} \inf_{y \in M} K(x,y,t) \int_M (f_0(y) - \bar{f}(x,t))^2\,dy\,dx$$

$$\geq \inf_{a \in \mathbb{R}} \int_M (f_0(y) - a)^2\,dy \int_{B_p(\rho/2)} \inf_{y \in M} K(x,y,t)\,dx. \quad (14.12)$$

Using (14.11) and (14.12), we conclude that

$$\lambda_1(M) \geq (2t)^{-1} \int_{B_p(\rho/2)} \inf_{y \in M} K(x,y,t)\,dx$$

for all $t > 0$. Note that the Harnack inequality (Corollary 12.3) of Li and Yau asserts that, for any $x, z \in B_p(\rho/2)$, we have

$$K(x,y,t) \geq K\left(z,y,\frac{t}{2}\right) C_4 \exp\left(-C_5\left(\frac{\rho^2}{t} + Rt + \frac{t}{\rho^2}\right)\right)$$

for some constants $C_4, C_5 > 0$ depending only on m. Hence

$$\int_{B_p(\rho/2)} \inf_{y \in M} K(x,y,t)\,dx$$

$$\geq V_p\left(\frac{\rho}{2}\right) \inf_{x \in B_p(\rho/2), y \in M} K(x,y,t)$$

$$\geq C_4 \exp\left(-C_5\left(\frac{\rho^2}{t} + Rt + \frac{t}{\rho^2}\right)\right) \inf_{y \in M} \int_{B_p(\rho/2)} K\left(z,y,\frac{t}{2}\right) dz,$$

and

$$\lambda_1(M) \geq \frac{C_4}{t} \exp\left(-C_5\left(\frac{\rho^2}{t} + Rt + \frac{t}{\rho^2}\right)\right) \inf_{y \in M} \int_{B_p(\rho/2)} K\left(z,y,\frac{t}{2}\right) dz. \quad (14.13)$$

To estimate the right-hand side, we apply Lemma 13.7 and obtain

$$\int_{B_p(\rho/2)} K\left(z,y,\frac{t}{2}\right) dz \geq \bar{\phi}\left(\bar{r}(\bar{y}), \frac{t}{2}\right), \quad (14.14)$$

where $\bar\phi$ is defined by

$$\bar\phi\left(\bar r(\bar y),\frac{t}{2}\right)=\int_{B_{\bar p}(\rho/2)}\bar K\left(\bar z,\bar y,\frac{t}{2}\right)d\bar z$$

with $\bar K$ being the Neumann heat kernel on $B_{\bar p}(\rho_0)$ with $\bar r(\bar y)=r(y)$. The proof of Lemma 13.7 asserts that $\bar\phi'\leq 0$, hence

$$\bar\phi\left(\bar r(\bar y),\frac{t}{2}\right)\geq\bar\phi\left(\rho_0,\frac{t}{2}\right). \tag{14.15}$$

On the other hand, since $B_{\bar p}(\rho_0)$ is convex, we can apply Theorem 12.6 to the function $\bar\phi$ and obtain

$$\bar\phi\left(\rho_0,\frac{t}{2}\right)\geq C_6\,\bar\phi\left(0,\frac{t}{4}\right)\exp\left(-C_7\left(\frac{\rho_0^2}{t}+Rt\right)\right) \tag{14.16}$$

for some constants C_6, $C_7>0$ depending only on m. Using $\bar\phi'\leq 0$ again, we conclude that $\bar p$ is a maximum point of $\bar\phi$ for any $t\in(0,\infty)$. Therefore, the heat equation implies that

$$\frac{\partial}{\partial t}\bar\phi=\bar\Delta\bar\phi(0,t)$$

$$\leq 0,$$

hence

$$\bar\phi\left(0,\frac{t}{4}\right)\geq\lim_{t\to\infty}\bar\phi(0,t)$$

$$=\int_{B_{\bar p}(\rho/2)}\bar V_{\bar p}^{-1}(\rho_0)\,d\bar z,$$

where $\bar V_{\bar p}(\rho_0)$ is the volume of $B_{\bar p}(\rho_0)$. Combining the above inequality with (14.13), (14.14), (14.15), and (14.16), we have

$$\lambda_1(M)\geq C_{21}\,(t)^{-1}\exp\left(-C_{22}\left(\frac{\rho^2}{t}+\frac{t}{\rho^2}+\frac{\rho_0^2}{t}+Rt\right)\right)\frac{\bar V_{\bar p}(\rho/2)}{\bar V_{\bar p}(\rho_0)},$$

for some constants C_{21}, $C_{22}>0$. Setting

$$t=\frac{\rho_0^2}{1+\rho_0\sqrt{R}}$$

and using the estimates

$$t^{-1}\geq\rho_0^{-2}$$

14 Inequalities of Sobolev, Poincaré and parabolic mean value

and

$$\frac{\rho_0^2 R}{1 + \rho_0 \sqrt{R}} \leq 1 + \rho_0 \sqrt{R},$$

we conclude that

$$\lambda_1(M) \geq \rho_0^{-2} \exp\left(-C_{23}\left(1 + \rho_0\sqrt{R}\right)\right) \frac{\bar{V}_{\bar{p}}(\rho/2)}{\bar{V}_{\bar{p}}(\rho_0)}$$

for some constant $C_{23} > 0$ depending only on m. The theorem now follows by observing that there exist constants $C_{24} > 0$ and $C_{25} > 0$, depending only on m, such that

$$\bar{r}^m \exp\left(C_{24}\left(1 + \bar{r}\sqrt{R}\right)\right) \leq \bar{V}_{\bar{p}}(\bar{r})$$

$$\leq \bar{r}^m \exp\left(C_{25}\left(1 + \bar{r}\sqrt{R}\right)\right).$$

This completes the proof of Theorem 14.5. □

Corollary 14.6 *Let $B_p(\rho)$ be a geodesic ball centered at a point $p \in M$ with radius $\rho > 0$ such that $B_p(\rho) \cap \partial M = \emptyset$. Suppose that the Ricci curvature of $B_p(\rho)$ is bounded from below by*

$$\mathcal{R}_{ij} \geq -(m-1)R$$

for some constant $R \geq 0$. Then there exists a constant $C_2 > 0$ depending only on m, such that the first nonzero Neumann eigenvalue $\lambda_1(B_p(\rho))$ has a lower bound given by

$$\lambda_1(B_p(\rho)) \geq \rho^{-2} \exp\left(-C_2\left(1 + \rho\sqrt{R}\right)\right).$$

The following theorem gives a parabolic mean value inequality for subsolutions of the heat equation. It was first proved by Li and Tam in [LT5].

Theorem 14.7 *Let M^m be a complete noncompact Riemannian manifold with boundary. Let $p \in M$ and $\rho > 0$ be such that $B_p(4\rho) \cap \partial M = \emptyset$. Suppose $g(x, t)$ is a nonnegative function defined on $B_p(4\rho) \times [0, T]$ for some $0 < T \leq \rho^2/4$ satisfying the differential inequality*

$$\Delta g - \frac{\partial g}{\partial t} \geq 0.$$

If the Ricci curvature of $B_p(4\rho)$ is bounded by

$$\mathcal{R}_{ij} \geq -(m-1)R$$

for some constant $R \geq 0$, then for any $q > 0$, there exist positive constants C_1 and C_2 depending only on m and q, such that for any $0 < \tau < T$, $0 < \delta < \frac{1}{2}$, and $0 < \eta < \frac{1}{2}$ we have

$$\sup_{B_p((1-\delta)\rho) \times [\tau, T]} g^q \leq C_1 \frac{\bar{V}(2\rho)}{V_p(\rho)} \left(\rho \sqrt{R} + 1\right) \exp\left(C_2 \sqrt{R}\, T\right)$$

$$\times \left(\frac{1}{\delta\rho} + \frac{1}{\sqrt{\eta\tau}}\right)^{m+2} \int_{(1-\eta)\tau}^{T} ds \int_{B_p(\rho)} g^q(y, s)\, dy,$$

where $\bar{V}(r)$ is the volume of the geodesic ball of radius r in the m-dimensional, simply connected space form with constant sectional curvature $-R$.

Proof For any $0 < \delta$, $\eta < \frac{1}{4}$, let $\tilde{\phi}$ and $\tilde{\psi}$ be two smooth functions on $[0, \infty)$ so that $0 \leq \tilde{\phi}, \tilde{\psi} \leq 1$, with

$$\tilde{\phi}(s) = \begin{cases} 1 & \text{for } 0 \leq s \leq 1 - \delta, \\ 0 & \text{for } s \geq 1, \end{cases}$$

and

$$\tilde{\psi}(s) = \begin{cases} 0 & \text{for } 0 \leq s \leq 1 - 2\eta, \\ 1 & \text{for } s \geq 1 - \eta. \end{cases}$$

We can choose $\tilde{\phi}$ and $\tilde{\psi}$ such that

$$-C\delta^{-1} \leq \tilde{\phi}' \tilde{\phi}^{-\frac{1}{2}} \leq 0,$$

$$\tilde{\phi}'' \geq -C\delta^{-2},$$

and

$$0 \leq \tilde{\psi}' \leq C\eta^{-1},$$

for some constant $C > 0$ which is independent of δ and η.

Let $r(x)$ denote the distance function from p, and we define $\phi(x) = \tilde{\phi}(r(x) R^{-1})$ and $\psi(t) = \tilde{\psi}(t \tau^{-1})$. Since g is a nonnegative subsolution of the heat equation, for $q > 1$ we have

$$\left(\Delta - \frac{\partial}{\partial t}\right)(\phi(x) \psi(t) g^q(x, t))$$

$$\geq (\Delta\phi) \psi g^q + 2q\psi g^{q-1} \langle \nabla\phi, \nabla g\rangle + q(q-1)\phi \psi g^{q-2} |\nabla g|^2 - \phi \psi' g^q$$

$$\geq (\Delta\phi) \psi g^q - \frac{q}{q-1} \psi g^q \frac{|\nabla\phi|^2}{\phi} - \phi \psi' g^q.$$

14 Inequalities of Sobolev, Poincaré and parabolic mean value

On the other hand, the Laplacian comparison theorem asserts that

$$\left(\Delta \phi - \frac{q}{q-1} \frac{|\nabla \phi|^2}{\phi}\right) \geq -C_3 \left(\rho \sqrt{R} + 1\right) \delta^{-2} \rho^{-2}, \qquad (14.17)$$

for some constant C_3 depending only on m and q.

If $H(x, y, t)$ is the Dirichlet heat kernel of M, by using (14.17) and the definition of ϕ and ψ, there exists a constant C_4 depending only on m and q so that

$$g^q(x, t) = g^q(x, t) \phi(x) \psi(t)$$

$$= -\int_0^t ds \int_M H(x, y, t-s) \left(\Delta_y - \frac{\partial}{\partial s}\right) (g^q(y, s) \phi(y) \psi(s)) \, dy$$

$$= -\int_{(1-2\eta)\tau}^t ds \int_{B_p(\rho)} H(x, y, t-s) \left(\Delta_y - \frac{\partial}{\partial s}\right)$$

$$\times (g^q(y, s) \phi(y) \psi(s)) \, dy$$

$$\leq C_4 \left\{ \left(\rho \sqrt{R} + 1\right) \delta^{-2} \rho^{-2} \int_{(1-2\eta)\tau}^T ds \int_{B_p(\rho)\backslash B_p((1-\delta)\rho)} \right.$$

$$\times H(x, y, t-s) g^q(y, s) \, dy$$

$$\left. + \frac{1}{\eta\tau} \int_{(1-2\eta)\tau}^{(1-\eta)\tau} ds \int_{B_p(\rho)} H(x, y, t-s) g^q(y, s) \, dy \right\}$$

$$(14.18)$$

for any $x \in B_p((1-2\delta)\rho)$ and $\tau < t < T$. Using the upper bound of the heat kernel given by Corollary 13.5, if $x, y \in B_p(R)$ and $t - s \leq R^2/4$, then we have

$$H(x, y, t-s) \leq C_5 V_x^{-\frac{1}{2}} \left(\sqrt{t-s}\right) V_y^{-\frac{1}{2}} \left(\sqrt{t-s}\right)$$

$$\times \exp\left(-C_6 \frac{r^2(x, y)}{t-s} + C_7 \sqrt{(\rho^{-2} + R)(t-s)}\right),$$

where C_5, C_6, and C_7 are constants depending only on m. On the other hand, the volume comparison theorem implies that for $x \in B_p(\rho)$ and $0 < s < t \leq T \leq \rho^2/4$

$$\frac{V_p(\rho)}{V_x \left(\sqrt{t-s}\right)} \leq \frac{V_x(2\rho)}{V_x \left(\sqrt{t-s}\right)} \leq \frac{\bar{V}(2\rho)}{\bar{V} \left(\sqrt{t-s}\right)} \leq C_8 (t-s)^{-m/2} \bar{V}(2\rho)$$

for some constant C_8 depending only on m. Hence

$$H(x, y, t-s) \leq C_9 \frac{\bar{V}(2\rho)}{V_p(\rho)} (t-s)^{-m/2}$$

$$\times \exp\left(-C_6 \frac{r^2(x,y)}{t-s} + C_7\sqrt{(\rho^{-2}+R)(t-s)}\right)$$

$$\leq C_{10} (\delta\rho)^{-m} \frac{\bar{V}(2\rho)}{V_p(\rho)} \exp\left(C_7\sqrt{(\rho^{-2}+R)(t-s)}\right) \quad (14.19)$$

for $x \in B_p((1-2\delta)\rho)$, $y \in B_p(\rho)\setminus B_p((1-\delta)\rho)$ and $0 < s \leq t \leq T \leq \rho^2/4$, where C_9 and C_{10} are constants depending only on m. Also for $t > \tau$, and $s \leq (1-\eta)\tau$, we can find a constant C_{11} depending only on m so that

$$H(x,y,t-s) \leq C_{11} (\eta\tau)^{-\frac{m}{2}} \frac{\bar{V}(2\rho)}{V_p(\rho)} \exp\left(C_7\sqrt{(\rho^{-2}+R)(t-s)}\right). \quad (14.20)$$

Combining (14.18), (14.19), and (14.20), we conclude that

$$\sup_{B((1-2\delta)\rho)\times[\tau,T]} g^q \leq C_{12} \frac{\bar{V}(2\rho)}{V_p(\rho)} \left(\rho\sqrt{R}+1\right) \exp\left(C_{13}\sqrt{(\rho^{-2}+R)T}\right)$$

$$\times \left((\delta\rho)^{-(m+2)} + (\eta\tau)^{-(m+2)/2}\right)$$

$$\times \int_{(1-2\eta)\tau}^{T} ds \int_{B_p(\rho)} g^q(y,s)\,dy \quad (14.21)$$

for $0 < \tau < T \leq \rho^2/4$, $0 < \delta$, $\eta \leq 1/4$, and $q > 1$, where C_{12} and C_{13} depend only on m and q. This completes the proof of the theorem for the case when $q > 1$ if we replace 2δ by δ and 2η by η.

In order to prove the case when $q \leq 1$, we can proceed as in [LS]. Taking $q = 2$ in (14.21), we have

$$\sup_{B_p((1-\delta)\rho)\times[\tau,T]} g^2 \leq C_{12} \frac{\bar{V}(2\rho)}{V_p(\rho)} \left(\rho\sqrt{R}+1\right) \exp\left(C_{13}\sqrt{(\rho^{-2}+R)T}\right)$$

$$\times \left((\delta\rho)^{-(m+2)} + (\eta\tau)^{-m+2/2}\right)$$

$$\times \int_{(1-\eta)\tau}^{T} ds \int_{B_p(\rho)} g^2(y,s)\,dy \quad (14.22)$$

14 Inequalities of Sobolev, Poincaré and parabolic mean value

for $0 < \delta$, $\eta < \frac{1}{2}$, and $T \leq \rho^2/4$. For $k \geq 0$ and $l \geq 1$, let us define

$$M(k,l) = \sup_{B_p((1-\delta_k)\rho) \times [\tau_l, T]} g^2,$$

where $\delta_k = 2^{-k}\delta$, and $\tau_l = \left(1 - \eta \sum_{i=1}^{l} 2^{-i}\right) \tau$. Inequality (14.22) implies that

$$M(k,l) \leq C_{12} \frac{\bar{V}(2\rho)}{V_p(\rho)} \left(\rho \sqrt{R} + 1\right) \exp\left(C_{13}\sqrt{(\rho^{-2} + R)T}\right)$$

$$\times \left(2^{(k+1)(m+2)}(\delta\rho)^{-(m+2)} + 2^{(l+1)(m+2)/2}(\eta\tau)^{-(m+2)/2}\right)$$

$$\times M^\lambda(k+1, l+1) \int_{(1-\eta)\tau}^{T} ds \int_{B_p(\rho)} g^q(y,s)\,dy,$$

where $\lambda = 1 - q/2$. Let $\tilde{M}(k) = M(k, 2k+1)$, then

$$\tilde{M}(k) \leq A(2^{m+2})^{k+1} \tilde{M}^\lambda(k+1),$$

where

$$A = C_{12} \frac{\bar{V}(2\rho)}{V_p(\rho)} \left(\rho\sqrt{R} + 1\right) \exp\left(C_{13}\sqrt{(\rho^{-2} + R)T}\right)$$

$$\times \left((\delta\rho)^{-(m+2)} + (\eta\tau)^{-(m+2)/2}\right) \int_{(1-\eta)\tau}^{T} ds \int_{B_p(\rho)} g^q(y,s)\,dy.$$

Iterating this inequality, we get

$$\tilde{M}(0) \leq C_{14} A^{1/(1-\lambda)},$$

where C_{14} is a constant depending only on m and q. From this, it is easy to conclude that the theorem is also true for $0 < q \leq 1$. □

Corollary 14.8 *Let M^m be a complete manifold with boundary and p be a fixed point in M. Suppose $\rho > 0$ is such that $B_p(4\rho) \cap \partial M = \emptyset$ and assume that the Ricci curvature of M on $B_p(4\rho)$ satisfies the bound*

$$\mathcal{R}_{ij} \geq -(m-1)R$$

for some constant $R \geq 0$. Let $0 < \delta < \frac{1}{2}$, $q > 0$, and $\lambda \geq 0$ be fixed constants. Then there exists a constant $C > 0$ depending only on δ, q, $\lambda \rho^2$, m, and $\rho \sqrt{R}$, such that for any nonnegative function f defined on $B_p(2\rho)$ satisfying the differential inequality

$$\Delta f \geq -\lambda f$$

we have

$$\sup_{B_p((1-\delta)\rho)} f^q \leq C\, \bar V_p^{-1}(\rho) \int_{B_p(\rho)} f^q(y)\, dy.$$

Proof Let us consider the function

$$g(x, t) = e^{-\lambda t} f(x)$$

which satisfies the differential inequality

$$\left(\Delta - \frac{\partial}{\partial t}\right) g(x, t) \geq 0$$

on $B_p(2\rho) \times [0, \infty)$. Applying Theorem 14.7 to $g(x, t)$ and setting $T = \rho^2/4$, $\tau = \rho^2/8$, and $\eta = \frac{1}{4}$, we obtain

$$\sup_{B_p((1-\delta)\rho) \times [\rho^2/8, \rho^2/4]} e^{-p\lambda t} f^q(x) \leq C_1\, \bar V(2\rho)\, \rho^{-(m+2)} \int_{3\rho^2/32}^{\rho^2/4} e^{-q\lambda s}\, ds$$

$$\times \left(\rho\sqrt{R} + 1\right) \exp\left(C_2 \rho \sqrt{R}\right) \bar V_p^{-1}(\rho)$$

$$\times \int_{B_p(\rho)} f^q(y)\, dy,$$

where $C_1 > 0$ depends on m, q, and δ. This implies that

$$\sup_{B_p((1-\delta)\rho)} f^q(x) \leq C_4 \exp\left(C_2 \rho \sqrt{R} + C_3 \lambda \rho^2\right) \bar V(2\rho)\, \rho^{-m}\, \bar V_p^{-1}(\rho)$$

$$\times \int_{B_p(\rho)} f^q(y)\, dy. \tag{14.23}$$

Obviously the corollary follows from this estimate. $\square$

Corollary 14.9 *Let M^m be a complete manifold with boundary and p be a fixed point in M. Suppose $\rho > 0$ be such that $B_p(4\rho) \cap \partial M = \emptyset$ and assume that the Ricci curvature of M is nonnegative. Let $0 < \delta < \frac{1}{2}$ and $q > 0$ be fixed constants. Then there exists a constant $C > 0$ depending only on δ, q, and m,*

14 Inequalities of Sobolev, Poincaré and parabolic mean value

such that for any nonnegative function f defined on $B_p(2\rho)$ satisfying the differential inequality

$$\Delta f \geq 0$$

we have

$$\sup_{B_p((1-\delta)\rho)} f^q \leq C\, V_p^{-1}(\rho) \int_{B_p(\rho)} f^q(y)\, dy.$$

Proof Applying the assumptions to (14.23) by setting

$$\bar{V}(2\rho) = \frac{\alpha_{m-1}}{m}\rho^m$$

and $\lambda = 0$, we obtain

$$\sup_{B_p((1-\delta)\rho)} f^q(x) \leq C_1\, \rho^{-2} \int_{3\rho^2/32}^{\rho^2/4} ds\, V_p^{-1}(\rho) \int_{B_p(\rho)} f^q(y)\, dy.$$

This proves the corollary. □

Another form of the mean value inequality similar to that of Theorem 14.7 is to allow information on the subsolution at $t = 0$ to be used. The proof is similar and slightly easier than the proof of Theorem 14.7 and is by choosing $\psi = 1$. We will only state the theorem, leaving the proof as an exercise.

Theorem 14.10 *Let M^m be a complete noncompact Riemannian manifold with boundary. Let $p \in M$ and $\rho > 0$ be such that $B_p(4\rho) \cap \partial M = \emptyset$. Suppose $g(x, t)$ is a nonnegative function defined on $B_p(4\rho) \times [0, T]$ for some $0 < T \leq \rho^2/4$ satisfying the differential inequality*

$$\Delta g - \frac{\partial g}{\partial t} \geq 0.$$

If the Ricci curvature of $B_p(4\rho)$ is bounded by

$$\mathcal{R}_{ij} \geq -(m-1)R$$

for some constant $R \geq 0$, then for any $q > 0$ and $\epsilon > 0$, there exist positive constants C_{15} and C_{16} depending on m and q, with C_{15} also depending on ϵ, such that for any $0 < \delta < \frac{1}{2}$ we have

$$\sup_{B_p((1-\delta)\rho)\times[0,T]} g^q \leq C_{15} \frac{\bar{V}(2\rho)}{V_p(\rho)} \left(\rho\sqrt{R}+1\right) \exp\left(C_2\sqrt{RT}\right)(\delta\rho)^{-(m+2)}$$

$$\times \int_0^T ds \int_{B_p(\rho)} g^q(y,s)\,dy + (1+\epsilon) \sup_{B_p(\rho)} g^q(\cdot,0),$$

where $\bar{V}(r)$ is the volume of the geodesic ball of radius r in the m-dimensional, simply connected space form with constant sectional curvature $-R$.

15

Uniqueness and the maximum principle for the heat equation

As alluded to in our discussion preceding Definition 10.3, if $H(x, y, t)$ is the minimal heat kernel on a complete manifold, constructed by Theorem 12.4, then the question is if

$$\int_M H(x, y, t)\, dy = 1 \tag{15.1}$$

is related to the uniqueness of the heat kernel. The validity of (15.1) implies that $H(x, y, t) = K(x, y, t)$ where $K(x, y, t)$ is a limiting heat kernel obtained from a sequence of Neumann heat kernels.

Note that the inequality

$$\int_M H(x, y, t) \leq 1 \tag{15.2}$$

implies that the (minimal) heat semi-group is contractive on L^∞, because

$$|f(x,t)| = \left| \int_M H(x, y, t)\, f_0(y)\, dy \right|$$

$$\leq \|f_0\|_\infty \int_M H(x, y, t)\, dy$$

$$\leq \|f_0\|_\infty,$$

implying $\|f(\cdot, t)\|_\infty \leq \|f_0\|_\infty$. Observe that the constant function 1 and $\int_M H(x, y, t)\, dy$ are solutions to the heat equation with the same initial data, hence if

$$\int_M H(x, y, t)\, dy < 1$$

for some $x \in M$, then there are two distinct L^∞ solutions to the heat equation with initial datum 1. In general, let $f(x, t)$ be any L^∞ solution to the heat equation with $f(x, 0) = f_0(x)$. If $|f(x, t)| < A$ on $M \times [0, T]$, then the functions $A + f$ and $A - f$ are both positive solutions to the heat equation with initial data given by $A + f_0$ and $A - f_0$, respectively. The minimality of the heat kernel H implies that

$$\int_M H(x, y, t)(A + f_0(y))\, dy \leq A + f(x, t)$$

and

$$\int_M H(x, y, t)(A - f_0(y))\, dy \leq A - f(x, t).$$

Therefore, we conclude that

$$\int_M H(x, y, t) f_0(y)\, dy - A\left(1 - \int_M H(x, y, t)\, dy\right) \leq f(x, t)$$
$$\leq \int_M H(x, y, t) f_0(y)\, dy + A\left(1 - \int_M H(x, y, t)\, dy\right).$$

In particular, the validity of (15.1) implies that $f(x, t) = \int_M H(x, y, t) f_0(y)\, dy$, asserting that the uniqueness of all bounded solutions is equivalent to (15.1).

Inequality (15.2) also implies that the semi-group is contractive on L^1, since

$$\int_M |f(x, t)|\, dy \leq \int_M \int_M H(x, y, t)|f_0|(y)\, dy\, dx$$
$$\leq \int_M |f_0|(y)\, dy.$$

When (15.1) is valid, then for nonnegative initial datum f_0, we have

$$\int_M f(x, t)\, dy = \int_M \int_M H(x, y, t) f_0(y)\, dy\, dx$$
$$= \int_M f_0(y)\, dy,$$

asserting that the heat semi-group preserves L^1-norm on nonnegative functions.

Also, note that the minimal heat semi-group is contractive on L^p for $1 < p < \infty$. To see this, let $f_0 \in L^p$ and let

$$f(x, t) = \int_M H(x, y, t) f_0(y)\, dy$$

be the solution of the heat equation with initial datum f_0. We will first assume that $f_0 \geq 0$. In this case, since $H(x, y, t)$ is the limit of a sequence of Dirichlet heat kernels $H_i(x, y, t)$ on a compact exhaustion $\{\Omega_i\}$, $f(x, t)$ is the limit of $f_i(x, t) = \int_M H_i(x, y, t) f_0 \, dy$, where we extend H_i to be 0 on $M \setminus \Omega_i$. However, since

$$\frac{\partial}{\partial t} \int_M f_i^p = p \int_M f_i^{p-1} \Delta f_i$$
$$= -p(p-1) \int_M f_i^{p-2} |\nabla f_i|^2$$
$$\leq 0,$$

we have

$$\|f_i(\cdot, t)\|_p \leq \|f_i(\cdot, 0)\|_p$$
$$\leq \|f_0\|_p$$

for all t. Taking the limit as $i \to \infty$, we conclude that

$$\|f(\cdot, t)\|_p \leq \|f_0\|_p,$$

as claimed. In general, we observe that

$$|f(x, t)| = \left| \int_M H(x, y, t) f_0(y) \, dy \right|$$
$$\leq \int_M H(x, y, t) |f_0(y)| \, dy.$$

Hence the contractivity of the minimal heat equation on L^p is established.

Proposition 15.1 *Let M^m be a complete manifold, and let $f(x, t)$ be a nonnegative subsolution of the heat equation. If $f(\cdot, t) \in L^p(M)$ for all $[0, T]$, for $1 < p < \infty$, then*

$$\|f(\cdot, t)\|_p \leq \|f(\cdot, 0)\|_p$$

for all $0 \leq t \leq T$. Moreover, the heat equation (not necessarily given by the minimal heat kernel) is contractive in L^p and it is uniquely determined by its initial condition and hence has to be given by

$$f(x, t) = \int_M H(x, y, t) f_0(y) \, dy.$$

Proof Using the fact that $f(x,t)$ is a nonnegative subsolution, if ϕ is a nonnegative cutoff function, we have

$$\frac{\partial}{\partial t} \int_M \phi^2(x) f^p(x,t)\,dx = p \int_M \phi^2(x) f^{p-1}(x,t) f_t(x,t)\,dx$$

$$\leq p \int_M \phi^2(x) f^{p-1}(x,t) \Delta f(x,t)\,dx$$

$$= -p(p-1) \int_M \phi^2(x) f^{p-2}(x,t) |\nabla f|^2(x,t)\,dx$$

$$- 2p \int_M \phi(x) f^{p-1}(x,t) \langle \nabla \phi(x), \nabla f(x,t) \rangle\,dx.$$
(15.3)

Schwarz inequality asserts that

$$-2p \int_M \phi f^{p-1} \langle \nabla \phi, \nabla f \rangle\,dx \leq p(p-1) \int_M \phi^2 f^{p-2} |\nabla f|^2\,dx$$

$$+ (p-1)^{-1} p \int_M |\nabla \phi|^2 f^p.$$

Combining this inequality with (15.3), we conclude that

$$\frac{\partial}{\partial t} \int_M \phi^2(x) f^p(x,t)\,dx \leq (p-1)^{-1} p \int_M |\nabla \phi|^2 f^p. \tag{15.4}$$

Choosing the cutoff function ϕ to satisfy

$$\phi = \begin{cases} 1 & \text{on} \quad B_q(\rho), \\ 0 & \text{on} \quad M \setminus B_q(2\rho), \end{cases}$$

and

$$|\nabla \phi|^2 \leq 2\rho^{-2},$$

(15.4) yields

$$\int_{B_q(\rho)} f^p(x,t)\,dx - \int_{B_q(2\rho)} f_0^p(x)\,dx$$

$$\leq \int_M \phi^2(x) f^p(x,t)\,dx - \int_M \phi^2(x) f_0^p(x)\,dx$$

$$\leq (p-1)^{-1} p \int_0^t \int_M |\nabla \phi|^2 f^p(x)\,dx\,dt$$

$$\leq 2p(p-1)^{-1} \rho^{-2} \int_0^t \int_M f^p(x)\,dx\,dt.$$

15 Uniqueness and the maximum principle for the heat equation

Using the assumption that $f(\cdot, t) \in L^p(M)$ and letting $\rho \to \infty$, we obtain

$$\int_M f^p(x, t) \, dx \leq \int_M f^p(x, 0) \, dx.$$

Observing that taking the absolute value of a solution yields a nonnegative subsolution, we conclude that the heat equation (not necessarily minimal) is contractive on nonnegative L^p functions. In particular, by taking the difference of two L^p solutions, this implies that L^p solutions are uniquely determined by their initial conditions. $\square$

We are now ready to address the issue of the uniqueness of L^∞ solutions. The theorem presented below was proved by Karp and Li [KL] in an unpublished paper. It was also independently proved by Grigor'yan [G1] with a slightly weaker assumption.

Theorem 15.2 (Karp–Li) *Let M^m be a complete manifold. Suppose $f(x, t)$ is a subsolution of the heat equation satisfying*

$$\left(\Delta - \frac{\partial}{\partial t}\right) f(x, t) \geq 0 \quad on \quad M \times (0, \infty)$$

with

$$f(x, 0) \leq 0 \quad on \quad M.$$

If there exist $p \in M$ and a constant $A > 0$ such that the L^2-norm of f over the geodesic ball centered at p of radius ρ satisfies

$$\int_{B_p(\rho)} f^2(x, t) \leq \exp(A \rho^2) \quad \text{for all} \quad t \in (0, T),$$

then $f(x, t) \leq 0$.

Proof Let us define the positive part of the function $f(x, t)$ by

$$f_+(x, t) = \max\{f(x, t), 0\},$$

which is a subsolution satisfying

$$\left(\Delta - \frac{\partial}{\partial t}\right) f_+(x, t) \geq 0 \quad on \quad M \times (0, \infty)$$

with

$$f_+(x, 0) = 0.$$

It suffices to show that $f_+ = 0$ on $M \times [0, \infty]$. For simplicity of notation, we will replace f_+ by f. Let us consider the function

$$g(x, t) = \frac{-r^2(x)}{4(2T - t)} \tag{15.5}$$

for $x \in M$ and $0 \le t < T$, where $r(x)$ is the distance function to the fixed point $p \in M$. One can verify that

$$|\nabla g|^2 + g_t = 0. \tag{15.6}$$

For $\epsilon > 0$ and $\rho > 0$, let ϕ be a nonnegative cutoff function with the properties that

$$\phi = \begin{cases} 1 & \text{on} \quad B_p(\rho), \\ 0 & \text{on} \quad M \setminus B_p(\rho + 1), \end{cases} \tag{15.7}$$

$0 \le \phi \le 1$ and $|\nabla \phi|^2 \le 2$. Using the fact that f is a nonnegative subsolution and its initial condition, we have

$$\begin{aligned}
0 \le{} & \int_0^\tau \int_M \phi^2 \exp(g) f \, \Delta f - \int_0^\tau \int_M \phi^2 \exp(g) f \, f_t \\
={} & -\int_0^\tau \int_M \phi^2 \exp(g) |\nabla f|^2 - 2 \int_0^\tau \int_M \phi \exp(g) f \, \langle \nabla \phi, \nabla f \rangle \\
& - \int_0^\tau \int_M \phi^2 \exp(g) f \, \langle \nabla g, \nabla f \rangle \\
& - \tfrac{1}{2} \int_M \phi^2 \exp(g) f^2 \big|_{t=\tau} + \tfrac{1}{2} \int_0^\tau \int_M \phi^2 \exp(g) f^2 \, g_t.
\end{aligned} \tag{15.8}$$

However, Schwarz inequality implies that

$$-\int_0^\tau \int_M \phi^2 \exp(g) f \, \langle \nabla g, \nabla f \rangle \le \tfrac{1}{2} \int_0^\tau \int_M \phi^2 \exp(g) f^2 |\nabla g|^2$$

$$+ \tfrac{1}{2} \int_0^\tau \int_M \phi^2 \exp(g) |\nabla f|^2$$

and

$$-\int_0^\tau \int_M \phi \exp(g) f \, \langle \nabla \phi, \nabla f \rangle \le \tfrac{1}{4} \int_0^\tau \int_M \phi^2 \exp(g) |\nabla f|^2$$

$$+ \int_0^\tau \int_M |\nabla \phi|^2 \exp(g) f^2,$$

hence (15.8) becomes

$$0 \le 2 \int_0^T \int_M |\nabla \phi|^2 \exp(g) f^2 + \frac{1}{2} \int_0^T \int_M \phi^2 \exp(g) f^2 |\nabla g|^2$$
$$- \frac{1}{2} \int_M \phi^2 \exp(g) f^2 |_{t=\tau} + \frac{1}{2} \int_0^T \int_M \phi^2 \exp(g) f^2 g_t.$$

Combining this inequality with (15.6) and (15.7), we obtain

$$\int_{B_p(\rho)} \exp(g) f^2 |_{t=\tau} \le \int_M \phi^2 \exp(g) f^2 |_{t=\tau}$$
$$\le 4 \int_0^T \int_{B_p(\rho+1) \setminus B_p(\rho)} \exp(g) f^2. \qquad (15.9)$$

Note that (15.5) implies that

$$\exp(g) \le \exp\left(-\frac{\rho^2}{8T}\right) \quad \text{on} \quad (B_p(\rho+1) \setminus B_p(\rho)) \times [0, T],$$

and (15.9) yields

$$\int_{B_p(\rho)} \exp(g) f^2(x, \tau) \, dx \le 4 \exp\left(-\frac{\rho^2}{8T}\right) \int_0^T \int_{B_p(\rho+1) \setminus B_p(\rho)} f^2(x, t) \, dx \, dt$$

for all $0 \le \tau \le T$. Applying the growth assumption on f, we conclude that

$$\int_{B_p(\rho)} \exp(g) f^2(x, \tau) \, dx \le 4\tau \exp\left(-\frac{\rho^2}{8T} + A(\rho+1)^2\right).$$

Choosing $T < 1/8A$ and letting $\rho \to \infty$, this implies that

$$\int_M \exp(g) f^2(x, \tau) \, dx \le 0 \quad \text{for all} \quad 0 < \tau \le T,$$

hence $f = 0$ on $M \times [0, T]$. Using $f(x, T)$ as the initial condition, we argue inductively that f is identically 0. □

Corollary 15.3 *Let M^m be a complete manifold. Suppose there exists a constant $A > 0$ such that the volume growth of M satisfies the estimate*

$$V_p(\rho) \le \exp(A\rho^2)$$

for all $\rho > 0$. Then any L^∞ solution of the heat equation is uniquely determined by its initial data. In particular,

$$\int_M H(x, y, t) = 1.$$

Proof By considering the difference of two L^∞ solutions, it suffices to prove that if $\|f(\cdot, t)\|_\infty \leq A$ and is a solution of the heat equation on M with $f(x, 0) = 0$, then f must be identically 0. This follows by applying Theorem 15.2 to the function $g = |f|$. □

We have yet to address the uniqueness of L^1 solutions. It turns out that with some mild curvature assumptions one can prove that the L^1 solution to the heat equation is also unique. We will delay our discussion until Chapter 31.

Note that Theorem 15.2 can be viewed as a form of the maximum principle on complete manifolds. Various generalizations based on the argument given in the proof of Theorem 15.2 have been developed. Interested readers should refer to [LoT] and [NT] for further discussion.

16
Large time behavior of the heat kernel

In this chapter, we will discuss the behavior of the heat equation when $t \to \infty$. When M is a compact manifold with (or without) boundary, due to the eigenfunction expansion, the Dirichlet and Neumann heat kernels are asymptotically given by the first eigenfunction and the first eigenvalue. In particular,

$$H(x, y, t) \sim e^{-\mu_1 t} \psi_1(x) \psi_1(y)$$

and

$$K(x, y, t) \sim V^{-1}(M)$$

as $t \to \infty$. However, if M is complete and noncompact, then the behavior of H as $t \to \infty$ is not so clear. If the bottom of the L^2 spectrum $\mu_1(M)$ is positive, it is still rather easy to see that

$$t^{-1} \log H(x, y, t) \sim -\mu_1(M).$$

The following discussion gives a sharp asymptotic behavior of H for manifolds with nonnegative Ricci curvature with maximal volume growth. As a corollary (Corollary 16.3), one concludes that such manifolds must have finite fundamental group. This was proved by the author in [L5].

Lemma 16.1 *Let M^m be a complete manifold with nonnegative Ricci curvature. If there exists a constant $\theta > 0$ such that*

$$\liminf_{\rho \to \infty} \frac{V_p(\rho)}{\rho^m} = \theta$$

for some point $p \in M$, then

$$\lim_{\rho \to \infty} \frac{A_p(\rho)}{m \rho^{m-1}} = \theta,$$

where $A_p(\rho)$ denotes the area of the geodesic sphere $\partial B_p(\rho)$.

Proof The volume comparison theorem asserts that if $\bar{A}(\rho)$ is the area of the sphere of radius ρ in $\mathbb{R}^m$, then

$$\frac{A_p(\rho)}{\bar{A}(\rho)}$$

is a nonincreasing function of ρ. In particular, the limit

$$\lim_{\rho \to \infty} \frac{A_p(\rho)}{m \rho^{m-1}}$$

exists. Note that since

$$\frac{V_p(\rho)}{\bar{V}(\rho)}$$

is also a nonincreasing function, we have the limit

$$\lim_{\rho \to \infty} \frac{V_p(\rho)}{\rho^m} = \theta.$$

L'Hôpital's rule implies that

$$\lim_{\rho \to \infty} \frac{V_p(\rho)}{\rho^m} = \lim_{\rho \to \infty} \frac{A_p(\rho)}{m \rho^{m-1}},$$

which proves the lemma. $\square$

Theorem 16.2 (Li) *Let M^m be a complete manifold with nonnegative Ricci curvature. If there exists a constant $\theta > 0$ such that*

$$\liminf_{\rho \to \infty} \frac{V_p(\rho)}{\rho^m} = \theta$$

for some point $p \in M$, then

$$\lim_{t \to \infty} V_p(\sqrt{t}) \, H(x, y, t) = m^{-1} \alpha_{m-1} (4\pi)^{-m/2},$$

where α_{m-1} denotes the area of the unit sphere $\mathbb{S}^{m-1} \subset \mathbb{R}^m$.

16 Large time behavior of the heat kernel

Proof The estimate of Corollary 12.5 asserts that

$$H(p, p, t_1) \leq H(p, p, t_2) \left(\frac{t_2}{t_1}\right)^{m/2}$$

for $t_1 \leq t_2$. This implies that

$$t^{m/2} H(p, p, t)$$

is a nondecreasing function of t. However, Theorem 13.4 asserts that

$$H(p, p, t) \leq C V_p^{-1}(\sqrt{t}).$$

From this and the assumption on the volume growth, we conclude that

$$t^{m/2} H(p, p, t)$$

is bounded from above. Hence the limit

$$\lim_{t \to \infty} t^{m/2} H(p, p, t) = \alpha$$

exists and

$$t^{m/2} H(p, p, t) \nearrow \alpha.$$

For $x, y \in M$, Corollary 12.5 asserts that

$$t^{m/2} H(x, y, t) \leq ((1+\delta)t)^{m/2} H(p, y, (1+\delta)t) \exp\left(\frac{r^2(x)}{4\delta t}\right)$$

$$\leq ((1+2\delta)t)^{m/2} H(p, p, (1+2\delta)t) \exp\left(\frac{r^2(x) + r^2(y)}{4\delta t}\right)$$

$$\leq \alpha \exp\left(\frac{r^2(x) + r^2(y)}{4\delta t}\right)$$

for $\delta > 0$, where $r(\cdot)$ denotes the distance function to the point p. This implies that

$$\limsup_{t \to \infty} t^{m/2} H(x, y, t) \leq \alpha.$$

The same argument also shows that

$$t^{\frac{m}{2}} H(x, y, t) \geq ((1-2\delta)t)^{m/2} H(p, p, (1-2\delta)t) \exp\left(-\frac{r^2(x) + r^2(y)}{4\delta t}\right)$$

implying that

$$\liminf_{t \to \infty} t^{m/2} H(x, y, t) \geq \alpha,$$

hence
$$\lim_{t\to\infty} t^{m/2} H(x, y, t) = \alpha. \qquad (16.1)$$

To compute the value of α, we use the heat kernel comparison in Lemma 13.6 and obtain

$$\int_{B_p(\rho)} H(p, y, t)\, dy \geq \int_{\bar{B}_{\bar{p}}(\rho)} \bar{H}(\bar{p}, \bar{y}, t)\, d\bar{y},$$

where $\bar{H}$ is the heat kernel of $\mathbb{R}^m$ with $r(y) = \bar{r}(\bar{y})$. In particular

$$\lim_{t\to\infty} t^{m/2} \int_{B_p(\rho)} H(p, y, t) dy \geq \lim_{t\to\infty} t^{m/2} \int_{\bar{B}_{\bar{p}}(\rho)} \bar{H}(\bar{p}, \bar{y}, t) d\bar{y}$$

$$= \lim_{t\to\infty} t^{m/2} \int_{\bar{B}_{\bar{p}}(\rho)} (4\pi t)^{-m/2} \exp\left(-\frac{\bar{r}^2(\bar{y})}{4t}\right) d\bar{y}$$

$$= (4\pi)^{-m/2} \bar{V}(\rho),$$

implying

$$\alpha V_p(\rho) \geq (4\pi)^{-m/2} \bar{V}(\rho).$$

Letting $\rho \to \infty$, we conclude that

$$\alpha \geq (4\pi)^{-m/2} m^{-1} \alpha_{m-1} \theta^{-1}. \qquad (16.2)$$

To obtain the upper bound for α, we apply Corollary 12.5 again and get

$$H^2(p, p, t) \exp\left(-\frac{r^2(y)}{2\delta t}\right) \leq (1+\delta)^m H^2(p, y, (1+\delta)t).$$

Integrating over $\partial B_p(\rho)$ yields

$$A_p(\rho) \exp\left(-\frac{\rho^2}{2\delta t}\right) H^2(p, p, t) \leq (1+\delta)^m \int_{\partial B_p(\rho)} H^2(p, y, (1+\delta)t)\, dy.$$

Integrating over $0 \leq \rho \leq \infty$ and using the semi-group property, we obtain

$$\int_0^\infty A_p(\rho) \exp\left(-\frac{\rho^2}{2\delta t}\right) H^2(p, p, t) d\rho \leq (1+\delta)^m \int_M H^2(p, y, (1+\delta)t) dy$$

$$= (1+\delta)^m H(p, p, 2(1+\delta)t).$$

$$(16.3)$$

However, Lemma 16.1 asserts that
$$\frac{A_p(\rho)}{m\,\rho^{m-1}} \searrow \theta,$$
hence
$$A_p(\rho) \geq \theta\, m\, \rho^{m-1}.$$
Observe that since $\bar{A}(\rho) = \alpha_{m-1}\,\rho^{m-1}$, when combined with (16.3), we conclude that
$$\theta\, m\, \alpha_{m-1}^{-1} \int_0^\infty \bar{A}(\rho) \exp\left(-\frac{\rho^2}{2\delta t}\right) H^2(p, p, t)\, d\rho$$
$$\leq (1+\delta)^m\, H(p, p, 2(1+\delta)t).$$
Multiplying both sides by $(2(1+\delta)t)^{m/2}$ and using (16.1), we get
$$\theta\, m\, \alpha_{m-1}^{-1}\, (2(1+\delta)t)^{m/2} \int_0^\infty \bar{A}(\rho) \exp\left(-\frac{\rho^2}{2\delta t}\right) H^2(p, p, t)\, d\rho \leq (1+\delta)^m\, \alpha.$$
(16.4)

Noting that
$$(8\pi\delta t)^{-m/2} = \bar{H}(\bar{p}, \bar{p}, 2\delta t)$$
$$= \int_{\mathbb{R}^m} \bar{H}^2(\bar{p}, \bar{y}, \delta t)\, d\bar{y}$$
$$= \int_0^\infty \bar{A}_{\bar{p}}(\rho)(4\pi\delta t)^{-m} \exp\left(-\frac{\rho^2}{2\delta t}\right) d\rho,$$
we can rewrite (16.4) as
$$\theta\, m\, \alpha_{m-1}^{-1}\, (2(1+\delta))^{m/2}\, (4\pi\delta t)^m\, H^2(p, p, t)\, (8\pi\delta)^{-m/2} \leq (1+\delta)^m\, \alpha.$$
Letting $t \to \infty$ and using (16.1), we obtain
$$\theta\, m\, \alpha_{m-1}^{-1}\, (1+\delta)^{-m/2}\, (4\pi\delta)^{m/2}\, \alpha^2 \leq \alpha,$$
hence
$$\alpha \leq \theta^{-1}\, m^{-1}\, \alpha_{m-1}\, (1+\delta)^{m/2}\, (4\pi\delta)^{-m/2}.$$
Letting $\delta \to \infty$ and combining with (16.2), we arrive at
$$\alpha = \theta^{-1}\, m^{-1}\, \alpha_{m-1}\, (4\pi)^{-m/2},$$
and the theorem follows by substituting the definition of θ. $\square$

Corollary 16.3 *Let M be a complete manifold with nonnegative Ricci curvature. Suppose there exists a constant $\theta > 0$ such that*

$$\liminf_{\rho \to \infty} \frac{V_p(\rho)}{\rho^m} = \theta$$

for some point $p \in M$. Then $|\pi_1(M)| < \infty$. Moreover,

$$|\pi_1(M)| = \lim_{\rho \to \infty} \frac{V_{\tilde{p}}(\rho)}{V_p(\rho)},$$

where $\tilde{p} \in \tilde{M}$ is a preimage point in the universal cover $\tilde{M}$ of M and $V_{\tilde{p}}(\rho)$ is the volume of a geodesic ball of radius ρ centered at $\tilde{p}$ in $\tilde{M}$.

Proof Let $\pi : \tilde{M} \to M$ be the projection map from the universal cover $\tilde{M}$ to M. We equip $\tilde{M}$ with the pull-back metric from M so that π is a local isometry and the elements of the fundamental group $\pi_1(M)$ act as isometries on $\tilde{M}$. If we let H and $\tilde{H}$ be minimal heat kernels on M and $\tilde{M}$, respectively, we then claim that

$$H(\pi(\tilde{x}), \pi(\tilde{y}), t) = \sum_{g \in \pi_1(M)} \tilde{H}(\tilde{x}, g(\tilde{y}), t)$$

for any $\tilde{x}, \tilde{y} \in \tilde{M}$. In fact, we first claim that

$$H_1(x, y, t) = \sum_{g \in \pi_1(M)} \tilde{H}(\tilde{x}, g(\tilde{y}), t) \tag{16.5}$$

for $x, y \in M$ with $\pi(\tilde{x}) = x$ and $\pi(\tilde{y}) = y$ is well defined. To see this, for any $x, y \in M$, let $B_y(\rho)$ be a sufficiently small geodesic ball around y such that their preimages $\{B_{g(\tilde{y})}(\rho) \mid g \in \pi_1(M)\}$ are all disjoint. In particular,

$$\sum_{g \in \pi_1(M)} \int_{B_{g(\tilde{y})}(\rho)} \tilde{H}(\tilde{x}, \tilde{z}, t) \, d\tilde{z} \leq \int_{\tilde{M}} \tilde{H}(\tilde{x}, \tilde{z}, t) \, d\tilde{z}$$

$$\leq 1. \tag{16.6}$$

On the other hand, Corollary 12.5 asserts that

$$V_{g(\tilde{y})}(\rho) \, \tilde{H}(\tilde{x}, \tilde{y}, t) \leq C \int_{B_{g(\tilde{y})}(\rho)} \tilde{H}(\tilde{x}, \tilde{z}, 2t) \, d\tilde{z} \, \exp\left(\frac{\rho^2}{4t}\right).$$

Combining this inequality with (16.6), we conclude that (16.5) is summable and hence well defined.

Obviously
$$\left(\Delta_y - \frac{\partial}{\partial t}\right) H_1(x, y, t) = 0 \quad \text{on} \quad M \times (0, \infty)$$
with
$$H_1(x, y, 0) = \delta_x(y).$$

The Duhamel principle now implies that for $y \in B_x(\rho)$,

$$H(x, y, t) - H_1(x, y, t)$$
$$= \int_0^t \frac{\partial}{\partial s} \int_M \phi^2(z) H_1(x, z, t-s) H(z, y, s) \, dz \, ds$$
$$= -\int_0^t \int_M \phi^2(z) \Delta H_1(x, z, t-s) H(z, y, s) \, dz \, ds$$
$$+ \int_0^t \int_M \phi^2(z) H_1(x, z, t-s) \Delta H(z, y, s) \, dz \, ds$$
$$= 2 \int_0^t \int_M \phi(z) H(z, y, s) \langle \nabla H_1(x, z, t-s), \nabla \phi(z) \rangle \, dz \, ds$$
$$- 2 \int_0^t \int_M \phi(z) H_1(x, z, t-s) \langle \nabla H(z, y, s), \nabla \phi(z) \rangle \, dz \, ds \quad (16.7)$$

for any nonnegative cutoff function ϕ satisfying

$$\phi = \begin{cases} 1 & \text{on} \quad B_x(\rho), \\ 0 & \text{on} \quad M \setminus B_x(2\rho), \end{cases}$$

and $|\nabla \phi|^2 \leq 2\rho^{-2}$. However, the Schwarz inequality asserts that

$$\left| \int_0^t \int_M \phi(z) H(z, y, s) \langle \nabla H_1(x, z, t-s), \nabla \phi(z) \rangle \, dz \, ds \right|$$
$$\leq \sqrt{2}\rho^{-1} \int_0^t \int_{B_x(2\rho) \setminus B_x(\rho)} H(z, y, s) |\nabla H_1(x, z, t-s)| \, dz \, ds$$
$$\leq \sqrt{2}\rho^{-1} \int_0^t \left(\int_{B_x(2\rho) \setminus B_x(\rho)} H^2(z, y, s) \, dz \right)^{1/2}$$
$$\times \left(\int_{B_x(2\rho) \setminus B_x(\rho)} |\nabla H_1(x, z, t-s)|^2 \, dz \right)^{1/2} ds. \quad (16.8)$$

Similarly, we also have

$$\left| \int_0^t \int_M \phi(z) H_1(x, z, t-s) \langle \nabla H(z, y, s), \nabla \phi(z) \rangle \, dz \, ds \right|$$

$$\leq \sqrt{2}\rho^{-1} \int_0^t \int_{B_x(2\rho) \setminus B_x(\rho)} H_1(x, z, t-s) |\nabla H(z, y, s)| \, dz \, ds$$

$$\leq \sqrt{2}\rho^{-1} \int_0^t \left(\int_{B_x(2\rho) \setminus B_x(\rho)} H_1^2(x, z, t-s) \, dz \right)^{1/2}$$

$$\times \left(\int_{B_x(2\rho) \setminus B_x(\rho)} |\nabla H(z, y, s)|^2 \, dz \right)^{1/2} ds. \qquad (16.9)$$

On the other hand, the semi-group property asserts that

$$\int_M H^2(z, y, s) \, dz = H(y, y, 2s),$$

hence $H \in L^2(M)$. Also, for the cutoff function ϕ defined above, we have

$$\int_M \phi^2(z) |\nabla H|^2(z, y, s) dz = -\int_M \phi^2 H(z, y, s) \Delta H(z, y, s) \, dz$$

$$- 2 \int_M \phi(z) H(z, y, s) \langle \nabla \phi(z), \nabla H(z, y, s) \rangle dz$$

$$\leq -\frac{1}{2} \frac{\partial}{\partial s} \left(\int_M \phi^2(z) H^2(z, y, s) \, dz \right)$$

$$+ \frac{1}{2} \int_M \phi^2(z) |\nabla H(z, y, s)|^2 \, dz$$

$$+ 2 \int_M |\nabla \phi|^2(z) H^2(z, y, s) \, dz,$$

implying that

$$\int_M \phi^2(z) |\nabla H(z, y, s)|^2 \, dz \leq -\frac{\partial}{\partial s} \left(\int_M \phi^2(z) H^2(z, y, s) \, dz \right)$$

$$+ 4 \int_M |\nabla \phi|^2(z) H^2(z, y, s) \, dz$$

$$\leq -\frac{\partial}{\partial s} \left(\int_M \phi^2(z) H^2(z, y, s) \, dz \right)$$

$$+ 2\rho^{-2} \int_M H^2(z, y, s) \, dz.$$

16 Large time behavior of the heat kernel

Letting $\rho \to \infty$ and using the fact that $H \in L^2$, we conclude that

$$\int_M |\nabla H|^2(z, y, s)\,dz \leq -\frac{\partial}{\partial s}\left(\int_M H^2(z, y, s)\,dz\right)$$

$$= \frac{\partial}{\partial s} H(y, y, 2s),$$

hence $|\nabla H| \in L^2(M)$. In the case of H_1, we also observe that if $\tilde{M}_1$ is a fixed fundamental domain of M in $\tilde{M}$ and use the fact that $\tilde{H}(g(\tilde{x}), g(\tilde{y}), t) = \tilde{H}(\tilde{x}, \tilde{y}, t)$ for any $g \in \pi_1(M)$, then

$$\int_M H_1^2(x, z, t-s)\,dz = \sum_{g, h \in \pi_1(M)} \int_{\tilde{M}_1} \tilde{H}(\tilde{x}, g(\tilde{z}), t-s)\,\tilde{H}(\tilde{x}, h(\tilde{z}))\,t-s)\,d\tilde{z}$$

$$= \sum_{g, h \in \pi_1(M)} \int_{\tilde{M}_1} \tilde{H}(\tilde{x}, g(\tilde{z}), t-s)$$

$$\times \tilde{H}(g \circ h^{-1}(\tilde{x}), g(\tilde{z})\,t-s)\,d\tilde{z}$$

$$= \sum_{g, h \in \pi_1(M)} \int_{g(\tilde{M}_1)} \tilde{H}(\tilde{x}, \tilde{z}, t-s)$$

$$\times \tilde{H}(g \circ h^{-1}(\tilde{x}), \tilde{z}, t-s)\,d\tilde{z}.$$

However, since

$$\sum_{h \in \pi_1(M)} \tilde{H}(g \circ h^{-1}(\tilde{x}), \tilde{z}, t-s) = \sum_{k \in \pi_1(M)} \tilde{H}(k(\tilde{x}), \tilde{z}, t-s),$$

we conclude that

$$\int_M H_1^2(x, z, t-s)\,dz = \sum_{g, k \in \pi_1(M)} \int_{g(\tilde{M}_1)} \tilde{H}(\tilde{x}, \tilde{z}, t-s)\,\tilde{H}(k(\tilde{x}), \tilde{z}, t-s)\,d\tilde{z}$$

$$= \sum_{k \in \pi_1(M)} \int_{\tilde{M}} \tilde{H}(\tilde{x}, \tilde{z}, t-s)\,\tilde{H}(k(\tilde{x}), \tilde{z}, t-s)\,d\tilde{z}$$

$$= \sum_{k \in \pi_1(M)} \tilde{H}(\tilde{x}, k(\tilde{x}), 2(t-s))$$

$$= H_1(\tilde{x}, \tilde{x}, 2(t-s)),$$

hence $H_1 \in L^2(M)$. The same argument we used before also implies that $|\nabla H_1| \in L^2(M)$. Therefore the right-hand sides of (16.8) and (16.9) tend to 0 as $\rho \to \infty$, and (16.7) yields

$$H(x, y, t) = H_1(x, y, t).$$

Applying Theorem 16.2, we have

$$m^{-1}\alpha_{m-1}(4\pi)^{-m/2} = \lim_{t\to\infty} V_p(\sqrt{t}) H(p, p, t)$$

$$= \lim_{t\to\infty} \sum_{g\in\pi_1(M)} V_p(\sqrt{t}) \tilde{H}(\tilde{p}, g(\tilde{p}), t)$$

$$= \lim_{t\to\infty} \sum_{g\in\pi_1(M)} \frac{V_p(\sqrt{t})}{V_{\tilde{p}}(\sqrt{t})} V_{\tilde{p}}(\sqrt{t}) \tilde{H}(\tilde{p}, g(\tilde{p}), t). \quad (16.10)$$

On the other hand, we observe that $\tilde{M}$ also has nonnegative Ricci curvature and $V_{\tilde{p}}(\sqrt{t}) \geq V_p(\sqrt{t})$, hence we can apply Theorem 16.2 to $\tilde{H}$ and obtain

$$\lim_{t\to\infty} \sum_{g\in\pi_1(M)} \frac{V_p(\sqrt{t})}{V_{\tilde{p}}(\sqrt{t})} V_{\tilde{p}}(\sqrt{t}) \tilde{H}(\tilde{p}, g(\tilde{p}), t) = |\pi_1(M)| m^{-1}\alpha_{m-1} (4\pi)^{-m/2}$$

$$\times \lim_{t\to\infty} \frac{V_p(\sqrt{t})}{V_{\tilde{p}}(\sqrt{t})}.$$

Combining this identity with (16.10), we conclude that

$$|\pi_1(M)| \lim_{t\to\infty} \frac{V_p(\sqrt{t})}{V_{\tilde{p}}(\sqrt{t})} = 1,$$

as claimed. In particular, since

$$\liminf_{\rho\to 0} \frac{V_p(\rho)}{\rho^m} = \theta$$

and

$$V_{\tilde{p}}(\rho) \leq m^{-1}\alpha_{m-1}\rho^m,$$

we deduce that

$$|\pi_1(M)| \leq m^{-1}\alpha_{m-1}\theta^{-1}$$

is finite. $\square$

16 Large time behavior of the heat kernel

The following lemma proved in [L5] asserts a mean value type property at infinity of a complete manifold with nonnegative Ricci curvature. This will be useful in Chapter 27 when we study linear growth harmonic functions.

Lemma 16.4 (Li) *Let M^m be a complete manifold with nonnegative Ricci curvature. Suppose f is a bounded subharmonic function defined on M, then*

$$\lim_{\rho \to \infty} V_p^{-1}(\rho) \int_{B_p(\rho)} f(x)\, dx = \sup_M f.$$

Proof Let us define $g_0(x) = \sup_M f - f(x)$, then g_0 satisfies

$$\Delta g_0 \leq 0$$

and

$$g_0 \geq 0.$$

It suffices to show that

$$\lim_{\rho \to \infty} V_p^{-1}(\rho) \int_{B_p(\rho)} g_0(x)\, dx = 0.$$

Let us consider the solution of the heat equation given by

$$g(x, t) = \int_M H(x, y, t)\, g_0(y)\, dy.$$

Observe that if $H_i(x, y, t)$ is a Dirichlet heat kernel on a compact subdomain $\Omega_i \subset M$, then

$$\frac{\partial}{\partial t} \int_{\Omega_i} H_i(x, y, t)\, g_0(y)\, dy = \int_{\Omega_i} \frac{\partial H_i}{\partial t}(x, y, t)\, g_0(y)\, dy$$

$$= \int_{\Omega_i} \Delta_y H_i(x, y, t)\, g_0(y)\, dy$$

$$= \int_{\Omega_i} H_i(x, y, t)\, \Delta_y g_0(y)\, dy$$

$$+ \int_{\partial \Omega_i} \frac{\partial H_i}{\partial \nu}(x, y, t)\, g_0(y)\, dy,$$

$$\leq 0,$$

implying the function $g_i(x, t) = \int_{\Omega_i} H_i(x, y, t)\, g_0(y)\, dy$ is nonincreasing in t. Since $H(x, y, t)$ is the limit of a sequence of Dirichlet heat kernels, we conclude that $g(x, t)$ is also nonincreasing in t. However, the maximum principle,

Theorem 15.2, asserts that $g(x, t) \geq 0$, hence $g(x, t)$ converges to a nonnegative, bounded function $g_\infty(x)$ as $t \to \infty$. Moreover, g_∞ is harmonic because

$$\Delta g(x, t) = \frac{\partial g(x, t)}{\partial t},$$

which tends to 0 as $t \to \infty$ by the fact that $\partial g/\partial t \leq 0$ and

$$\int_0^\infty \frac{\partial g}{\partial t}(x, t) \, dt = g_\infty(x) - g_0(x).$$

On the other hand, since M has nonnegative Ricci curvature, a bounded harmonic function must be constant. The facts that $g_\infty \leq g_0$ and $\inf_M g_0 = 0$ imply that $g_\infty = 0$, hence $g(x, t) \to 0$.

Note that for $p \in M$, we have

$$g(p, t) = \int_M H(p, y, t) g_0(y) \, dy$$

$$\geq \int_{B_p(\sqrt{t})} H(p, y, t) g_0(y) \, dy. \qquad (16.11)$$

Applying the lower bound (Corollary 13.9), we obtain

$$H(p, y, t) \geq C \, V_p^{-1}(\sqrt{t})$$

for all $y \in B_p(\sqrt{t})$. Hence combining this inequality with (16.11) yields

$$g(p, t) \geq C \, V_p^{-1}(\sqrt{t}) \int_{B_p(\sqrt{t})} g_0(y) \, dy,$$

and the lemma follows. $\square$

17
Green's function

Let M^m be a compact manifold of dimension m with boundary ∂M. Let Δ be the Laplacian defined on functions with a Dirichlet boundary condition on M. Then standard elliptic theory asserts that there exists a Green's function $G(x, y)$ defined on $M \times M \setminus D$, where $D = \{(x, x) \mid x \in M\}$, so that

$$\int_M G(x, y) \Delta f(y) \, dy = -f(x) \tag{17.1}$$

for all functions f satisfying the Dirichlet boundary condition

$$f|_{\partial M} = 0.$$

Moreover, $G(x, y) = 0$ for $y \in \partial M$ and $x \in M \setminus \partial M$. Since both G and f satisfy the Dirichlet boundary condition, after integration by parts, (17.1) becomes

$$\int_M \Delta_y G(x, y) f(y) \, dy = -f(x), \tag{17.2}$$

which is equivalent to saying that

$$\Delta_y G(x, y) = -\delta_x(y), \tag{17.3}$$

where $\delta_x(y)$ is the delta function at x. If we let $f(x) = G(z, x)$, then (17.2) yields

$$G(z, x) = -\int_M \Delta_y G(x, y) G(z, y) \, dy.$$

However, applying (17.1) to the right-hand side, we obtain

$$G(z, x) = G(x, z). \tag{17.4}$$

189

This shows that $G(x, y)$ must be symmetric in the variables x and y. We also observe that by letting $f(x) = \Delta_x G(z, x)$ in (17.2), we have

$$\Delta_x G(z, x) = -\int_M \Delta_y G(x, y) \Delta_y G(z, y) \, dy.$$

Since the right-hand side is symmetric in x and z, we conclude that

$$\Delta_x G(z, x) = \Delta_z G(x, z).$$

On the other hand,

$$\Delta_z G(x, z) = \Delta_z G(z, x)$$

by the symmetry of G, and we conclude that

$$\Delta_x G(z, x) = \Delta_z G(z, x).$$

Therefore (17.2) can be written as

$$\Delta_x \int_M G(x, y) f(y) \, dy = \int_M \Delta_x G(x, y) f(y) \, dy$$
$$= -f(x). \tag{17.5}$$

Hence if we define the operator Δ^{-1} by

$$\Delta^{-1} f(x) = -\int_M G(x, y) f(y) \, dy,$$

then (17.1) and (17.5) give

$$\Delta \Delta^{-1} = I$$

and

$$\Delta^{-1} \Delta = I,$$

respectively.

When M^m is a complete, noncompact manifold without boundary, one would like to obtain a Green's function $G(x, y)$ that also satisfies the properties (17.1), (17.5) and (17.4) for any compactly supported function $f \in C_c^\infty(M)$.

For the case $M = \mathbb{R}^m$, it is well known that the function

$$G(x, y) = \begin{cases} ((m-2)\alpha_{m-1})^{-1} r(x, y)^{2-m} & \text{for } m \geq 3, \\ -(\alpha_1)^{-1} \log(r(x, y)) & \text{for } m = 2, \end{cases}$$

17 Green's function

where α_{m-1} is the area of the unit $(m-1)$-sphere in $\mathbb{R}^m$, satisfies these properties. In fact, these formulas, when interpreted appropriately, reflect the ideal situations for manifolds also.

Let us now assume that M^m has a point p around which the metric is rotationally symmetric. This is equivalent to saying that if we take polar coordinates $(r, \theta_2, \ldots, \theta_m)$ around p, then

$$ds_M^2 = dr^2 + g_{\alpha\beta}\, d\theta_\alpha\, d\theta_\beta$$

with $g_{\alpha\beta}(r)$ being functions of r alone. The Laplacian with respect to this coordinate system will take the form

$$\Delta = \frac{\partial^2}{\partial r^2} + \frac{1}{\sqrt{g}} \frac{\partial \sqrt{g}}{\partial r} \frac{\partial}{\partial r} + \Delta_{\partial B_p(r)},$$

where $g = \det(g_{\alpha\beta})$ and $\Delta_{\partial B_p(r)}$ denotes the Laplacian defined on the sphere, $\partial B_p(r)$, of radius r centered at p. If we let $A_p(r)$ be the area of $\partial B_p(r)$, then

$$\int \sqrt{g(r)}\, d\theta = A_p(r).$$

Since $\sqrt{g}$ is independent of the θ_αs,

$$\frac{1}{\sqrt{g}} \frac{\partial \sqrt{g}}{\partial r} = \frac{A'_p(r)}{A_p(r)},$$

and we have

$$\Delta = \frac{\partial^2}{\partial r^2} + \frac{A'_p(r)}{A_p(r)} \frac{\partial}{\partial r} + \Delta_{\partial B_p(r)}.$$

Let us now consider the function

$$G(y) = -\int_1^{r(y)} \frac{dt}{A_p(t)},$$

where $r(y)$ is the distance from p to y. Direct computation gives

$$\Delta G = \frac{\partial^2 G}{\partial r^2} + \frac{A'_p(r)}{A_p(r)} \frac{\partial G}{\partial r}$$
$$= 0$$

for $y \neq p$. Moreover, if $f \in C_c^\infty(M)$ is a compactly supported smooth function, then

$$\int_{M\setminus B_p(\epsilon)} G(y)\,\Delta f(y)\,dy = \int_{M\setminus B_p(\epsilon)} \Delta G(y)\,f(y)\,dy$$

$$- \int_{\partial B_p(\epsilon)} G(y) \frac{\partial f}{\partial r}\,d\theta + \int_{\partial B_p(\epsilon)} \frac{\partial G}{\partial r}\,f\,d\theta$$

$$= \int_1^\epsilon \frac{dt}{A_p(t)} \int_{\partial B_p(\epsilon)} \frac{\partial f}{\partial r}\,d\theta - \frac{1}{A_p(\epsilon)} \int_{\partial B_p(\epsilon)} f\,d\theta.$$

(17.6)

Note that the continuity of f implies that

$$\frac{1}{A_p(\epsilon)} \int_{\partial B_p(\epsilon)} f\,d\theta \to f(p)$$

as $\epsilon \to 0$. Also the continuity of Δf and the divergence theorem imply that

$$\frac{1}{V_p(\epsilon)} \int_{\partial B_p(\epsilon)} \frac{\partial f}{\partial r}\,d\theta = \frac{1}{V_p(\epsilon)} \int_{B_p(\epsilon)} \Delta f\,dy$$

$$\to \Delta f(p)$$

as $\epsilon \to 0$, where $V_p(r)$ is the volume of $B_p(r)$. This implies that

$$\int_{\partial B_p(\epsilon)} \frac{\partial f}{\partial r}\,d\theta \sim \Delta f(p)\,V_p(\epsilon).$$

On the other hand, since

$$A_p(t) \sim \alpha_{m-1}\,t^{m-1},$$

where α_{m-1} is the area of the unit $(m-1)$-sphere in $\mathbb{R}^m$, we have

$$\int_1^\epsilon \frac{dt}{A_p(t)} \sim \begin{cases} -\dfrac{1}{(m-2)\alpha_{m-1}} \epsilon^{2-m} & \text{for} \quad m \geq 3, \\ \dfrac{1}{\alpha_1} \log(\epsilon) & \text{for} \quad m = 2. \end{cases}$$

Therefore, by letting $\epsilon \to 0$, the right-hand side of (17.6) becomes

$$\lim_{\epsilon \to 0} \left(\int_1^\epsilon \frac{dt}{A_p(t)} \int_{\partial B_p(\epsilon)} \frac{\partial f}{\partial r}\,d\theta - \frac{1}{A_p(\epsilon)} \int_{\partial B_p(\epsilon)} f\,d\theta \right) = -f(p),$$

17 Green's function

hence
$$\int_M G(y) \Delta f(y)\, dy = -f(p).$$

In particular, this implies that one can take
$$G(p, y) = -\int_1^{r(y)} \frac{dt}{A_p(t)} \tag{17.7}$$

to be the Green's function. Note that if
$$\int_1^\infty \frac{dt}{A_p(t)} < \infty,$$

then by adding this constant to G, we can use the formula
$$G(p, y) = \int_{r(y)}^\infty \frac{dt}{A_p(t)}. \tag{17.8}$$

The Green's functions in $\mathbb{R}^m$ are given in the form of (17.7) and (17.8) for dimensions $m = 2$ and $m \geq 3$, respectively.

We are now ready to construct a Green's function on a complete manifold. We should point out that an existence proof was first given by Malgrange [M]. However, for the purpose of application, a constructive argument is a key step in getting the appropriate estimates. The first constructive proof was published in [LT2].

Let $\{\Omega_i\}_{i=1}^\infty$ be a compact exhaustion of M, such that
$$\Omega_i \subsetneq \Omega_j \quad \text{for} \quad i < j,$$
$$\cup_{i=1}^\infty \Omega_i = M,$$

and each Ω_i is a sufficiently smooth compact subdomain of M. For each i, let $G_i(x, y)$ be the symmetric, positive, Green's function with a Dirichlet boundary condition of Ω_i. Moreover,
$$G_i(p, y) \sim \begin{cases} ((m-2)\alpha_{m-1})^{-1} r^{2-m}(p, y) & \text{for } m \geq 3, \\ -(\alpha_1)^{-1} \log r(p, y) & \text{for } m = 2, \end{cases}$$

as $y \to p$.

Lemma 17.1 *Let $\Omega \subset M$ be a connected open subset of a complete manifold M. Suppose $\{f_i\}$ is a sequence of positive harmonic functions defined on Ω. If there exists a point $p \in \Omega$ such that the sequence $f_i(p)$ is bounded, then after passing through a subsequence f_i converges uniformly on compact subsets of Ω to a harmonic function f.*

Proof For any point $x \in \Omega$, let $2\rho(x)$ be the distance from x to $\partial \Omega$, then Theorem 6.1 asserts that

$$|\nabla \log f_i| \leq C(m, K, \rho(x)) \quad \text{on} \quad B_x(\rho(x))$$

for some constant $C(m, K, \rho(x))$ depending only on m, K, and $\rho(x)$. Let D be any compact, connected subset of Ω containing p. By compactness, D can be covered by a finite number of geodesic balls of the form $B_{x_i}(\rho(x_i))$. Hence the functions $|\nabla \log f_i|$ are uniformly bounded on D. Integrating this along a curve γ in D joining any point $x \in D$ to p yields

$$\log f_i(x) - \log f_i(p) = \int_\gamma \frac{d \log f_i}{ds} ds$$

$$\leq \int_\gamma |\nabla \log f_i| ds$$

$$\leq C \ell(\gamma).$$

This implies that

$$f_i(x) \leq f_i(p) \exp(C \tilde{d}(p, x)),$$

where $\tilde{d}(p, x)$ is the distance from p to x within D. The compactness of D and the fact that the sequence $\{f_i(p)\}$ is uniformly bounded for all i implies that the sequence of functions $\{f_i\}$ is uniformly bounded on D. Hence by passing through a subsequence, $f_i \to f$ uniformly on compact subsets of Ω. Since

$$\frac{|\nabla f_i|}{f_i} = |\nabla \log f_i|$$

are bounded on D, $f_i \to f$ must converge in the $C^{0,1}$ sense and f is Lipschitz. Moreover, f_i being harmonic implies that

$$0 = \int_\Omega \phi \, \Delta f_i$$

$$= - \int_\Omega \langle \nabla \phi, \nabla f_i \rangle$$

for any compactly supported function ϕ defined on Ω. Passing to the limit, we conclude that

$$0 = - \int_\Omega \langle \nabla \phi, \nabla f \rangle$$

and f is weakly harmonic. Regularity theory then implies that f is smooth and harmonic. □

17 Green's function

Lemma 17.2 *Let $p \in M$ be a fixed point and D be any compact subset of $M \setminus \{p\}$. For sufficiently large i, so that p and D are properly contained in Ω_i, the sequence of Green's functions $G_i(p, y)$ as functions of y must have uniformly bounded oscillations on D.*

Proof It suffices to show that the oscillations of the G_is are uniformly bounded on sets of the form $B_p(\rho_2) \setminus B_p(\rho_1)$, where $0 < \rho_1 < \rho_2 < \infty$. Let us denote the oscillation of $G_i(p, y)$ by

$$\omega_i(\rho_1, \rho_2) = \operatorname{osc}_{y \in B_p(\rho_2) \setminus B_p(\rho_1)} G_i(p, y)$$

$$= \sup_{y \in B_p(\rho_2) \setminus B_p(\rho_1)} G_i(p, y) - \inf_{y \in B_p(\rho_2) \setminus B_p(\rho_1)} G_i(p, y).$$

We will prove the uniform oscillation bound by contradiction.

Let us assume that there is a subsequence of the $\{\omega_i\}$ such that $\omega_i \to \infty$. We define the set of functions $g_i(y)$ by

$$g_i(y) = \omega_i^{-1} G_i(p, y) - \omega_i^{-1} \inf_{z \in B_p(\rho_2)} G_i(p, z).$$

Each g_i is a harmonic function defined on $\Omega_i \setminus \{p\}$ with the properties that

$$\operatorname{osc}_{y \in B_p(\rho_2) \setminus B_p(\rho_1)} g_i(y) = 1, \tag{17.9}$$

$$g_i(y) \sim \omega_i^{-1} G_i(p, y) \quad \text{as} \quad y \to p, \tag{17.10}$$

and

$$\inf_{y \in B_p(\rho_2)} g_i(y) = 0.$$

Note that since the function $G_i(p, y)$ is harmonic when restricted to $B_p(\rho_2) \setminus \{p\}$ and

$$G_i(p, y) \to \infty$$

as $y \to p$, the maximum principle asserts that its minimum must occur on $\partial B_p(\rho_2)$, hence

$$\inf_{y \in \partial B_p(\rho_2)} g_i(y) = 0. \tag{17.11}$$

It is also convenient to define

$$i_i(r) = \inf_{y \in \partial B_p(r)} G_i(p, y)$$

and
$$s_i(r) = \sup_{y \in \partial B_p(r)} G_i(p, y).$$

Applying the maximum principle to the domain $\Omega_i \setminus B_p(r)$, we conclude that $s_i(r)$ is a montonically decreasing function since $G_i(p, y) = 0$ on $\partial \Omega_i$.

If we denote the Dirichlet Green's function on $B_p(\rho_2)$ by $\bar{G}(p, y)$, then we claim that g_i must satisfy the estimates

$$\omega_i^{-1} \bar{G}(p, y) \leq g_i(y)$$
$$\leq \omega_i^{-1} \bar{G}(p, y) + 1. \qquad (17.12)$$

To see this, let $\alpha > 0$ be any positive constant, then the function
$$(1 + \alpha) g_i(y) - \omega_i^{-1} \bar{G}(p, y)$$
tends to ∞ as $y \to p$ because of the asymptotic behavior of $\bar{G}$ and (17.10). It is also nonnegative on $\partial B_p(\rho_2)$. Hence the maximum principle implies that it must be nonnegative on $B_p(\rho_2)$. Letting $\alpha \to 0$, we obtain the lower bound for g_i. The upper bound follows similarly by considering the function
$$(1 - \alpha) g_i(y) - \omega_i^{-1} \bar{G}(p, y) - 1$$
and using (17.9) and (17.11). The assumption that $\omega_i \to \infty$, (17.12), and Lemma 17.1 imply that there is a subsequence of $g_i(y)$ that converges uniformly on compact subsets of $B_p(\rho_2) \setminus \{p\}$. We also denote this subsequence by $g_i(y)$. Moreover, the limiting function g is harmonic and bounded between 0 and 1. However, the removable singularity theorem for harmonic functions asserts that g can be extended to a harmonic function defined on $B_p(\rho_2)$. On the other hand, the property that $s_i(r)$ is monotonically decreasing implies that
$$\sup_{y \in \partial B_p(r)} g_i(y)$$
is decreasing in r. Passing to the limit, we conclude that
$$\sup_{y \in \partial B_p(r)} g(y)$$
is nonincreasing in r. In particular, $g(p) \geq \sup_{y \in \partial B_p(r)} g(y)$ for all $r > 0$, and p must be a local maximum for g. However, this violates the maximum principle unless g is identically constant.

On the other hand, the fact that $s_i(r)$ is monotonically decreasing implies that the supremum of g_i on $M \setminus B_p(\rho_1)$ is achieved on $\partial B_p(\rho_1)$. Using (17.9) and (17.11), we conclude that $1 - g_i(y)$ is nonnegative on $M \setminus B_p(\rho_1)$. Using

Lemma 17.1, we conclude that $1 - g_i$ must have a uniformly convergent subsequence since $g_i \to g$ on $B_p(\rho_2) \setminus \{p\}$. Hence $g_i \to g$, which is a constant function on M. However, after passing to the limit (17.9) implies that

$$\operatorname{osc}_{y \in B_p(\rho_2) \setminus B_p(\rho_1)} g(y) = 1,$$

which is a contradiction. Hence the oscillation of $G(p, y)$ must be bounded on $B_p(\rho_2) \setminus B_p(\rho_1)$. $\square$

Theorem 17.3 *Let M^m be a complete manifold without boundary. There exists a Green's function $G(x, y)$ which is smooth on $(M \times M) \setminus D$ satisfying properties (17.1), (17.4), and (17.5). Moreover, $G(x, y)$ can be taken to be positive if and only if there exists a positive superharmonic function f on $M \setminus B_p(\rho)$ with the property that*

$$\liminf_{x \to \infty} f(x) < \inf_{x \in \partial B_p(\rho)} f(x).$$

Proof Let us first observe that if $G(x, y)$ is a positive symmetric Green's function, then the maximum principle asserts that the infimum of $G(p, y)$ must be achieved at infinity of M. The function $f(y)$ given by the restriction of $G(p, y)$ on $M \setminus B_p(\rho)$ is a positive harmonic function with the desired property.

We will first construct $G(p, y)$ assuming that there is a nonconstant, positive, superharmonic function f defined on $M \setminus B_p(\rho)$ whose infimum is achieved at infinity of M. Let $\{\Omega_i\}_{i=1}^{\infty}$ be a compact exhaustion of M with the property that $B_p(\rho) \subset \Omega_i$ for all i. Let $G_i(p, y)$ be the Dirichlet Green's function defined on Ω_i. Note that $G_i(p, y)$ are monotonically increasing. In fact, the maximum principle and the asymptotic behavior of $G_i(p, y)$ as $y \to p$ assert that

$$(1 + \alpha) G_j(p, y) \geq G_i(p, y)$$

for $\alpha > 0$ and $i < j$, and the monotonicity follows by taking $\alpha \to 0$. As before, let

$$s_i(\rho) = \sup_{y \in \partial B_p(\rho)} G_i(p, y),$$

then $s_i(\rho)$ is a monotonically increasing sequence in i. We now claim that $s_i(\rho)$ is bounded from above. To see this, let us consider the sequence of functions

$$s_i^{-1}(\rho) G_i(p, y).$$

By addition and multiplication by constants, we may assume that

$$\inf_{x \in \partial B_p(\rho)} f(x) = 1$$

and

$$\liminf_{x \to \infty} f(x) = 0. \tag{17.13}$$

The maximum principle asserts that

$$s_i^{-1}(\rho) G_i(p, y) \le f(y) \quad \text{on} \quad M \setminus B_p(\rho).$$

By passing to a subsequence, the sequence of functions $\{s_i^{-1}(\rho) G_i(p, y)\}$ converges uniformly on compact subsets of $M \setminus B_p(\rho)$ to a harmonic function g satisfying the properties that

$$g(y) \le f(y) \quad \text{on} \quad M \setminus B_p(\rho)$$

and

$$\sup_{x \in \partial B_p(\rho)} g(x) = 1. \tag{17.14}$$

Lemma 17.1 asserts that the subsequence, in fact, converges on a compact subset of $M \setminus \{p\}$ and g is a harmonic function on $M \setminus \{p\}$. However, the maximum principle implies that

$$s_i^{-1}(\rho) \bar{G}(p, y) \le s_i^{-1}(\rho) G_i(p, y) \le s_i^{-1}(\rho) \bar{G}(p, y) + 1,$$

where $\bar{G}$ is the Dirichlet Green's function on $B_p(\rho)$. If $s_i(\rho) \to \infty$, then similarly to the argument of Lemma 17.2 we conclude that g must be identically 0 due to (17.13), hence contradicting (17.14). Therefore, $s_i(\rho)$ must be bounded.

Let $s(\rho) = \lim_{i \to \infty} s_i(\rho)$. The maximum principle again implies that

$$G_i(p, y) \le s(\rho) f(y) \quad \text{on} \quad M \setminus B_p(\rho).$$

Arguing as before, we conclude that the monotonic sequence $G_i(p, y)$ converges uniformly on compact subsets of $M \setminus \{p\}$ to a harmonic function $G(p, y)$. In particular, given any $\epsilon > 0$, for sufficiently large i, we have

$$0 \le G(p, y) - G_i(p, y) \le \epsilon \quad \text{on} \quad \partial B_p(1).$$

The maximum principle then asserts that

$$G_i(p, y) \le G(p, y) \le G_i(p, y) + \epsilon \quad \text{on} \quad B_p(1) \setminus \{p\}.$$

In particular, if $f \in C_c^\infty(M)$, then

$$\left| \int_M G(p, y) \Delta f(y) \, dy - \int_M G_i(p, y) \Delta f(y) \right|$$

$$\leq \int_M |G(p, y) - G_i(p, y)| \, |\Delta f|(y) \, dy$$

$$\leq \epsilon \int_{B_p(1)} |\Delta f|(y) \, dy + \int_{M \setminus B_p(1)} |G(p, y) - G_i(p, y)| \, |\Delta f|(y) \, dy.$$

Note that the second term on the right-hand side tends to 0 as $i \to \infty$, and using

$$\int_M G_i(p, y) \Delta f(y) \, dy = -f(p)$$

we conclude that

$$\left| \int_M G(p, y) \Delta f(y) \, dy + f(p) \right| \leq \epsilon \int_{B_p(1)} |\Delta f|(y) \, dy.$$

Letting $\epsilon \to 0$, this yields property (17.1) for $G(p, y)$.

Observe that the above argument shows that

$$G_i(q, y) \to G(q, y)$$

as long as $q \in B_p(\rho/2)$. For arbitrary $q \in M$, we just need to take ρ sufficiently large such that $q \in B_p(\rho/2)$. Hence $G_i(q, y) \to G(q, y)$ for any $q \in M$. Since each G_i satisfies (17.4) and (17.5), G will also inherit the same properties.

Note that the function G obtained in this manner will have the minimal property, namely,

$$\tilde{G}(x, y) \geq G(x, y),$$

for any positive Green's function $\tilde{G}$. Indeed, since the maximum principle implies that

$$\tilde{G}(x, y) \geq G_i(x, y) \quad \text{on} \quad \Omega_i,$$

the minimal property of G follows by taking the limit as $i \to \infty$.

We now assume that f does not exist on $M \setminus B_p(\rho)$. This is equivalent to saying that

$$s_i(p) \to \infty$$

as $i \to \infty$ for all fixed p. Let

$$a_i = \inf_{y \in \partial B_p(1)} G_i(p, y).$$

The sequence of functions $\{h_i(p, y) = G_i(p, y) - a_i\}$ then has the property that

$$\inf_{y \in \partial B_p(1)} h_i(p, y) = 0.$$

Also, Lemma 17.2 asserts that there is a constant ω such that the oscillations of $G_i(p, y)$ over the set $B_p(2) \setminus B_p(\frac{1}{2})$ is bounded by

$$\omega_i(1, 2) \leq \omega.$$

In particular,

$$\sup_{y \in \partial B_p(1)} h_i(p, y) \leq \omega$$

for all i. Hence the sequence of positive harmonic functions $\{h_i + \omega\}$ when restricted to $B_p(2) \setminus \{p\}$ has a convergent subsequence, also denoted by $\{h_i + \omega\}$, such that $\{h_i(p, y) + \omega\}$ converges uniformly on compact subsets of $B_p(2) \setminus \{p\}$ to a harmonic function $G(p, y) + \omega$. Similarly, the sequence of positive harmonic functions $\{\omega - h_i\}$ when restricted to $M \setminus B_p(\frac{1}{2})$ has a subsequence converging to $\omega - G_1(p, y)$ on compact subsets of $M \setminus B_p(\frac{1}{2})$. Since $G_1(p, y) = G(p, y)$ on $B_p(2) \setminus B_p(\frac{1}{2})$, we can denote G_1 by G also, and $G(p, y)$ is the uniform limit of $h_i(p, y)$ on compact subsets of $M \setminus \{p\}$. Moreover, since

$$\int_M h_i(p, y) \Delta f(y) \, dy = \int_M G_i(p, y) \Delta f(y) \, dy - a_i \int_M \Delta f(y) \, dy$$
$$= -f(p)$$

for $f \in C_c^\infty(M)$, the previous argument implies that $G(p, y)$ also satisfies property (17.1). We also observe that if $\bar{G}(p, y)$ is the Dirichlet Green's function on $B_p(1)$, then the maximum principle implies that

$$\bar{G}(p, y) \leq h_i(p, y) \leq \bar{G}(p, y) + \omega.$$

Hence by passing through the limit, we conclude that

$$\bar{G}(p, y) \leq G(p, y) \leq \bar{G}(p, y) + \omega. \tag{17.15}$$

17 Green's function

For another point $q \neq p$, the same argument implies that there is a subsequence i_j of i and a sequence of constants b_{i_j} such that

$$G_{i_j}(q, y) - b_{i_j}$$

converges uniformly on compact subset of $M \setminus \{q\}$. Setting $y = p$, we observe that since

$$(G_{i_j}(q, p) - b_{i_j}) = (G_{i_j}(p, q) - a_{i_j}) + (a_{i_j} - b_{i_j}),$$

by taking the limit as $j \to \infty$, we conclude that

$$\lim_{j \to \infty} G_{i_j}(q, p) = G(p, q) + \lim_{j \to \infty} (a_{i_j} - b_{i_j}).$$

Hence the limit

$$\lim_{j \to \infty} (a_{i_j} - b_{i_j}) = c$$

must exist and the sequence of functions

$$G_{i_j}(q, y) - a_{i_j} = (G_{i_j}(q, y) - b_{i_j}) - (a_{i_j} - b_{i_j})$$

also converges. Let

$$\lim_{j \to \infty} G_{i_j}(q, y) - a_{i_j} = J(q, y).$$

Again, setting $y = p$, we conclude that

$$G(p, q) = J(q, p).$$

We now claim that the original sequence $G_i(q, y) - a_i$ must converge to $J(q, y)$ also. To see this, it suffices to show that if

$$G_{i_k}(q, y) - a_{i_k}$$

is any other convergent subsequence of i, then it must converge to $J(q, y)$. Let

$$\lim_{k \to \infty} (G_{i_k}(q, y) - a_{i_k}) = K(q, y)$$

denote its limit. The same argument as above shows that

$$K(q, p) = G(p, q),$$

hence

$$K(q, p) = J(q, p). \tag{17.16}$$

On the other hand, if we let ρ be sufficiently large so that $B_q(1) \subset B_p(\rho)$, then

$$G(p, y) - K(q, y) = \lim_{k \to \infty} (G_{i_k}(p, y) - a_{i_k}) - \lim_{k \to \infty} (G_{i_k}(q, y) - a_{i_k})$$

$$= \lim_{k \to \infty} (G_{i_k}(p, y) - G_{i_k}(q, y)).$$

Since G_{i_k} satisfies the Dirichlet boundary condition on Ω_{i_k}, for sufficiently large k such that $B_p(\rho) \subset \Omega_{i_k}$, the maximum principle asserts that the harmonic function

$$G_{i_k}(p, y) - G_{i_k}(q, y)$$

when restricted on $\Omega_{i_k} \setminus B_p(\rho)$ must have its maximum achieved on $\partial B_p(\rho)$. Passing to the limit, we conclude that

$$\sup_{y \in M \setminus B_p(\rho)} (G(p, y) - K(q, y)) = \sup_{y \in \partial B_p(\rho)} (G(p, y) - K(q, y))$$

and $G(p, y) - K(q, y)$ is bounded on $M \setminus B_p(\rho)$. A similar argument implies that $G(p, y) - J(q, y)$ is also bounded on $M \setminus B_p(\rho)$ and therefore $K(q, y) - J(q, y)$ is bounded on $M \setminus B_p(\rho)$. This together with estimate (17.15), when applied to both K and J at the point q, implies that $K(q, y) - J(q, y)$ is bounded on $M \setminus \{q\}$. Again the removable singularity theorem asserts that $K(q, y) - J(q, y)$ is a bounded harmonic function in y. However, this will imply that M admits a harmonic function on $M \setminus B_p(\rho)$ with its infimum achieved at infinity unless $K(q, y) - J(q, y)$ is identically constant. The identity (17.16) implies that $K(q, y) = J(q, y)$. In particular, we write this as $G(q, y)$. Properties (17.4) and (17.5) obviously follow since G is the limit of functions satisfying these properties. $\square$

18
Measured Neumann Poincaré inequality and measured Sobolev inequality

In this chapter, we will adapt an argument by Hajłasz and Koskela [HK] that proved the validity of the Neumann Poincaré inequality which when combined with a volume doubling property implies the validity of a Sobolev type inequality. In particular, we will use this argument to establish a measured Sobolev inequality to be used in the next chapter.

Definition 18.1 We say that a subdomain Ω of a manifold satisfies the volume doubling property ($\mathcal{VD}$) if there exists a constant $C_{\mathcal{VD}}(\Omega) > 0$ such that

$$V_p(2\rho) \leq C_{\mathcal{VD}}(\Omega)\, V_p(\rho)$$

for all $p \in \Omega$ and for all $0 < \rho \leq d(y, \partial\Omega)/2$, where $d(y, \partial\Omega)$ is the distance from y to $\partial\Omega$. In particular, we call the smallest such constant,

$$C_{\mathcal{VD}}(\Omega) = \sup \frac{V_p(2\rho)}{V_p(\rho)},$$

the volume doubling constant on Ω.

In addition to the volume doubling property, we also assume that M admits a uniform lower bound for the scaled first nonzero Neumann eigenvalue for geodesic balls. Again, one can localize the argument by introducing the following local λ_1 bound.

Definition 18.2 We say that $C_{\mathcal{P}}(\Omega) > 0$ is a scaled local λ_1 bound for the first nonzero Neumann eigenvalues over the smooth subdomain Ω if

$$\inf(\rho^2 \lambda_1(B_y(\rho))) = C_{\mathcal{P}}(\Omega),$$

where the infimum is taken over all $y \in \Omega$ and all $\rho \leq d(y, \partial\Omega)/2$, where $d(y, \partial\Omega)$ is the distance from y to $\partial\Omega$.

We will now state the main theorem of this chapter.

Theorem 18.3 *Let M be a complete manifold (possibly with boundary) and $B_p(\rho)$ be a geodesic ball in M satisfying $B_p(\rho) \cap \partial M = \emptyset$. For any fixed $0 < \delta < 1$, let $\phi(r)$ be a function defined on $B_p(\rho)$ given by*

$$\phi(r) = \begin{cases} 1 & \text{if } r \leq \delta\rho, \\ \dfrac{\rho - r}{(1-\delta)\rho} & \text{if } \delta\rho \leq r \leq \rho, \\ 0 & \text{if } \rho \leq r. \end{cases}$$

Then there exists a universal constant $C_1 > 0$ depending on $C_\mathcal{P}(B_p(\rho))$ and $C_{\mathcal{VD}}(B_p(\rho))$ such that the measured Neumann Poincaré inequality of the form

$$C_1 \rho^{-2} \int_{B_p(\rho)} |f(x) - \bar{f}|^2 \phi^2(x)\, dV \leq \int_{B_p(\rho)} |\nabla f|^2(x) \phi^2(x)\, dV$$

is valid for all $f \in C^\infty(B_p(\rho))$, where $\bar{f} = \int_{B_p(\rho)} f(x) \phi^2(x)\, dV$.

To prove the theorem, we will first establish a local version of the measured Neumann Poincaré inequality.

Lemma 18.4 *Let $B_p(\rho) \subset M$ be a geodesic ball satisfying $B_p(\rho) \cap \partial M = \emptyset$. Assume that $B_p(\rho)$ has a scaled local λ_1 bound given by $C_\mathcal{P}(B_p(\rho)) > 0$. Then the measured Neumann Poincaré inequality*

$$\frac{C_\mathcal{P}(B_p(\rho))}{9\tau^2} \int_{B_y(\tau)} |f(x) - \tilde{f}|^2 \phi^2(x)\, dV \leq \int_{B_y(\tau)} |\nabla f|^2(x) \phi^2(x)\, dV \tag{18.1}$$

is valid for all $y \in B_p(\rho)$ and for all $0 \leq \tau \leq d(y, \partial B_p(\rho))/2$, where $\tilde{f} = \int_{B_y(\tau)} f(x) \phi^2(x)\, dV$.

Proof The local λ_1 bound asserts that

$$\int_{B_y(\tau)} |f(x) - \tilde{f}|^2 \phi^2 \, dV = \inf_{k \in \mathbb{R}} \int_{B_y(\tau)} |f(x) - k|^2 \phi^2 \, dV$$

$$\leq \sup_{B_y(\tau)} \phi^2 \inf_{k \in \mathbb{R}} \int_{B_y(\tau)} |f(x) - k|^2 \, dV$$

$$\leq C_{\mathcal{P}}^{-1}(B_p(\rho)) \tau^2 \sup_{B_y(\tau)} \phi^2 \int_{B_y(\tau)} |\nabla f|^2(x) \, dV$$

$$\leq C_{\mathcal{P}}^{-1}(B_p(\rho)) \tau^2 \frac{\sup_{B_y(\tau)} \phi^2}{\inf_{B_y(\tau)} \phi^2}$$

$$\times \int_{B_y(\tau)} |\nabla f|^2(x) \phi^2(x) \, dV. \qquad (18.2)$$

We observe that

$$\sup_{B_y(\tau)} \phi^2 = \begin{cases} 1 & \text{if} \quad r(p, y) - \tau \leq \delta\rho, \\ \left(\dfrac{\rho - r(p, y) + \tau}{(1-\delta)\rho}\right)^2 & \text{if} \quad r(p, y) - \tau \geq \delta\rho, \end{cases}$$

and

$$\inf_{B_y(\tau)} \phi^2 = \begin{cases} \left(\dfrac{\rho - r(p, y) - \tau}{(1-\delta)\rho}\right)^2 & \text{if} \quad r(p, y) + \tau \geq \delta\rho, \\ 1 & \text{if} \quad r(p, y) + \tau \leq \delta\rho, \end{cases} \qquad (18.3)$$

implying that

$$\frac{\sup_{B_y(\tau)} \phi^2}{\inf_{B_y(\tau)} \phi^2}$$

$$= \begin{cases} 1 & \text{if} \quad r(p, y) + \tau \leq \delta\rho, \\ \left(\dfrac{(1-\delta)\rho}{\rho - r(p, y) - \tau}\right)^2 & \text{if} \quad r(p, y) - \tau \leq \delta\rho \leq r(p, y) + \tau, \\ \left(\dfrac{\rho - r(p, y) + \tau}{\rho - r(p, y) - \tau}\right)^2 & \text{if} \quad \delta\rho \leq r(p, y) - \tau. \end{cases}$$

In the case when $r(p, y) - \tau \leq \delta\rho \leq r(p, y) + \tau$, we have

$$\frac{(1-\delta)\rho}{\rho - r(p, y) - \tau} \leq \frac{\rho - r(p, y) + \tau}{\rho - r(p, y) - \tau}$$
$$\leq 3$$

since $(\rho - r(p, y) + \tau)/(\rho - r(p, y) - \tau)$ is an increasing function of $0 \leq \tau \leq \frac{1}{2}(\rho - r(p, y))$. The same estimate also applies to the case when $\delta\rho \leq r(p, y) - \tau$. Inequality (18.1) follows by combining these estimates with (18.2). $\square$

We will also need the following volume doubling property for the measure $\phi^2 \, dV$, which follows as a consequence of the volume doubling property ($\mathcal{VD}$) for the background metric.

Lemma 18.5 *Let $B_p(\rho) \subset M$ be a geodesic ball satisfying $B_p(\rho) \cap \partial M = \emptyset$. Suppose $C_{\mathcal{VD}}(B_p(\rho))$ is the volume doubling constant for $B_p(\rho)$. Then the inequality*

$$\int_{B_y(2\tau)} \phi^2(x) \, dV \leq 16 C_{\mathcal{VD}}(B_p(\rho)) \int_{B_y(\tau)} \phi^2(x) \, dV$$

is valid for all $y \in B_p(\rho)$ and $\tau \leq d(y, \partial B_p(\rho))/2$.

Proof If $y \in B_p(\rho)$ and $\tau \leq d(y, \partial B_p(\rho))/2$, then the volume doubling property for the background metric asserts that

$$\int_{B_y(2\tau)} \phi^2(x) \, dV \leq V_y(2\tau) \sup_{B_y(2\tau)} \phi^2$$

$$\leq C_{\mathcal{VD}}(B_p(\rho)) V_y(\tau) \sup_{B_y(2\tau)} \phi^2$$

$$\leq C_{\mathcal{VD}}(B_p(\rho)) \frac{\sup_{B_y(2\tau)} \phi^2}{\inf_{B_y(\tau)} \phi^2} \int_{B_y(\tau)} \phi^2(x) \, dV. \quad (18.4)$$

Since

$$\sup_{B_y(2\tau)} \phi^2 = \begin{cases} 1 & \text{if } r(p, y) - 2\tau \leq \delta\rho, \\ \left(\dfrac{\rho - r(p, y) + 2\tau}{(1-\delta)\rho}\right)^2 & \text{if } r(p, y) - 2\tau \geq \delta\rho, \end{cases}$$

18 Measured Neumann Poincaré and Sobolev inequalities

when combining this with (18.3), we have

$$\frac{\sup_{B_y(2\tau)} \phi^2}{\inf_{B_y(\tau)} \phi^2} = \begin{cases} 1 & \text{if } r(p,y) + \tau \leq \delta\rho, \\ \left(\dfrac{(1-\delta)\rho}{\rho - r(p,y) - \tau}\right)^2 & \text{if } r(p,y) - 2\tau \leq \delta\rho \leq r(p,y) + \tau, \\ \left(\dfrac{\rho - r(p,y) + 2\tau}{\rho - r(p,y) - \tau}\right)^2 & \text{if } \delta\rho \leq r(p,y) - 2\tau \end{cases}$$

Estimating the same way as in the proof of Lemma 18.4 by using the assumption that $\tau \leq d(y, \partial B_p(\rho))/2$, we conclude that

$$\frac{\sup_{B_y(2\tau)} \phi^2}{\inf_{B_y(\tau)} \phi^2} \leq 16$$

and the lemma follows from (18.4). □

Theorem 18.3 now follows by applying Lemma 18.4, Lemma 18.5, and a measured Hajłasz–Koskela [HK] theorem.

Theorem 18.6 *Let $B_p(\rho)$ be a geodesic ball in a complete manifold satisfying $B_p(\rho) \cap \partial M = \emptyset$. Let ϕ be a nonnegative function defined on $B_p(\rho)$. Suppose $C_{\mathcal{VD},\phi}(B_p(\rho)) > 0$ is a constant such that the volume doubling property with respect to the measure $\phi^2 \, dV$ of the form*

$$\int_{B_y(2\tau)} \phi^2 \, dV \leq C_{\mathcal{VD},\phi}(B_p(\rho)) \int_{B_y(\tau)} \phi^2 \, dV$$

is valid for all $y \in B_p(\rho)$ and for all $\tau \leq d(y, \partial B_p(\rho))/2$, where $d(y, \partial B_p(\rho)) = \rho - r(p, y)$. We also assume that there exists a constant $C_{\mathcal{P},\phi}(B_p(\rho)) > 0$ such that the measured Neumann Poincaré inequality

$$C_{\mathcal{P},\phi}(B_p(\rho)) \, \tau^{-2} \inf_{k \in \mathbb{R}} \int_{B_y(\tau)} |f(x) - k|^2 \, \phi^2 \, dV \leq \int_{B_y(\tau)} |\nabla f|^2 \, \phi^2 \, dV$$

is valid for all $y \in B_p(\rho)$ and for all $\tau \leq d(y, \partial B_p(\rho))/2$. Then the measured Neumann Poincaré inequality is valid on $B_p(\rho)$. In particular, there exists a constant $C_\phi > 0$ depending only on $C_{\mathcal{P},\phi}(B_p(\rho))$ and $C_{\mathcal{VD},\phi}(B_p(\rho))$ such that

$$C_\phi \rho^{-2} \int_{B_p(\rho)} |f - \bar{f}|^2 \phi^2 \, dV \leq \int_{B_p(\rho)} |\nabla f|^2 \phi^2 \, dV$$

for all $f \in C^\infty(B_p(\rho))$, where $\bar{f} = \int_{B_p(\rho)} f \phi^2 \, dV$. Moreover, there exists $\alpha > 2$ depending on $C_{\mathcal{VD},\phi}(B_p(\rho))$ such that the measured Neumann Sobolev inequality of the form

$$C_\phi \rho^{-2} V_\phi^{2/\alpha}(B_p(\rho)) \left(\int_{B_p(\rho)} |f - \bar{f}|^{\frac{2\alpha}{\alpha-2}} \phi^2 \, dV \right)^{(\alpha-2)/\alpha} \leq \int_{B_p(\rho)} |\nabla f|^2 \phi^2 \, dV$$

is valid, where $V_\phi(B_p(\rho)) = \int_{B_p(\rho)} \phi^2 \, dV$ denotes the volume of $B_p(\rho)$ with respect to the measure $\phi^2 \, dV$.

Proof In this proof, we will abbreviate our notation by using $C_{\mathcal{VD},\phi}$ and $C_{\mathcal{P},\phi}$ for $C_{\mathcal{VD},\phi}(B_p(\rho))$ and $C_{\mathcal{P},\phi}(B_p(\rho))$, respectively. For any $y \in B_p(\rho)$, let $\gamma(t)$ be a normal geodesic with $\gamma(0) = p$ and $\gamma(\ell) = y$. Let us define the sequence of points $y_0 = \gamma(0) = p$, $y_1 = \gamma(\rho/2)$, and $y_i = \gamma\left(\sum_{j=1}^i \rho/2^j\right)$. Let i_0 be the largest i such that

$$\sum_{j=1}^i \frac{\rho}{2^j} < \ell.$$

Let us define

$$B_i = \begin{cases} B_{y_i}\left(\dfrac{\rho}{2^{i+1}}\right) & \text{for } i < i_0, \\ B_y\left(\dfrac{\rho}{2^{i+1}}\right) & \text{for } i \geq i_0. \end{cases}$$

For any $f \in C^\infty(B_p(\rho))$, we define

$$f_{B_i} = V_\phi^{-1}(B_i) \int_{B_i} f \phi^2 \, dV,$$

where

$$V_\phi(B_i) = \int_{B_i} \phi^2 \, dV$$

denotes the volume of B_i with respect to the measure $\phi^2 \, dV$. We observe that $f_{B_i} \to f(y)$ as $i \to \infty$, and

$$|f_{B_0} - f(y)| \leq \sum_{i=0}^\infty |f_{B_i} - f_{B_{i+1}}|$$

$$\leq \sum_{i=0}^\infty (|f_{B_i} - f_{D_i}| + |f_{D_i} - f_{B_{i+1}}|), \qquad (18.5)$$

18 Measured Neumann Poincaré and Sobolev inequalities

where

$$D_i = B_{z_i}\left(\frac{\rho}{2^{i+3}}\right) \subset B_i \cap B_{i+1}$$

with $z_i = \gamma\left(\sum_{j=1}^{i} \rho/2^j + 3\rho/2^{i+3}\right)$. However, since $D_i \subset B_i$, we have

$$|f_{B_i} - f_{D_i}| \leq V_\phi^{-1}(D_i) \int_{D_i} |f - f_{B_i}| \phi^2 \, dV$$

$$\leq V_\phi^{-1}(D_i) \int_{B_i} |f - f_{B_i}| \phi^2 \, dV$$

$$\leq C_{\mathcal{VD},\phi}^3 V_\phi^{-1}(B_i) \int_{B_i} |f - f_{B_i}| \phi^2 \, dV$$

and similarly

$$|f_{B_{i+1}} - f_{D_i}| \leq C_{\mathcal{VD},\phi}^3 V_\phi^{-1}(B_{i+1}) \int_{B_{i+1}} |f - f_{B_{i+1}}| \phi^2 \, dV.$$

Therefore combining this inequality with (18.5), we conclude that

$$|f_{B_0} - f(y)| \leq 2C_{\mathcal{VD},\phi}^3 \sum_{i=0}^{\infty} V_\phi^{-1}(B_i) \int_{B_i} |f - f_{B_i}| \phi^2 \, dV$$

$$\leq 2C_{\mathcal{VD},\phi}^3 \sum_{i=0}^{\infty} \left(V_\phi^{-1}(B_i) \int_{B_i} |f - f_{B_i}|^2 \phi^2 \, dV\right)^{1/2}$$

$$\leq 2C_{\mathcal{VD},\phi}^3 C_{\mathcal{P},\phi}^{-1/2} \sum_{i=0}^{\infty} \frac{\rho}{2^{i+1}} \left(V_\phi^{-1}(B_i) \int_{B_i} |\nabla f|^2 \phi^2 \, dV\right)^{1/2}.$$
(18.6)

Note that for $\epsilon > 0$, since

$$\sum_{i=0}^{\infty} 2^{-\epsilon(i+1)} < \infty,$$

(18.6) implies that

$$\sum_{i=0}^{\infty} \rho_i (V_\phi^{-1}(B_i) \int_{B_i} |\nabla f|^2 \phi^2 \, dV)^{1/2}$$

$$\geq C_1 C_{\mathcal{VD},\phi}^{-3} C_{\mathcal{P},\phi}^{1/2} \rho^{-\epsilon} |f_{B_0} - f(y)| \sum_{i=0}^{\infty} \rho_i^\epsilon,$$

where $\rho_i = 2^{-(i+1)}\rho$ and $C_1 > 0$ is a universal constant. In particular, we deduce that there exists $j \geq 0$ such that

$$V_\phi^{-1}(B_j) \int_{B_j} |\nabla f|^2 \phi^2 \, dV \geq C_1^2 C_{\mathcal{VD},\phi}^{-6} C_{\mathcal{P},\phi} \rho^{-2\epsilon} |f_{B_0} - f(y)|^2 \rho_j^{2\epsilon-2}. \tag{18.7}$$

Note that since

$$r(y, y_i) \leq \rho - r(p, y_i)$$

$$\leq \rho - \sum_{j=1}^{i} \frac{\rho}{2^j}$$

$$= \frac{\rho}{2^i},$$

we have

$$B_i \subset B_y(3\rho_i) \quad \text{for all} \quad i.$$

Therefore taking this together with (18.7), we obtain

$$\int_{B_y(3\rho_j) \cap B_p(\rho)} |\nabla f|^2 \phi^2 \, dV$$

$$\geq C_1^2 C_{\mathcal{VD},\phi}^{-6} C_{\mathcal{P},\phi} V_\phi(B_j) \rho^{-2\epsilon} |f_{B_0} - f(y)|^2 \rho_j^{2\epsilon-2}$$

$$\geq C_1^2 C_{\mathcal{VD},\phi}^{-8} C_{\mathcal{P},\phi} \rho^{-2\epsilon} |f_{B_0} - f(y)|^2 \rho_j^{2\epsilon-2} V_\phi(B_y(\rho_j)).$$

In particular, we conclude that for all $y \in B_p(\rho)$, there exists $\rho_y > 0$ such that

$$\int_{B_y(\rho_y) \cap B_p(\rho)} |\nabla f|^2 \phi^2 \, dV \geq C_\phi \rho^{-2\epsilon} |f_{B_0} - f(y)|^2 \rho_y^{2\epsilon-2} V_\phi(B_y(\rho_y)), \tag{18.8}$$

where $C_\phi = C_1^2 C_{\mathcal{VD},\phi}^{-10} C_{\mathcal{P},\phi}$.

We now claim that the volume doubling property with respect to the measure $\phi^2 \, dV$ implies that there exists a constant $\beta > 0$ depending only on $C_{\mathcal{VD},\phi}$ such that

$$V_\phi(B_p(\rho)) \leq C_{\mathcal{VD},\phi} \left(\frac{\rho}{\rho_y}\right)^\beta V_\phi(B_y(\rho_y)) \tag{18.9}$$

for all $y \in B_p(\rho)$. Indeed, let γ be the geodesic joining y to p and we choose a sequence of points $w_0, w_1, \ldots, w_k$ with $w_0 = y$ and $w_i \in \gamma$ such that $r(w_i, w_{i+1}) = 2^i \rho_y$, where k is the first integer such that $p \in B_{w_k}(2^k \rho_y)$. In particular, $\rho_y + \sum_{i=2}^{k} 2^i \rho_y \leq r(p, y)$, hence

$$k \leq \log_2 \left(\frac{\rho}{\rho_y} - 1 \right).$$

Note that the volume doubling property asserts that

$$C_{\mathcal{VD},\phi}^3 V_\phi(B_{w_i}(2^i \rho_y)) \geq V_\phi(B_{w_i}(5(2^i \rho_y)))$$
$$\geq V_\phi(B_{w_{i+1}}(2^{i+1} \rho_y))$$

and

$$C_{\mathcal{VD},\phi} V_\phi(B_{w_k}(2^k \rho_y)) \geq V_\phi(B_p(2^k \rho_y)),$$

hence we have

$$C_{\mathcal{VD},\phi}^{3k+1} V_\phi(B_y(\rho_y))\phi^2 \, dV \geq V_\phi(B_p(2^k \rho_y)). \quad (18.10)$$

On the other hand, if we set k' to be the smallest integer such that

$$2^{k'}(2^k \rho_y) \geq \rho,$$

then

$$k' + k \leq \log_2 \left(\frac{\rho}{\rho_y} \right) + 1$$

and

$$C_{\mathcal{VD},\phi}^{k'} V_\phi(B_p(2^k \rho_y)) \geq V_\phi(B_p(\rho)).$$

Combining this volume estimate with (18.10), we obtain

$$C_{\mathcal{VD},\phi}^{3\log_2(\rho/\rho_y)+1} V_\phi(B_{w_k}(2^k \rho_y)) \geq V_\phi(B_p(\rho)),$$

and the claim (18.9) is established with

$$\beta = 3 \log_2 C_{\mathcal{VD},\phi}.$$

Let us now define the sublevel set by

$$A_t = \{y \in B_p(\rho) \mid |f_{B_0} - f(y)| \geq t\}.$$

Applying (18.9) to (18.8), we conclude that for all $y \in A_t$, there exists $\rho_y > 0$ such that

$$\rho^2 V_\phi^{2(\epsilon-1)\beta^{-1}}(B_p(\rho)) \int_{B_y(\rho_y) \cap B_p(\rho)} |\nabla f|^2 \phi^2 \, dV$$

$$\geq C_\phi t^2 \left(\frac{\rho}{\rho_y}\right)^{2-2\epsilon} V_\phi^{2(\epsilon-1)\beta^{-1}}(B_p(\rho)) V_\phi(B_y(\rho_y))$$

$$\geq C_\phi t^2 V_\phi^{1+2(\epsilon-1)\beta^{-1}}(B_y(\rho_y)).$$

Without loss of generality, we may assume that $\beta \geq 2$. The Vitali covering lemma asserts that there is a countable (possibly finite), mutually disjoint, subcollection $\{B_1(\rho_1), B_2(\rho_2), \ldots, B_i(\rho_i), \ldots\}$ of the collection $\{B_y(\rho_y) \mid y \in A_t\}$ with the properties that $\rho_1 \geq \rho_2 \geq \cdots \geq \rho_i \geq \cdots$. Moreover, for any $y \in A_t$, there exists an i such that $B_i(\rho_i) \cap B_y(\rho_y) = \emptyset$ and $B_y(\rho_y) \subset B_i(3\rho_i)$. In particular, we have

$$V_\phi^{1-2(1-\epsilon)\beta^{-1}}(A_t) \leq C_{\mathcal{VD},\phi}^2 \sum_i V_\phi^{1-2(1-\epsilon)\beta^{-1}}(B_i(\rho_i))$$

$$\leq C_\phi^{-1} t^{-2} \rho^2 V_\phi^{-2(1-\epsilon)\beta^{-1}}(B_p(\rho))$$

$$\times \sum_i \int_{B_i(\rho_i) \cap B_p(\rho)} |\nabla f|^2 \phi^2 \, dV$$

$$\leq C_\phi^{-1} t^{-2} \rho^2 V_\phi^{-2(1-\epsilon)\beta^{-1}}(B_p(\rho)) \int_{B_p(\rho)} |\nabla f|^2 \phi^2 \, dV,$$

where $C_\phi > 0$ is a constant depending only on $C_{\mathcal{VD},\phi}$ and $C_{\mathcal{P},\phi}$. Setting

$$B = \left(C_\phi^{-1} \rho^2 V_\phi^{-2(1-\epsilon)\beta^{-1}}(B_p(\rho)) \int_{B_p(\rho)} |\nabla f|^2 \phi^2 \, dV\right)^{1/(1-2(1-\epsilon)\beta^{-1})},$$

then if q is a constant satisfying $q < 2/(1 - 2(1 - \epsilon)\beta^{-1})$ the coarea formula implies that

$$\int_{B_p(\rho)} |f - f_{B_0}|^q \phi^2 \, dV = q \int_0^\infty t^{q-1} V_\phi(A_t) \, dt$$

$$= q \int_0^T t^{q-1} V_\phi(A_t) \, dt + q \int_T^\infty t^{q-1} V_\phi(A_t) \, dt$$

$$\leq T^q V_\phi(B_p(\rho)) + q B \int_T^\infty t^{q-1-2/(1-2(1-\epsilon)\beta^{-1})} \, dt$$

$$= T^q V_\phi(B_p(\rho)) + \frac{q B}{q - \frac{2}{1-2(1-\epsilon)\beta^{-1}}}$$

$$\times T^{q-2/(1-2(1-\epsilon)\beta^{-1})}.$$

Optimizing by choosing

$$T = \rho V_\phi^{-1/2}(B_p(\rho)) \left(\int_{B_p(\rho)} |\nabla f|^2 \phi^2 \, dV \right)^{1/2},$$

we obtain

$$\left(\int_{B_p(\rho)} |f - f_{B_0}|^q \phi^2 \, dV \right)^{1/q} \leq C_\phi^{-1} \rho V_\phi^{1/q - 1/2}(B_p(\rho))$$

$$\times \left(\int_{B_p(\rho)} |\nabla f|^2 \phi^2 \, dV \right)^{1/2}. \quad (18.11)$$

Since $2/(1 + 2(\epsilon - 1)\beta^{-1}) > 2$, we can choose $q = 2$ and obtain

$$\inf_{k \in \mathbb{R}} \int_{B_p(\rho)} |f - k|^2 \phi^2 \, dV \leq \int_{B_p(\rho)} |f - f_{B_0}|^2 \phi^2 \, dV$$

$$\leq C_\phi^{-1} \rho^2 \int_{B_p(\rho)} |\nabla f|^2 \phi^2 \, dV, \quad (18.12)$$

establishing the measured Neumann Poincaré inequality.

In fact, since we can take $2 < q < 2/(1 + 2(\epsilon - 1)\beta^{-1})$, a measured Neumann Sobolev type inequality also follows. To see this, we apply (18.11) and obtain

$$\left(\int_{B_p(\rho)} |f - \bar{f}|^q \phi^2 \, dV\right)^{1/q} \leq \left(\int_{B_p(\rho)} |f - f_{B_0}|^q \phi^2 \, dV\right)^{1/q}$$

$$+ \left(\int_{B_p(\rho)} |f_{B_0} - \bar{f}|^q \phi^2 \, dV\right)^{1/q}$$

$$\leq C_\phi^{-1} \rho \, V_\phi^{1/q - 1/2}(B_p(\rho))$$

$$\times \left(\int_{B_p(\rho)} |\nabla f|^2 \phi^2 \, dV\right)^{1/2}$$

$$+ V_\phi^{1/q}(B_p(\rho)) |f_{B_0} - \bar{f}|. \qquad (18.13)$$

However,

$$V_\phi^{1/q}(B_p(\rho)) |f_{B_0} - \bar{f}| \leq V_\phi^{1/q}(B_p(\rho)) V_\phi^{-1}(B_0) \int_{B_0} |f - \bar{f}| \phi^2 \, dV$$

$$\leq V_\phi^{1/q}(B_p(\rho)) V_\phi^{-1/2}(B_0) \left(\int_{B_p(\rho)} |f - \bar{f}|^2 \phi^2 dV\right)^{1/2}$$

$$\leq C_{\mathcal{VD},\phi} V_\phi^{1/q - 1/2}(B_p(\rho)) \left(\int_{B_p(\rho)} |f - \bar{f}|^2 \phi^2 dV\right)^{1/2},$$

hence combining this inequality with (18.12) and (18.13), we conclude that

$$\left(\int_{B_p(\rho)} |f - \bar{f}|^{2\alpha/(\alpha-2)} \phi^2 \, dV\right)^{(\alpha-2)/\alpha} \leq C_\phi^{-1} \rho^2 \, V_\phi^{-2/\alpha}(B_p(\rho))$$

$$\times \int_{B_p(\rho)} |\nabla f|^2 \phi^2 \, dV$$

with $\alpha = 2q/(q-2)$. $\square$

Obviously, Theorem 18.3 follows from Lemma 18.4, Lemma 18.5, and Theorem 18.6.

Let us also point out that if we set $\phi \equiv 1$, then Theorem 18.6 asserts that a local λ_1 bound and volume doubling implies a Neumann Sobolev inequality.

Corollary 18.7 *Let $B_p(\rho)$ be a geodesic ball in a complete manifold satisfying $B_p(\rho) \cap \partial M = \emptyset$. Suppose $C_{\mathcal{VD}}(B_p(\rho)) > 0$ is the volume doubling constant on $B_p(\rho)$ and $C_{\mathcal{P}}(B_p(\rho)) > 0$ is the scaled local λ_1 bound as given by Definition 18.1 and Definition 18.2, respectively. Then there exist $\alpha > 2$ and a* constant $C_1 > 0$ depending only on $C_\mathcal{P}(B_p(\rho))$ and $C_{\mathcal{VD}}(B_p(\rho))$ such that, the Neumann Sobolev inequality of the form

$$\left(\int_{B_p(\rho)} |f - \bar{f}|^{2\alpha/(\alpha-2)} dV\right)^{(\alpha-2)/\alpha} \leq C_1 \rho^2 V^{-2/\alpha}(B_p(\rho)) \int_{B_p(\rho)} |\nabla f|^2 dV$$

is valid, for all $f \in C^\infty(B_p(\rho))$, where $\bar{f} = \fint_{B_p(\rho)} f\, dV$.

We would like to point out that both constants $C_{\mathcal{VD}}(\Omega)$ and $C_\mathcal{P}(\Omega)$ are monotonic in the sense that

$$C_{\mathcal{VD}}(\Omega_1) \leq C_{\mathcal{VD}}(\Omega_2)$$

and

$$C_\mathcal{P}(\Omega_1) \geq C_\mathcal{P}(\Omega_2)$$

if $\Omega_1 \subset \Omega_2$. In particular, if M has nonnegative Ricci curvature, then the volume comparison theorem asserts that

$$C_{\mathcal{VD}}(\Omega) = 2^m$$

for all $\Omega \subset M$. Moreover, Corollary 14.6 asserts that

$$C_\mathcal{P}(B_p(\rho)) \geq C_0$$

for all $p \in M$ and $\rho > 0$, where $C_0 > 0$ is a constant depending only on m.

19
Parabolic Harnack inequality and regularity theory

In this chapter, we will present Moser's version of the Nash–Mosers Harnack inequality for parabolic equations. The elliptic version (also proved by De Giorgi) can be considered as a special case when the solution is time independent. The iteration procedure of Moser was particularly useful in the theory of geometric analysis. We will attempt to cover this in as much generality as possible while keeping explicit account of the dependency of various geometric and analytic constants. In applying this type of argument in the study of geometric partial differential equations, often the explicit geometric dependency is crucial. As a result of these estimates, one derives a mean value inequality for nonnegative subsolutions and a Harnack inequality for positive solutions of a fairly general class of parabolic operators. In particular, it gives a C^α estimate for solutions of any second order parabolic (elliptic) operators of divergence form with only measurable coefficients. This regularity result was the original motivation for the development of this theory. We shall point out that the mean value inequality and the Harnack inequality derived from this argument are applicable to a more general class of equations, while the ones given in earlier chapters yield stronger results but require more smoothness from the operator. Both approaches are important in the theory of geometric analysis, but they are suited to different types of situation. The following account is a slightly modified version of Moser's argument that has been adapted to a more geometrical setting.

Let us define the average value of a function f on a geodesic ball $B_p(\rho)$ by

$$\fint_{B_p(\rho)} f\, dV = V_p(\rho)^{-1} \int_{B_p(\rho)} f\, dV.$$

19 Parabolic Harnack inequality and regularity theory

When the point p is fixed, the average L^q-norm of f over $B_p(\rho)$ is denoted by

$$\#f\|_{q,\rho} = \left(\fint_{B_p(\rho)} f^q \, dV\right)^{1/q}$$

and the regular L^q-norm is denoted by

$$\|f\|_{q,\rho} = \left(\int_{B_p(\rho)} f^q \, dV\right)^{1/q}.$$

When the function is defined on $B_p(\rho) \times [T_0, T_1]$, we denote its L^q-norm over the set $B_p(\rho) \times [T_0, T_1]$ by

$$\|f\|_{q,\rho,[T_0,T_1]} = \left(\int_{T_0}^{T_1} \int_{B_p(\rho)} f^q \, dV \, dt\right)^{1/q}$$

and the average L^q-norm by

$$\#f\|_{q,\rho,[T_0,T_1]} = \left((T_1 - T_0)^{-1} \int_{T_0}^{T_1} \fint_{B_p(\rho)} f^q \, dV\right)^{1/q}.$$

For the sake of further generalization to Riemannian manifolds, we will introduce a normalized Sobolev inequality on a geodesic ball centered at a point $p \in M$ of radius ρ. The inequality asserts that for a fixed constant $\mu \leq m/(m-2)$, there exists a constant $C_{\mathcal{SD}} > 0$, depending on $B_p(\rho)$, such that

$$\fint_{B_p(\rho)} |\nabla \phi|^2 \geq \frac{C_{\mathcal{SD}}}{\rho^2} \left(\fint_{B_p(\rho)} \phi^{2\mu}\right)^{1/\mu}, \tag{19.1}$$

for all $\phi \in H^c_{1,2}(B_p(\rho))$ that has boundary condition $\phi = 0$ on $\partial B_p(\rho)$. Typically, the constant μ is given by $\mu = m/(m-2)$ when $m \geq 3$, and $2 < \mu < \infty$ when $m = 2$. However, in the following context, we are not restricting the value of μ.

Lemma 19.1 *Let M be a complete manifold of dimension m. Let us assume that the geodesic ball $B_p(\rho)$ centered at p with radius ρ satisfies $B_p(\rho) \cap \partial M = \emptyset$. Suppose that $u(x,t)$ is a nonnegative function defined on $B_p(\rho) \times [T_0, T_1]$ such that*

$$\left(\Delta - \frac{\partial}{\partial t}\right) u \geq -fu$$

in the weak sense. Assume that the function f is nonnegative on $B_p(\rho) \times [T_0, T_1]$ and its L^q-norm on $B_p(\rho)$ is finite for some $\mu/(\mu-1) < q \le \infty$ on $[T_0, T_1]$, with

$$A = \sup_{[T_0,T_1]} \#\|f\|_{q,\rho} = \sup_{[T_0,T_1]} \left(\fint_{B_p(\rho)} f^q(x,t)\,dV \right)^{1/q}$$

for $\mu/(\mu-1) < q < \infty$ and

$$A = \|f\|_{\infty,\rho,[T_0,T_1]} = \sup_{B_p(\rho) \times [T_0,T_1]} f$$

for $q = \infty$. If α and ν are defined by $\alpha = q(\mu-1)/(\mu(q-1)-q) > 0$ and $\nu = (2\mu-1)/\mu > 1$, then for any $k > 1$, there exists constant $C_2 > 0$, depending only on k, μ, and q, such that

$$\|u\|_{\infty,\theta\rho,[T,T_1]} \le C_2 \left((kA\rho^2 C_{SD}^{-1})^\alpha \left(\frac{C_{SD}}{\rho^2}\right)^{(\nu-1)/\nu} \right.$$

$$\left. + ((1-\theta)^{-2}\rho^{-2} + (T-T_0)^{-1})\left(\frac{\rho^2}{C_{SD}}\right)^{1/\nu} \right)^{\nu/k(\nu-1)}$$

$$\times (T_1 - T_0)^{\frac{1}{k}} \#\|u\|_{k,\rho,[T_0,T_1]}$$

for any $\theta \in (0,1)$ and $T \in (T_0, T_1)$. For $0 < k \le 1$, there exists constant $C_3 > 0$, depending only on k, μ, and q, such that

$$\|u\|_{\infty,\theta\rho,[T,T_1]} \le C_3 \left(\left(A\frac{\rho^2}{C_{SD}}\right)^\alpha \left(\frac{C_{SD}}{\rho^2}\right)^{(\nu-1)/\nu} + C_{SD}^{-1/\nu}(1-\theta)^{-2} \right.$$

$$\left. \times \rho^{-2(\nu-1)/\nu} + C_{SD}^{-1/\nu}\rho^{2/\nu}(T-T_0)^{-1} \right)^{\nu/k(\nu-1)}$$

$$\times V_p^{-1/k}\left(\frac{\rho}{2}\right) \|u\|_{k,\rho,[T_0,T_1]}.$$

Proof Note that the normalized Sobolev inequality in the form described is scale invariant. Hence, without loss of generality, we may assume that $V_p(\rho) = 1$. For any arbitrary constant $a \ge 1$, the assumption on u implies that

$$\int \phi^2 f u^{2a} - \int \phi^2 u^{2a-1} u_t \ge -\int \phi^2 u^{2a-1} \Delta u,$$

19 Parabolic Harnack inequality and regularity theory

for any nonnegative, compactly supported, Lipschitz function ϕ on $B_p(\rho)$. The right-hand side after integration by parts yields

$$-\int \phi^2 u^{2a-1} \Delta u = 2\int \phi u^{2a-1} \langle \nabla \phi, \nabla u \rangle + (2a-1) \int \phi^2 u^{2a-2} |\nabla u|^2. \tag{19.2}$$

However, using the identity

$$\int |\nabla(\phi u^a)|^2 = \int |\nabla \phi|^2 u^{2a} + 2a \int \phi u^{2a-1} \langle \nabla \phi, \nabla u \rangle + a^2 \int \phi^2 u^{2a-2} |\nabla u|^2 \tag{19.3}$$

and the assumption that $a \geq 1$, we have

$$a \int \phi^2 f u^{2a} + \int |\nabla \phi|^2 u^{2a} \geq \int |\nabla(\phi u^a)|^2 + \frac{1}{2} \int \phi^2 (u^{2a})_t$$

$$\geq \frac{C_{SD}}{\rho^2} \left(\int (\phi u^a)^{2\mu} \right)^{1/\mu} + \frac{1}{2} \int \phi^2 (u^{2a})_t. \tag{19.4}$$

When $q = \infty$, (19.4) implies that

$$aA \int \phi^2 u^{2a} + \int |\nabla \phi|^2 u^{2a} \geq \frac{C_{SD}}{\rho^2} \left(\int (\phi u^a)^{2\mu} \right)^{1/\mu} + \frac{1}{2} \int \phi^2 (u^{2a})_t. \tag{19.5}$$

When $\mu/(\mu-1) < q < \infty$, by Hölder's inequality, we have

$$a \int \phi^2 f u^{2a} \leq aA \left(\int (\phi^2 u^{2a})^{q/(q-1)} \right)^{(q-1)/q}$$

$$\leq aA \left(\int \phi^2 u^{2a} \right)^{(\mu(q-1)-q)/q(\mu-1)} \left(\int (\phi^2 u^{2a})^{\mu} \right)^{1/q(\mu-1)}. \tag{19.6}$$

However, applying the inequality

$$x^\epsilon \leq \delta^{(\epsilon-1)/\epsilon} x + \delta \epsilon^{1/(1-\epsilon)} \left(\frac{1}{\epsilon} - 1 \right)$$

by setting $\epsilon = (\mu(q-1) - q)/q(\mu-1)$ and

$$x = (aA)^{q(\mu-1)/(\mu(q-1)-q)} \left(\int \phi^2 u^{2a} \right) \left(\int (\phi^2 u^{2a})^\mu \right)^{-1/\mu},$$

we have

$$aA \left(\int \phi^2 u^{2a} \right)^{(\mu(q-1)-q)/q(\mu-1)} \left(\int (\phi^2 u^{2a})^\mu \right)^{(q-\mu(q-1))/q\mu(\mu-1)}$$

$$\leq \delta^{(\epsilon-1)/\epsilon} (aA)^{q(\mu-1)/(\mu(q-1)-q)} \left(\int \phi^2 u^{2a} \right) \left(\int (\phi^2 u^{2a})^\mu \right)^{-1/\mu}$$

$$+ \delta\epsilon^{1/(1-\epsilon)} \left(\frac{1}{\epsilon} - 1 \right). \tag{19.7}$$

Multiplying through by

$$\left(\int (\phi^2 u^{2a})^\mu \right)^{1/\mu}$$

and choosing δ so that

$$\delta\epsilon^{1/(1-\epsilon)} \left(\frac{1}{\epsilon} - 1 \right) = \frac{C_{SD}}{2\rho^2},$$

(19.6) and (19.7) yield

$$a \int \phi^2 f u^{2a} \leq C_1 \left(\frac{C_{SD}}{\rho^2} \right)^{-\mu/(\mu(q-1)-q)} (aA)^{q(\mu-1)/(\mu(q-1)-q)} \left(\int \phi^2 u^{2a} \right)$$

$$+ \frac{C_{SD}}{2\rho^2} \left(\int (\phi^2 u^{2a})^\mu \right)^{1/\mu}$$

for some constant $C_1 > 0$ depending only on μ and q. Hence applying the above inequality to the first term of (19.4), we have

$$C_1 \left(\frac{C_{SD}}{\rho^2} \right)^{-\mu/(\mu(q-1)-q)} (aA)^{q(\mu-1)/(\mu(q-1)-q)} \left(\int \phi^2 u^{2a} \right) + \int |\nabla\phi|^2 u^{2a}$$

$$\geq \frac{C_{SD}}{2\rho^2} \left(\int (\phi u^a)^{2\mu} \right)^{1/\mu} + \frac{1}{2} \int \phi^2 (u^{2a})_t. \tag{19.8}$$

19 Parabolic Harnack inequality and regularity theory

In any event, (19.5) and (19.8) imply the inequality

$$2C_1 \left(\frac{C_{SD}}{\rho^2}\right)^{1-\alpha} (aA)^\alpha \int \phi^2 u^{2a} + 2 \int |\nabla \phi|^2 u^{2a}$$

$$\geq \frac{C_{SD}}{\rho^2} \left(\int (\phi u^a)^{2\mu}\right)^{1/\mu} + \int \phi^2 (u^{2a})_t. \qquad (19.9)$$

with $\alpha = q(\mu - 1)/(\mu(q-1) - q) > 0$ for all $\mu/(\mu - 1) < q \leq \infty$.
If $\psi(t)$ is a Lipschitz function given by

$$\psi(t) = \begin{cases} 0 & \text{for } T_0 \leq t \leq s, \\ \dfrac{t-s}{v} & \text{for } s \leq t \leq s+v, \\ 1 & \text{for } s+v \leq t \leq T_1, \end{cases}$$

then (19.9) implies that

$$2C_1 \left(\frac{C_{SD}}{\rho^2}\right)^{1-\alpha} (aA)^\alpha \int_{T_0}^{t'} \int_{B_p(\rho)} \psi^2(t) \phi^2(x) u^{2a}(x,t) \, dx \, dt$$

$$+ 2 \int_{T_0}^{t'} \int_{B_p(\rho)} \psi^2(t) |\nabla \phi|^2(x) u^{2a}(x,t) \, dx \, dt$$

$$+ 2 \int_{T_0}^{t'} \int_{B_p(\rho)} \psi(t) \psi_t(t) \phi^2(x) u^{2a}(x,t) \, dx \, dt$$

$$\geq \frac{C_{SD}}{\rho^2} \int_{T_0}^{t'} \psi^2(t) \left(\int_{B_p(\rho)} (\phi(x) u^a(x,t))^{2\mu} \, dx\right)^{1/\mu} dt$$

$$+ \int_{T_0}^{t'} \frac{\partial}{\partial t} \left(\int_{B_p(\rho)} \psi^2(t) \phi^2(x) u^{2a}(x,t) \, dx\right) dt$$

$$= \frac{C_{SD}}{\rho^2} \int_{T_0}^{t'} \psi^2(t) \left(\int_{B_p(\rho)} (\phi(x) u^a(x,t))^{2\mu} \, dx\right)^{1/\mu} dt$$

$$+ \int_{B_p(\rho)} \psi^2(t') \phi^2(x) u^{2a}(x,t') \, dx \qquad (19.10)$$

for all $t' \in (T_0, T_1]$. First observe that since the first term on the right-hand side is nonnegative, we conclude that

$$2C_1 \left(\frac{C_{SD}}{\rho^2}\right)^{1-\alpha} (aA)^\alpha \int_{T_0}^{T_1} \int_{B_p(\rho)} \psi^2(t) \phi^2(x) u^{2a}(x,t) \, dx \, dt$$

$$+ 2\int_{T_0}^{T_1} \int_{B_p(\rho)} \psi^2(t) |\nabla\phi|^2(x) u^{2a}(x,t) \, dx \, dt$$

$$+ 2\int_{T_0}^{T_1} \int_{B_p(\rho)} \psi(t) \psi_t(t) \phi^2(x) u^{2a}(x,t) \, dx \, dt$$

$$\geq \sup_{s+v \leq t' \leq T_1} \int_{B_p(\rho)} \phi^2(x) u^{2a}(x,t') \, dx. \tag{19.11}$$

By setting $t' = T_1$ in (19.10), we also have

$$2C_1 \left(\frac{C_{SD}}{\rho^2}\right)^{1-\alpha} (aA)^\alpha \int_{T_0}^{T_1} \int_{B_p(\rho)} \psi^2(t) \phi^2(x) u^{2a}(x,t) \, dx \, dt$$

$$+ 2\int_{T_0}^{T_1} \int_{B_p(\rho)} \psi^2(t) |\nabla\phi|^2(x) u^{2a}(x,t) \, dx \, dt$$

$$+ 2\int_{T_0}^{T_1} \int_{B_p(\rho)} \psi(t) \psi_t(t) \phi^2(x) u^{2a}(x,t) \, dx \, dt$$

$$\geq \frac{C_{SD}}{\rho^2} \int_{s+v}^{T_1} \left(\int_{B_p(\rho)} (\phi(x) u^a(x,t))^{2\mu} \, dx\right)^{1/\mu} dt. \tag{19.12}$$

On the other hand, Schwarz inequality implies that

$$\left(\int (\phi u^a)^2\right)^{(\mu-1)/\mu} \left(\int (\phi u^a)^{2\mu}\right)^{1/\mu} \geq \int (\phi u^a)^{2(2\mu-1)/\mu},$$

hence

$$\left(\sup_{s+v \leq t \leq T_1} \int_{B_p(\rho)} (\phi u^a)^2 dx\right)^{(\mu-1)/\mu} \int_{s+v}^{T_1} \left(\int_{B_p(\rho)} (\phi(x) u^a(x,t))^{2\mu} dx\right)^{1/\mu} dt$$

$$\geq \int_{s+v}^{T_1} \int_{B_p(\rho)} (\phi u^a)^{2(2\mu-1)/\mu}.$$

19 Parabolic Harnack inequality and regularity theory

Combining this inequality with (19.11) and (19.12), we obtain

$$2C_1 \left(\frac{C_{SD}}{\rho^2}\right)^{1-\alpha} (aA)^\alpha \int_{T_0}^{T_1} \int_{B_p(\rho)} \psi^2(t) \phi^2(x) u^{2a}(x,t) \, dx \, dt$$

$$+ 2 \int_{T_0}^{T_1} \int_{B_p(\rho)} \psi^2(t) |\nabla \phi|^2(x) u^{2a}(x,t) \, dx \, dt$$

$$+ 2 \int_{T_0}^{T_1} \int_{B_p(\rho)} \psi(t) \psi_t(t) \phi^2(x) u^{2a}(x,t) \, dx \, dt$$

$$\geq \left(\frac{C_{SD}}{\rho^2} \int_{s+v}^{T_1} \int_{B_p(\rho)} (\phi(x) u^a(x,t))^{2(2\mu-1)/\mu} \, dx \, dt\right)^{\mu/(2\mu-1)}. \qquad (19.13)$$

Let us now choose $\phi(r(x))$ to be a function of the distance to p, given by

$$\phi(r) = \begin{cases} 0 & \text{on} \quad B_p(\rho) \setminus B_p(\tau + \sigma), \\ \dfrac{\tau + \sigma - r}{\sigma} & \text{on} \quad B_p(\tau + \sigma) \setminus B_p(\tau), \\ 1 & \text{on} \quad B_p(\tau), \end{cases}$$

and (19.13) then yields

$$\left(2C_1(\rho^2 C_{SD}^{-1})^{\alpha-1}(aA)^\alpha + 2\sigma^{-2} + 2v^{-1}\right) (\rho^2 C_{SD}^{-1})^{\mu/(2\mu-1)}$$

$$\times \int_s^{T_1} \int_{B_p(\tau+\sigma)} u^{2a} \geq \left(\int_{s+v}^{T_1} \int_{B_p(\tau)} u^{2a(2\mu-1)/\mu}\right)^{\mu/(2\mu-1)} \qquad (19.14)$$

with $\alpha = q(\mu-1)/(\mu(q-1)-q)$ for all $m/2 < q \leq \infty$.

Let us now choose the sequences of a_i, τ_i, and σ_i such that

$$a_0 = \frac{k}{2}, \quad a_1 = \frac{k\nu}{2}, \quad \ldots, a_i = \frac{k\nu^i}{2}, \quad \ldots,$$

$$\sigma_0 = 2^{-1}(1-\theta)\rho, \quad \sigma_1 = 2^{-2}(1-\theta)\rho, \quad \ldots, \quad \sigma_i = 2^{-(1+i)}(1-\theta)\rho, \quad \ldots,$$

and

$$\tau_0 = \rho, \quad \tau_1 = \rho - \sigma_0, \quad \ldots, \quad \tau_i = \rho - \sum_{j=0}^{i-1} \sigma_j, \quad \ldots.$$

with $v = (2\mu - 1)/\mu$. We also choose the sequences of s_i and v_i, such that

$$s_0 = T_0, \quad s_1 = T_0 + 2^{-1}(T - T_0), \quad \ldots \quad s_i = T_0 + \sum_{j=1}^{i} 2^{-i}(T - T_0), \quad \ldots,$$

and

$$v_0 = 2^{-1}(T - T_0), \quad v_1 = 2^{-2}(T - T_0), \quad \ldots, \quad v_i = 2^{-(1+i)}(T - T_0), \quad \ldots.$$

Observe that $\lim_{i \to \infty} \rho_i = \theta\rho$ and $\lim_{i \to \infty} s_i = T$. Applying (19.14) to $a = a_i$, $\tau = \tau_{i+1}$, $\sigma = \sigma_i$, $s = s_i$, and $v = v_i$, we have

$$\|u\|_{2a_{i+1}, \tau_{i+1}, [s_{i+1}, T_1]} \leq \left(2^{1-\alpha}(kA)^\alpha C_1(\rho^2 C_{SD}^{-1})^{\alpha - 1 + 1/v} v^{i\alpha} \right.$$
$$+ (2^{3+2i}(1-\theta)^{-2} \rho^{-2}$$
$$\left. + 2^{2+i}(T-T_0)^{-1})(\rho^2 C_{SD}^{-1})^{1/v} \right)^{1/kv^i} \|u\|_{2a_i, \tau_i, [s_i, T_1]}$$

with

$$\|u\|_{2a_i, \tau_i, [s_i, T_1]} = \left(\int_{s_i}^{T_1} \int_{B_\rho(\tau_i)} u^{2a_i} \right)^{1/2a_i}.$$

Iterating this inequality, we conclude that

$$\|u\|_{2a_{i+1}, \tau_{i+1}, [s_{i+1}, T_1]} \leq \prod_{j=0}^{i} \left(2^{1-\alpha}(kA)^\alpha C_1(\rho^2 C_{SD}^{-1})^{\alpha - 1 + 1/v} v^{j\alpha} \right.$$
$$+ (2^{3+2j}(1-\theta)^{-2} \rho^{-2}$$
$$\left. + 2^{2+j}(T-T_0)^{-1})(\rho^2 C_{SD}^{-1})^{1/v} \right)^{1/kv^j}$$
$$\times \|u\|_{2a_0, \tau_0, [s_0, T_1]}.$$

On the other hand, we have the inequality

$$\lim_{i \to \infty} ((T_1 - T) V(\theta\rho))^{-1/2a_{i+1}} \|u\|_{2a_{i+1}, \tau_{i+1}, [s_{i+1}, T_1]}$$
$$\geq \lim_{i \to \infty} ((T_1 - T) V(\theta\rho))^{-1/2a_{i+1}} \|u\|_{2a_{i+1}, \theta\rho, [T, T_1]}$$
$$= \|u\|_{\infty, \theta\rho, [T, T_1]},$$

where the right-hand side is defined to be the supremum of u over the set $B_p(\theta\rho) \times [T, T_1]$. Therefore, letting $i \to \infty$, we conclude that

$$\|u\|_{\infty,\theta\rho,[T,T_1]} \leq \prod_{j=0}^{\infty} \left(2^{1-\alpha}(kA)^{\alpha} C_1 (\rho^2 C_{SD}^{-1})^{\alpha-1+1/\nu} \nu^{j\alpha}\right.$$

$$+ (2^{3+2j}(1-\theta)^{-2}\rho^{-2}$$

$$\left. + 2^{2+j}(T-T_0)^{-1})(\rho^2 C_{SD}^{-1})^{1/\nu}\right)^{1/k\nu^j} \|u\|_{k,\rho,[T_0,T_1]}.$$

(19.15)

The product can be estimated by using the fact that $\nu = (2\mu - 1)/\mu > 1$, hence we obtain the identity

$$\prod_{j=0}^{\infty} B^{\nu^{-j}} = B^{\nu/(\nu-1)},$$

and the fact that $\sum_{j=0}^{\infty} j\mu^{-j}$ is finite. Therefore we have

$$\prod_{j=0}^{\infty} \left(2^{1-\alpha}(kA)^{\alpha} C_1 (\rho^2 C_{SD}^{-1})^{\alpha-1+1/\nu} \nu^{j\alpha} + (2^{3+2j}(1-\theta)^{-2}\rho^{-2}\right.$$

$$\left. + 2^{2+j}(T-T_0)^{-1}) \left(\rho^2 C_{SD}^{-1}\right)^{1/\nu}\right)^{1/k\nu^j}$$

$$\leq \prod_{j=0}^{\infty} \left(2^{1-\alpha}(kA)^{\alpha} C_1 (\rho^2 C_{SD}^{-1})^{\alpha-1+1/\nu} + ((1-\theta)^{-2}\rho^2\right.$$

$$\left. + (T-T_0)^{-1})(\rho^2 C_{SD}^{-1})^{1/\nu}\right)^{1/k\nu^j} \max\{\nu^{\alpha}, 4\}^{j/k\nu^j}$$

$$\leq C_2 \left((kA\rho^2 C_{SD}^{-1})^{\alpha} (\rho^2 C_{SD}^{-1})^{-(\nu-1)/\nu} + ((1-\theta)^{-2}\rho^{-2}\right.$$

$$\left. + (T-T_0)^{-1})(\rho^2 C_{SD}^{-1})^{1/\nu}\right)^{\nu/k(\nu-1)},$$

where $C_2 \geq 0$ depends only on k, μ, and q. Inequality (19.15) now takes the form

$$\|u\|_{\infty,\theta\rho,[T,T_1]} \leq C_2 \left((kA\rho^2 C_{\mathcal{SD}}^{-1})^\alpha \left(\frac{C_{\mathcal{SD}}}{\rho^2} \right)^{(\nu-1)/\nu} + ((1-\theta)^{-2}\rho^{-2} \right.$$

$$+ (T-T_0)^{-1}) \left(\frac{\rho^2}{C_{\mathcal{SD}}} \right)^{1/\nu} \right)^{\nu/k(\nu-1)}$$

$$\times (T_1 - T_0)^{1/k} \#u\|_{k,\rho,[T_0,T_1]} \qquad (19.16)$$

for $k \geq 2$, where

$$\#u\|_{k,\rho,[T_0,T_1]} = \left((T_1 - T_0)^{-1} V_p^{-1}(\rho) \int_{T_0}^{T_1} \int_{B_p(\rho)} u^k \right)^{1/k}.$$

We now claim that (19.16) holds for $1 < k < 2$ also. To see this, we observe that the iteration process begins with $a = k/2$, and hence for $k > 1$, we have $2a - 1 > 0$ in (19.2). Using the inequality

$$2\int \phi u^{2a-1} \langle \nabla\phi, \nabla u \rangle \geq -a^{-1}(2a-1)^{-1} \int |\nabla\phi|^2 u^{2a}$$

$$- a(2a-1) \int \phi^2 u^{2a-2} |\nabla u|^2$$

and combining it with (19.2), we conclude that

$$-\int \phi^2 u^{2a-1} \Delta u = 2\delta \int \phi u^{2a-1} \langle \nabla\phi, \nabla u \rangle + 2(1-\delta) \int \phi u^{2a-1} \langle \nabla\phi, \nabla u \rangle$$

$$+ (2a-1) \int \phi^2 u^{2a-2} |\nabla u|^2$$

$$\geq -2\delta a^{-1}(2a-1)^{-1} \int |\nabla\phi|^2 u^{2a}$$

$$+ 2(1-\delta) \int \phi u^{2a-1} \langle \nabla\phi, \nabla u \rangle$$

$$+ (2a - 1 - \delta a(2a-1)) \int \phi^2 u^{2a-2} |\nabla u|^2.$$

Setting $\delta = (2a)^{-1}$ and using (19.3), we have

$$-\int \phi^2 u^{2a-1} \Delta u \geq -\left(\frac{1}{a^2(2a-1)} + \frac{2a-1}{2a^2}\right)\int |\nabla \phi|^2 u^{2a}$$
$$+ \frac{2a-1}{2a^2}\int |\nabla(\phi u^a)|^2,$$

hence

$$\int \phi^2 f u^{2a} + \left(\frac{1}{a^2(2a-1)} + \frac{2a-1}{2a^2}\right)\int |\nabla \phi|^2 u^{2a}$$
$$\geq \frac{2a-1}{2a^2}\int |\nabla(\phi u^a)|^2 + \frac{1}{2a}\int \phi^2 (u^{2a})_t$$
$$\geq \frac{(2a-1)C_{SD}}{2a^2 \rho^2}\left(\int (\phi u^a)^{2\mu}\right)^{1/\mu} + \frac{1}{2a}\int \phi^2 (u^{2a})_t.$$

Using this inequality instead of (19.4) for values of $\frac{1}{2} < a < 1$, the same argument as above yields (19.16) for $k > 1$.

For $k \leq 1$, we begin with the inequality case $k = 2$. In that case, (19.16) yields

$$\|u\|_{\infty,\gamma\tau,[S,T_1]} \leq C_2 \left(\left(2A\frac{\tau^2}{C_{SD}}\right)^\alpha \left(\frac{C_{SD}}{\tau^2}\right)^{(\nu-1)/\nu}\right.$$
$$+ \left(\frac{\tau^2}{C_{SD}}\right)^{1/\nu}\left.((1-\gamma)^{-2}\tau^{-2} + (1-\xi)^{-1}S^{-1})\right)^{\nu/2(\nu-1)}$$
$$\times (T_1 - \xi S)^{1/2} \#\|u\|_{2,\tau,[\xi S,T_1]}$$
$$\leq C_2 \left(\left(2A\frac{\rho^2}{C_{SD}}\right)^\alpha \left(\frac{C_{SD}}{\rho^2}\right)^{(\nu-1)/\nu}\right.$$
$$+ C_{SD}^{-1/\nu}(1-\gamma)^{-2}\tau^{-2(\nu-1)/\nu}$$
$$+ \left(\frac{\rho^2}{C_{SD}}\right)^{1/\nu}(1-\xi)^{-1}S^{-1}\right)^{\nu/2(\nu-1)}$$
$$\times (T_1 - \xi S)^{1/2} \#\|u\|_{k,\tau,[\xi S,T_1]}^{k/2} \|u\|_{\infty,\tau,[\xi S,T_1]}^{1-k/2}, \quad (19.17)$$

for any $\theta\rho \leq \tau \leq \rho$, $0 < \gamma < 1$, $T_0 \leq S \leq T$ and $0 < \xi < 1$. Let us choose the sequences of τ_i and γ_i to be

$$\tau_{-1} = \theta\rho, \quad \tau_0 = \theta\rho + 2^{-1}(1-\theta)\rho, \ldots, \quad \tau_{i-1} = \theta\rho + (1-\theta)\rho \sum_{j=1}^{i} 2^{-j}, \ldots$$

and

$$\gamma_i \tau_i = \tau_{i-1}.$$

Also, the sequences S_i and ξ_i are chosen to be

$$S_0 = T, \quad S_1 = (1 - 2^{-1})(T - T_0) + T_0, \ldots,$$

$$S_i = \left(1 - \sum_{j=1}^{i} 2^{-j}\right)(T - T_0) + T_0, \ldots$$

and

$$\xi_i S_i = S_{i+1}.$$

Applying (19.17) by setting $\tau = \tau_i$, $\gamma = \gamma_i$, $S = S_i$, and $\xi = \xi_i$ and iterating the inequality yields

$$\|u\|_{\infty,\theta\rho,[T,T_1]} \leq \prod_{i=0}^{j} C_2 \left(\left(2A\frac{\rho^2}{C_{SD}}\right)^{\alpha} \left(\frac{C_{SD}}{\rho^2}\right)^{(\nu-1)/\nu}\right.$$

$$+ C_{SD}^{-1/\nu}(1-\gamma_i)^{-2}\tau_i^{-2(\nu-1)/\nu}$$

$$+ \left(\frac{\rho^2}{C_{SD}}\right)^{1/\nu}(S_i - S_{i+1})^{-1}\right)^{\nu/2(\nu-1)(1-k/2)^i}$$

$$\times (T_1 - S_{i+1})^{(1-k/2)^i/2}$$

$$\times \|u\|_{\infty,\rho,[T_0,T_1]}^{(1-k/2)^{j+1}} \prod_{i=0}^{j} \left(\#u\|_{k,\tau_i,[S_{i+1},T_1]}^{k/2}\right)^{(1-k/2)^i}. \quad (19.18)$$

We observe that since

$$(1 - \gamma_i) = \frac{\tau_i - \tau_{i-1}}{\tau_i}$$

$$\geq 2^{-(i+1)}(1-\theta),$$

we have
$$(1-\gamma_i)^{-2}\tau_i^{-2(\nu-1)/\nu} \leq 4^{(i+1)}(1-\theta)^{-2}\rho^{-2(\nu-1)/\nu}.$$

Also, since
$$S_i - S_{i+1} = 2^{-(i+1)}(T-T_0),$$

(19.18) becomes
$$\|u\|_{\infty,\theta\rho,[T,T_1]} \leq \prod_{i=0}^{j} C_2 \left(\left(2A\frac{\rho^2}{C_{SD}}\right)^\alpha \left(\frac{C_{SD}}{\rho^2}\right)^{(\nu-1)/\nu} \right.$$
$$+ C_{SD}^{-1/\nu}(1-\theta)^{-2}\rho^{-2(\nu-1)/\nu}$$
$$\left. + C_{SD}^{-1/\nu}\rho^{2/\nu}(T-T_0)^{-1} \right)^{(\nu/2(\nu-1))(1-k/2)^i}$$
$$\times \prod_{i=0}^{j} 2^{((i+1)\nu/\nu-1)(1-k/2)^i}(T_1-T_0)^{1/k}\|u\|_{\infty,\rho,[T_0,T_1]}^{(1-k/2)^{j+1}}$$
$$\times \prod_{i=0}^{j} \left(\#u\|_{k,\tau_i,[S_{i+1},T_1]}^{k/2}\right)^{(1-k/2)^i}.$$

Letting $j \to \infty$ and using the facts that
$$\prod_{i=0}^{\infty} 2^{((i+1)\nu/\nu-1)(1-k/2)^i} < \infty,$$
$$\|u\|_{\infty,\rho,[T_0,T_1]}^{(1-k/2)^{j+1}} \to 1 \text{ as } j \to \infty,$$

and
$$\#u\|_{k,\tau_i,[S_{i+1},T_1]}^{k/2} \leq ((T_1-S_1)V_p(\tau_0))^{-1/2}\|u\|_{k,\rho,[T_0,T_1]}^{k/2}$$
$$\leq \left(\frac{T_1-T_0}{2}V_p\left(\frac{\rho}{2}\right)\right)^{-1/2}\|u\|_{k,\rho,[T_0,T_1]}^{k/2},$$

we obtain

$$\|u\|_{\infty,\theta\rho,|T,T_1|} \leq C_3 \left(\left(2A \frac{\rho^2}{C_{SD}} \right)^{\alpha} \left(\frac{C_{SD}}{\rho^2} \right)^{(\nu-1)/\nu} \right.$$

$$+ C_{SD}^{-1/\nu} (1-\theta)^{-2} \rho^{-2(\nu-1)/\nu}$$

$$\left. + C_{SD}^{-1/\nu} \rho^{2/\nu} (T-T_0)^{-1} \right)^{\nu/k(\nu-1)}$$

$$\times V_\rho^{-1/k} \left(\frac{\rho}{2} \right) \|u\|_{k,\rho,|T_0,T_1|},$$

proving the desired inequality for $k \leq 1$. □

To complete the proof of the Harnack inequality, we will need the following lemma.

Lemma 19.2 *Let g be a positive measurable function defined on a set $D \subset M$. Suppose $\{D_\sigma \mid 0 < \sigma \leq 1\}$ is a family of measurable subsets of D with the properties that $D_{\sigma'} \subset D_\sigma$ for $0 < \sigma' \leq \sigma \leq 1$ and $D_1 = D$. Let us assume that there are constants $C_4 > 0$, $\gamma > 0$, $0 < \delta < 1$, and $0 < k_0 < \infty$ so that the inequality*

$$\left(\int_{D_{\sigma'}} g^{k_0} \right)^{1/k_0} \leq \left(C_4 (\sigma - \sigma')^{-\gamma} V^{-1}(D) \right)^{1/k - 1/k_0} \left(\int_{D_\sigma} g^k \right)^{1/k}$$

is valid for all $\delta \leq \sigma' \leq \sigma \leq 1$ and $0 < k \leq k_0$. If there exists a constant $C_5 > 0$ such that

$$m(\{x \in D \mid \ln g > \lambda\}) \leq C_5 V(D) \lambda^{-1}$$

for all $\lambda > 0$, then there exists a constant $C_6 > 0$ depending only on γ, δ, C_4 and C_5, such that

$$\left(\int_{D_\delta} g^{k_0} \right)^{1/k_0} \leq C_6 V^{1/k_0}(D).$$

Proof Let us normalize the volume of D to be $V(D) = 1$ and define the function

$$\psi(\sigma) = \ln \left(\left(\int_{D_\sigma} g^{k_0} \right)^{1/k_0} \right)$$

19 Parabolic Harnack inequality and regularity theory

for $\delta \le \sigma < 1$. We may assume that

$$\psi(\sigma) \ge 2C_5,$$

otherwise the lemma follows because of the monotonicity of ψ and by choosing $C_6 = \exp(2C_5)$.

We first claim that

$$\psi(\sigma') \le \tfrac{3}{4}\psi(\sigma) + 8C_4^2 C_5 (\sigma - \sigma')^{-2\gamma} \quad (19.19)$$

for all $\sigma' < \sigma$. To see this, we only need to verify (19.19) for those σ' such that

$$\psi(\sigma') \ge 8C_4^2 C_5 (\sigma - \sigma')^{-2\gamma}. \quad (19.20)$$

Let us consider

$$\left(\int_{D_\sigma} g^k\right)^{1/k} = \left(\int_{D_\sigma \cap \{\ln g > \psi(\sigma)/2\}} g^k\right)^{1/k} + \left(\int_{D_\sigma \cap \{\ln g \le \psi(\sigma)/2\}} g^k\right)^{1/k}$$

$$\le \left(\int_{D_\sigma} g^{k_0}\right)^{1/k_0} m\left(D_\sigma \cap \left\{\ln g > \frac{\psi(\sigma)}{2}\right\}\right)^{1/k - 1/k_0}$$

$$+ \exp\left(\frac{\psi(\sigma)}{2}\right)$$

$$\le \exp(\psi(\sigma))(2C_5 \psi(\sigma)^{-1})^{1/k - 1/k_0} + \exp\left(\frac{\psi(\sigma)}{2}\right), \quad (19.21)$$

where we have used the fact that $V(D_\sigma) \le 1$. Note that we may choose $k \le k_0$ to satisfy

$$\left(\frac{1}{k} - \frac{1}{k_0}\right)^{-1} = \frac{2}{\psi(\sigma)} \ln\left(\frac{\psi(\sigma)}{2C_5}\right),$$

hence combining this identity with (19.21), we obtain

$$\left(\int_{D_\sigma} g^k\right)^{1/k} \le 2\exp\left(\frac{\psi(\sigma)}{2}\right).$$

On the other hand, the assumption of the lemma asserts that

$$\psi(\sigma') \leq \ln\left((C_4 (\sigma - \sigma')^{-\gamma})^{1/k - 1/k_0} \left(\int_{D_\sigma} g^k\right)^{1/k}\right)$$

$$\leq \left(\frac{1}{k} - \frac{1}{k_0}\right) \ln(2C_4(\sigma - \sigma')^{-\gamma}) + \frac{\psi(\sigma)}{2}$$

$$\leq \frac{\psi(\sigma)}{2} \left(\frac{\ln(2C_4(\sigma - \sigma')^{-\gamma})}{\ln\left(\frac{\psi(\sigma)}{2C_5}\right)} + 1\right)$$

$$\leq \frac{3\psi(\sigma)}{4}$$

because of (19.20) and the monotonicity of ψ. The claim (19.19) is then validated.

For some $\beta > 1$, let us now choose a sequence of $\{\sigma_i\}$ with $\sigma_0 = \delta$, and $\sigma_{i+1} = \sigma_i + \beta^{-i-1}(1 - \delta)(\beta - 1)$ for all $i \geq 0$. Inequality (19.19) asserts that

$$\psi(\sigma_i) \leq \tfrac{3}{4} \psi(\sigma_{i+1}) + 8C_4^2 C_5 (1 - \delta)^{-2\gamma} (\beta - 1)^{-2\gamma} \beta^{2\gamma(i+1)}$$

for all $i \geq 0$. Iterating this inequality yields

$$\psi(\sigma_0) \leq \left(\tfrac{3}{4}\right)^j \psi(\sigma_j) + 8C_4^2 C_5 (1 - \delta)^{-2\gamma} (\beta - 1)^{-2\gamma} \sum_{i=1}^{j} \left(\tfrac{3}{4}\right)^{i-1} \beta^{2\gamma i}.$$

Using the fact that $\sigma_j = (1 - \delta)(\beta - 1) \sum_{i=1}^{j} \beta^{-i} \leq (1 - \delta)$, and letting $j \to \infty$, we conclude that

$$\psi(\sigma_0) \leq 8C_4^2 C_5 (1 - \delta)^{-2\gamma} (\beta - 1)^{-2\gamma} \beta^{2\gamma} \sum_{i=0}^{\infty} \left(\frac{3\beta^{2\gamma}}{4}\right)^i.$$

By choosing $1 < \beta$ and $3\beta^{2\gamma}/4 < 1$, we deduce that the right-hand side is finite and the lemma is proved. □

Theorem 19.3 *Let M be a complete manifold. Suppose that the geodesic ball $B_p(\rho)$ centered at p with radius ρ satisfies $B_p(\rho) \cap \partial M = \emptyset$. Let $u \geq 0$ be a function defined on $B_p(\rho) \times [0, T]$, satisfying the equation*

$$\left(\Delta - \frac{\partial}{\partial t}\right) u = fu$$

19 Parabolic Harnack inequality and regularity theory

in the weak sense for some function f on $B_p(\rho)$ whose supremum norm is given by

$$A = \|f\|_{\infty, \rho, [0, T]}.$$

Let $\nu = m/2$ for $m > 2$, and let $1 < \nu < \infty$ be arbitrary when $m = 2$. If $T \geq \rho^2$, then for any $0 < \theta < 1$, there exists a constant $C_{16} > 0$, depending only on θ, $C_{\mathcal{SD}}$, $C_{\mathcal{VD}}$, and $C_{\mathcal{P}}$, such that

$$\sup_{B_p(\theta\rho) \times [T - 3\rho^2/4, T - \rho^2/2]} u \leq C_{16} \exp\left(\frac{A\rho^2}{2}\right) \inf_{B_p(\theta\rho) \times [T - \rho^2/4, T]} u.$$

Proof Let us first prove the theorem for the case $A = 0$. We define the function $w = -\ln u$, then it satisfies the equation

$$\Delta w = \frac{\partial w}{\partial t} + |\nabla w|^2.$$

Let us point out that in the argument that follows we only need the inequality

$$\Delta w \geq \frac{\partial w}{\partial t} + |\nabla w|^2,$$

hence for any compactly supported nonnegative smooth function $\phi \in C_c^\infty(B_p(\rho))$, we have

$$\frac{\partial}{\partial t} \int_{B_p(\rho)} \phi^2 w \leq \int_{B_p(\rho)} \phi^2 \Delta w - \int_{B_p(\rho)} \phi^2 |\nabla w|^2$$

$$= -2 \int_{B_p(\rho)} \phi \langle \nabla \phi, \nabla w \rangle - \int_{B_p(\rho)} \phi^2 |\nabla w|^2$$

$$\leq -\frac{1}{2} \int_{B_p(\rho)} \phi^2 |\nabla w|^2 + 2 \int_{B_p(\rho)} |\nabla \phi|^2.$$

In particular, choosing ϕ to be

$$\phi(r) = \begin{cases} 1 & \text{if } r \leq \delta\rho, \\ \dfrac{\rho - r}{(1 - \delta)\rho} & \text{if } \delta\rho \leq r \leq \rho, \\ 0 & \text{if } \rho \leq r, \end{cases}$$

we obtain

$$\frac{\partial}{\partial t} \int_{B_p(\rho)} \phi^2 w + \frac{1}{2} \int_{B_p(\rho)} \phi^2 |\nabla w|^2 \leq (1 - \delta)^{-2} \rho^{-2} V_p(\rho). \quad (19.22)$$

However, Theorem 18.3 asserts that there exists a constant $C_7 > 0$, depending on $C_\mathcal{P}$, $C_{\mathcal{VD}}$, and δ, such that

$$\int_{B_p(\rho)} |\nabla w|^2 \phi^2 \geq C_7 \rho^{-2} \int_{B_p(\rho)} |w - \bar{w}|^2 \phi^2,$$

where $\bar{w} = V_\phi^{-1}(B_p(\rho)) \int_{B_p(\rho)} w \phi^2$ denotes the average of w over $B_p(\rho)$ with respect to the measure $\phi^2 \, dV$. Taking the above inequality together with (19.22), we have

$$\frac{\partial \bar{w}}{\partial t} = V_\phi^{-1}(B_p(\rho)) \frac{\partial}{\partial t} \int_{B_p(\rho)} w \phi^2$$

$$\leq -C_7 \rho^{-2} V_\phi^{-1}(B_p(\rho)) \int_{B_p(\rho)} |w - \bar{w}|^2 \phi^2$$

$$+ (1 - \delta)^{-2} \rho^{-2} V_p(\rho) V_\phi^{-1}(B_p(\rho))$$

$$\leq -C_7 \rho^{-2} V_p^{-1}(\rho) \int_{B_p(\delta\rho)} |w - \bar{w}|^2 + C_8 \rho^{-2}, \qquad (19.23)$$

because of

$$V_p(\delta\rho) \leq V_\phi(B_p(\rho)) \leq V_p(\rho).$$

Let us define

$$v(x, t) = w(x, t) - C_8(t - s') \rho^{-2}$$

and

$$\bar{v}(t) = \bar{w}(t) - C_8(t - s') \rho^{-2}$$

for a fixed $s' \in [T - \rho^2/2, T - \rho^2/4]$. We can rewrite (19.23) as

$$\frac{\partial \bar{v}}{\partial t} + C_7 \rho^{-2} V_p^{-1}(\rho) \int_{B_p(\delta\rho)} |v - \bar{v}|^2 \leq 0. \qquad (19.24)$$

For any $\lambda > 0$, we define

$$\Omega_\lambda^+(t) = \{x \in B_p(\delta\rho) \mid v(x, t) > a + \lambda\}$$

and

$$\Omega_\lambda^-(t) = \{x \in B_p(\delta\rho) \mid v(x, t) < a - \lambda\},$$

19 Parabolic Harnack inequality and regularity theory

where $a = \bar{v}(s') = \bar{w}(s')$. Since $(\partial/\partial t)\bar{v} \leq 0$, we have

$$v(x,t) - \bar{v}(t) > a + \lambda - \bar{v}(t)$$
$$\geq \lambda$$

when $x \in \Omega_\lambda^+(t)$ and $t > s'$. Combining this inequality with (19.24), this implies that

$$\frac{\partial}{\partial t}\bar{v}(t) + C_7 \rho^{-2} V_p^{-1}(\rho) (a + \lambda - \bar{v}(t))^2 V\left(\Omega_\lambda^+(t)\right) \leq 0$$

and

$$-C_7^{-1} \rho^2 V_p(\rho) \frac{\partial}{\partial t}\left(|a + \lambda - \bar{v}|^{-1}\right) \geq V\left(\Omega_\lambda^+(t)\right)$$

for $t > s'$. Integrating over the interval (s', T) yields

$$C_7^{-1} \rho^2 V_p(\rho)(\lambda^{-1} - |a + \lambda - \bar{v}(T)|^{-1})$$

$$\geq \int_{s'}^{T} V(\Omega_\lambda^+(t))$$

$$= m\{(x,t) \in B_p(\delta\rho) \times (s', T) \mid w(x,t) > a + \lambda + C_8 (t - s') \rho^{-2}\},$$

(19.25)

where $m\{D\}$ denotes the measure with respect to the product metric of the set $D \subset M \times \mathbb{R}$. If $\lambda \geq C_8 (T - s') \rho^{-2}$, then (19.25) implies

$$C_7^{-1} \rho^2 V_p(\rho) \lambda^{-1}$$
$$\geq m\{(x,t) \in B_p(\delta\rho) \times (s', T) \mid w(x,t) > a + \lambda + C_8 (t - s') \rho^{-2}\}$$
$$\geq m\{(x,t) \in B_p(\delta\rho) \times (s', T) \mid w(x,t) > a + 2\lambda\}.$$

(19.26)

On the other hand, if $0 < \lambda < C_8 (T - s') \rho^{-2}$, then we can estimate

$$m\{(x,t) \in B_p(\delta\rho) \times (s', T) \mid w(x,t) > a + 2\lambda\} \leq (T - s') V_p(\rho)$$
$$\leq C_8(T - s')^2 \rho^{-2} \lambda^{-1} V_p(\rho).$$

However, using the fact that

$$T - s' \leq \frac{\rho^2}{2},$$

together with (19.26), we conclude that

$$m\{(x,t) \in B_p(\delta\rho) \times (s', T) \mid w(x,t) > a + \lambda\} \leq C_9 \rho^2 V_p(\rho) \lambda^{-1}.$$
(19.27)

When $t \leq s'$, again using (19.24), we have

$$v(x,t) - \bar{v}(t) \leq a - \lambda - \bar{v}(t)$$
$$= -\lambda$$

on $\Omega_\lambda^-(t)$, hence

$$\frac{\partial}{\partial t}\bar{v} + C_7 \rho^{-2} V_p^{-1}(\rho) (a - \lambda - \bar{v}(t))^2 V\left(\Omega_\lambda^-(t)\right) \leq 0$$

and

$$V\left(\Omega_\lambda^-(t)\right) \leq -C_7 \rho^2 V_p(\rho) \frac{\partial}{\partial t}(|a - \lambda - \bar{v}|^{-1}).$$

Integrating from $T - \rho^2$ to s' we obtain

$$m\{(x,t) \in B_p(\delta\rho) \times (T - \rho^2, s') \mid w(x,t) < a - \lambda + C_8 (t - s') \rho^{-2}\}$$
$$= \int_{T-\rho^2}^{s'} V\left(\Omega_\lambda^-(t)\right) dt$$
$$\leq C_7 \rho^2 V_p(\rho) (\lambda^{-1} - |a - \lambda - \bar{v}(T - \rho^2)|^{-1}$$
$$\leq C_7 \rho^2 V_p(\rho) \lambda^{-1},$$

hence

$$m\{(x,t) \in B_p(\delta\rho) \times (T - \rho^2, s') \mid w(x,t) < a - \lambda\} \leq C_7 \rho^2 V_p(\rho) \lambda^{-1}.$$
(19.28)

We will now apply Lemma 19.2 to the function

$$g = e^a u.$$

In particular, (19.28) asserts that

$$m\{(x,t) \in D \mid \ln g > \lambda\} \leq C_9 \rho^2 V_p(\rho) \lambda^{-1}$$
$$\leq C_{10} V(D) \lambda^{-1}$$

with $D = B_p(\delta\rho) \times (T - \rho^2, s')$, where C_{10} depends on C_9, $C_{\mathcal{VD}}$, and δ. Let us define the subdomains $D_\sigma \subset D$ by

$$D_\sigma = B_p(\sigma\delta\rho) \times (T - \sigma\rho^2), s').$$

Obviously, $D_{\sigma'} \subset D_\sigma$ for $\frac{1}{4} < \sigma' \leq \sigma \leq 1$ with $D_1 = D$.

Applying Lemma 19.1 to the domain $D_{\sigma'}$ and D_σ we have

$$\|g\|_{\infty,\sigma'\delta\rho,[T-\sigma'\rho^2,s']} \leq C_{11} ((\sigma - \sigma')\delta\rho)^{-2} + (\sigma - \sigma')^{-1}\rho^{-2})^{\nu/k(\nu-1)}$$
$$\times (\sigma\delta\rho)^{2/k(\nu-1)} V_p^{-1/k}(\sigma\delta\rho) \|g\|_{k,\sigma\delta\rho,[T-\sigma\rho^2,s']}$$
$$\leq C_{11} (\sigma - \sigma')^{-2\nu/k(\nu-1)} (\delta\rho)^{-2/k}$$
$$\times V_p^{-1/k}(\sigma\delta\rho) \|g\|_{k,\sigma\delta\rho,[T-\sigma\rho^2,s']}, \quad (19.29)$$

where $C_{11} > 0$ depends on k, $\nu = (2\mu - 1)/\mu$, δ, $C_{\mathcal{SD}}$, $C_{\mathcal{P}}$ and $C_{\mathcal{VD}}$. Note that for $\sigma \geq \frac{1}{4}$, the volume doubling property implies that

$$V(D) = \rho^2 V_p(\delta\rho)$$
$$\leq C_{\mathcal{VD}}^2 \rho^2 V_p(\sigma\delta\rho)$$

for $k_0 \geq k$, and (19.29) asserts that

$$\left(\fint_{D_{\sigma'}} g^{k_0}\right)^{1/k_0} \leq C_{12} \left((\sigma - \sigma')^{-2k_0\nu/(k_0-k)(\nu-1)} V(D)^{-1}\right)^{1/k - 1/k_0}$$
$$\times \left(\fint_{D_\sigma} g^k\right)^{1/k},$$

where $C_{12} > 0$ depends on k, ν, δ, $C_{\mathcal{SD}}$, $C_{\mathcal{P}}$ and $C_{\mathcal{VD}}$. Lemma 19.1 now asserts that

$$\left(\fint_{D_{\frac{3}{4}}} g^{k_0}\right)^{1/k_0} \leq C_{13}, \quad (19.30)$$

where $C_{13} > 0$ depends on k, ν, δ, $C_{\mathcal{SD}}$, and $C_{\mathcal{VD}}$. On the other hand, applying (19.29) again, we have

$$\|g\|_{\infty,\delta\rho/2,[T-\rho^2/4,s']} \leq C_{14} (\delta\rho)^{-2/k_0} V_p^{-1/k_0}\left(\frac{3\delta\rho}{4}\right) \|g\|_{k_0,3\delta\rho/4,[T-3\rho^2/4,s']},$$

which when combined with (19.30) yields

$$\|g\|_{\infty,\delta\rho/2,[T-\rho^2/4,s']} \leq C_{15}\, e^{-a}. \tag{19.31}$$

where $C_{15} > 0$ depends on ν, δ, $C_{\mathcal{SD}}$, C_P and C_V.

Similarly, using (19.27), we apply Lemma 19.2 to the function

$$g = e^{-a} u^{-1}.$$

As above, we use Lemma 19.1 to verify

$$\left(\int_{D_{\sigma'}} g^{-k_0}\right)^{1/k_0} \leq C_{12}\left((\sigma-\sigma')^{-2k_0\nu/(k_0-k)(\nu-1)} V(D)^{-1}\right)^{1/k-1/k_0}$$
$$\times \left(\int_{D_\sigma} g^{-k}\right)^{1/k},$$

hence the same argument implies that

$$\|u^{-1}\|_{\infty,\delta\rho/2,[s'+\rho^2/4,T]} \leq C_{15}\, e^{a},$$

which can be rewritten as

$$e^{-a} \leq C_{15} \inf_{B_p(\delta\rho/2)\times[s'+\rho^2/4,T]} u. \tag{19.32}$$

The Harnack inequality follows by combining this with (19.31) and choosing $\theta = \delta/2$ and $s' = T - \rho^2/2$.

When f is not identically 0, we observe that

$$\left(\Delta - \frac{\partial}{\partial t}\right)(e^{-At} u) = f e^{-At} u + A e^{-At} u$$
$$\geq 0$$

and

$$\left(\Delta - \frac{\partial}{\partial t}\right)(e^{-At} u^{-1}) = f e^{-At} u^{-1} - e^{-At}|\nabla u|^2 u^{-2} + A e^{-At} u^{-1}$$
$$\leq 0.$$

Hence we can apply (19.31) and (19.32) to the functions $e^{-At} u$ and $e^{-At} u^{-1}$ and obtain

$$e^{-AT}\|u\|_{\infty,\delta\rho/2,[T-\rho^2/2,T]} \leq C_{15}\, e^{-a}$$

and
$$e^{-a} \le C_{15} \, e^{-A\left(T-\frac{\rho^2}{2}\right)} \inf_{B_p(\delta\rho/2) \times [T-\rho^2/2, T]} u,$$

respectively. In particular, we conclude that

$$\sup_{B_p(\delta\rho/2) \times [T-\rho^2/2, T]} u \le C_{15}^2 \, e^{A\rho^2/2} \inf_{B_p(\delta\rho/2) \times [T-\rho^2/2, T]} u. \qquad \square$$

The following corollary follows directly from Theorem 19.3. The case when M is quasi-isometric to $\mathbb{R}^m$ follows readily from the Nash–Moser theory. The general case when M is quasi-isometric to a complete manifold with nonnegative Ricci curvature was first proved by Saloff-Coste [SC] and Grigor'yan [G3].

Corollary 19.4 *Let M be a manifold of dimension m. Suppose that ds^2 is a complete metric on M such that there is a point $p \in M$ and, for all ρ, the quantities $C_\mathcal{P}, C_{\mathcal{SD}},$ and $C_{\mathcal{VD}}$ are all bounded and positive independent of ρ. Then for any metric $\overline{ds}^2$ (not necessarily continuous) on M which is uniformly equivalent to ds^2, there does not exist any nonconstant positive harmonic functions for the Laplacian with respect to $\overline{ds}^2$. In particular, any manifold which is quasi-isometric to a complete manifold with nonnegative Ricci curvature does not admit a nonconstant positive harmonic function.*

Proof To see this, we first observe that $C_\mathcal{P}, C_{\mathcal{SD}},$ and $C_{\mathcal{VD}}$ are quasi-isometric invariants. Since a harmonic function is a stationary solution to the heat equation, Theorem 19.3 implies that any positive harmonic function u defined on M must satisfies the Harnack inequality

$$\sup_{B_p(\rho/4)} u \le C \inf_{B_p(\rho/4)} u$$

for any $\rho > 0$. On the other hand, since u is positive, by translation, we may assume that $\inf_M u = 0$. Hence, by taking $\rho \to \infty$, we conclude that

$$\sup_M u \le C \inf_M u = 0.$$

Therefore, u must be identically 0. $\qquad \square$

The next corollary yields a Hölder regularity result for weak solutions. One important point of this is that the metric $\overline{ds}^2$ need not be continuous as long as it is uniformly bounded with respect to some background metric ds^2.

Corollary 19.5 *Let M be a complete manifold with metric ds^2. Suppose $\overline{ds}^2$ is uniformly equivalent to ds^2. Suppose u is a weak solution to the equation*

$$\left(\bar{\Delta} - \frac{\partial}{\partial t}\right) u = fu$$

defined on $B_p(1) \times [T-1, T]$ with respect to the metric $\overline{ds}^2$. Assume that f is bounded on $B_p(1) \times [T-1, T]$, then u must be Hölder continuous at the point (p, T_0) for all $T - 1 < T_0 \leq T$.

Proof Let ρ be a constant such that $0 < \rho \leq T_0 - T + 1$. We define $s(\rho) = \sup_{B_p(\rho) \times [T_0 - \rho^2, T_0]} u$ and $i(\rho) = \inf_{B_p(\rho) \times [T_0 - \rho^2, T_0]} u$. Applying Theorem 19.3 to the functions $s(\rho) - u$ and $u - i(\rho)$, we have

$$s(\rho) - \inf_{B_p(\rho/2) \times [T_0 - 3\rho^2/4, T_0 - \rho^2/2]} u \leq C \left(s(\rho) - s\left(\frac{\rho}{2}\right) \right)$$

and

$$\sup_{B_p(\rho/2) \times [T_0 - 3\rho^2/4, T_0 - \rho^2/2]} u - i(\rho) \leq C \left(i\left(\frac{\rho}{2}\right) - i(\rho) \right).$$

Adding the two inequalities yields

$$\operatorname{osc}(\rho) + \operatorname{osc}\left(B_p\left(\frac{\rho}{2}\right) \times [T_0 - \frac{3\rho^2}{4}, T_0 - \frac{\rho^2}{2}] \right) \leq C \left(\operatorname{osc}(\rho) - \operatorname{osc}\left(\frac{\rho}{2}\right) \right),$$

where $\operatorname{osc}(\rho) = s(\rho) - i(\rho)$ denotes the oscillation of u on $B_p(\rho) \times [T_0 - \rho^2, T_0]$. This implies that

$$\operatorname{osc}\left(\frac{\rho}{2}\right) \leq \gamma \operatorname{osc}(\rho)$$

for $\gamma = (C-1)/C < 1$. Iterating this inequality gives

$$\operatorname{osc}(2^{-k}\rho) \leq \gamma^k \operatorname{osc}(\rho).$$

This implies that u is Hölder continuous of exponent $-\log \gamma / \log 2$ in the space variable and Hölder continuous of exponent $-\log \gamma / \log 4$ in the time variable. □

20
Parabolicity

In view of the construction in Chapter 17, the existence and nonexistence of a positive Green's function divides the class of complete manifolds into two categories. In general, the methods in dealing with function theory on these manifolds are different, hence it is important to understand the difference between the two categories.

Definition 20.1 A complete manifold is said to be parabolic if it does not admit a positive Green's function. Otherwise it is said to be nonparabolic.

As pointed out in Theorem 17.3, a manifold is nonparabolic if and only if there exists a positive superharmonic function whose infimum is achieved at infinity. This property can be localized at any unbounded component at infinity.

Definition 20.2 An end E with respect to a compact subset $\Omega \subset M$ is an unbounded connected component of $M \setminus \Omega$. The *number of ends* with respect to Ω, denoted by $N_\Omega(M)$, is the number of unbounded connected components of $M \setminus \Omega$.

It is obvious that if $\Omega_1 \subset \Omega_2$, then $N_{\Omega_1}(M) \leq N_{\Omega_2}(M)$. Hence if $\{\Omega_i\}$ is a compact exhaustion of M, then $N_{\Omega_i}(M)$ is a monotonically nondecreasing sequence. If this sequence is bounded, then we say that M has finitely many ends. In this case, we denote the number of ends of M by

$$N(M) = \max_{i \to \infty} N_{\Omega_i}(M).$$

One can readily check that this is independent of the compact exhaustion $\{\Omega_i\}$. In fact, it is also easy to see that there must be an i such that $N(M) = N_{\Omega_i}(M)$. Moreover, for any compact set Ω containing Ω_i, $N_\Omega(M) = N(M)$. Hence for all practical purposes, we may assume that $M \backslash B_p(\rho_0)$ has $N(M)$ unbounded connect components, for some ρ_0.

In general, when we say that E is an end we mean that it is an end with respect to some compact subset Ω. In particular, its boundary ∂E is given by $\partial \Omega \cap \bar{E}$.

Definition 20.3 An end E is said to be parabolic if it does not admit a positive harmonic function f satisfying

$$f = 1 \quad \text{on} \quad \partial E$$

and

$$\liminf_{y \to E(\infty)} f(y) < 1,$$

where $E(\infty)$ denotes the infinity of E. Otherwise, E is said to be nonparabolic and the function f is said to be a barrier function of E.

Observe that by subtraction and multiplication of constants, we may assume that the barrier function satisfies

$$\liminf_{y \to E(\infty)} f(y) = 0.$$

With this notion, we can also count the number of nonparabolic (parabolic) ends as in the definition of $N(M)$.

Definition 20.4 Let $\{\Omega_i\}$ be a compact exhaustion of M. We let $N_i^0(M)$ be the number of nonparabolic ends with respect to the compact set Ω_i. We say that M has finitely many nonparabolic ends if the sequence $\{N_i^0(M)\}$ is bounded. In this case, we say that $N^0(M) = \lim_{i \to \infty} N_i^0(M)$ is the number of nonparabolic ends of M.

Definition 20.5 Let $\{\Omega_i\}$ be a compact exhaustion of M. We let $N_i'(M)$ be the number of parabolic ends with respect to the compact set Ω_i. We say that M has finitely many parabolic ends if the sequence $\{N_i'(M)\}$ is bounded. In this case, we say that $N'(M) = \lim_{i \to \infty} N_i'(M)$ is the number of parabolic ends of M.

Note that if E is a nonparabolic end of M, then by extending f to be identically 1 on $(M \backslash \Omega) \backslash E$, it can be used to construct a positive Green's function on M. Hence, M is nonparabolic if and only if M has a nonparabolic end. Of course, it is possible for a nonparabolic manifold to have many parabolic ends. Let us also point out that E being nonparabolic is equivalent to saying that E has a positive Green's function with the Neumann boundary condition. This

20 Parabolicity

can be seen by repeating the construction of the Green's function in the previous chapter on a manifold with boundary by taking the Neumann boundary condition. One needs to check that the maximum principle arguments are still valid in this case because of the presence of ∂E. However, the Hopf boundary lemma asserts that at a boundary maximum point of a harmonic function f, the outward pointing normal derivative must be positive, violating the Neumann boundary condition.

For the purpose of geometric application, it is often important to obtain appropriate estimates on the barrier function. We will give a canonical method of constructing barrier functions which will help in obtaining estimates later [LT6].

Lemma 20.6 *An end E with respect to the compact set $B_p(\rho_0)$ is nonparabolic if and only if the sequence of positive harmonic functions $\{f_i\}$, defined on $E_p(\rho_i) = E \cap B_p(\rho_i)$ for $\rho_0 < \rho_1 < \rho_2 < \cdots \to \infty$, satisfying*

$$f_i = 1 \quad on \quad \partial E$$

and

$$f_i = 0 \quad on \quad \partial B_p(\rho_i) \cap E,$$

converges uniformly on compact subsets of $E \cup \partial E$ to a barrier function f. The barrier constructed in this manner is minimal among all barrier functions, and f has finite Dirichlet integral on E.

Proof If E is nonparabolic, then there exists a barrier function g satisfying

$$g = 1 \quad on \quad \partial E$$

and

$$\liminf_{y \to E(\infty)} g(y) = 0.$$

For each i, the maximum principle asserts that

$$f_i \leq g \quad on \quad B_p(\rho_i) \cap E$$

and the sequence of functions $\{f_i\}$ is monotonically increasing in i. Lemma 17.1 implies that f_i must converge uniformly on compact subsets of $E \cup \partial E$ to a nonnegative function f satisfying

$$f \leq g \quad on \quad E \tag{20.1}$$

with the boundary conditions

$$f = 1 \quad \text{on} \quad \partial E$$

and

$$\liminf_{y \to E(\infty)} f(y) = 0.$$

Moreover, f is harmonic on E and we claim that the convergence is Lipschitz up to the boundary ∂E. To see this, it suffices to show that $|\nabla \log f_i|^2$ is also uniformly bounded on compact subsets of $E \cup \partial E$. Following the argument for the gradient estimate in Chapter 6, we can consider the function

$$\phi |\nabla \log f_i|^2,$$

where $\phi(r)$ is the cutoff function satisfying the properties

$$\phi(r) = \begin{cases} 1 & \text{for} \quad r \leq R_0, \\ 0 & \text{for} \quad r \geq 2R_0, \end{cases}$$

$$-C R_0^{-1} \leq \phi' \leq 0,$$

and

$$|\phi''| \leq C R_0^{-2}.$$

If the maximum point of this function occurs in the interior of E, then the same argument as in Theorem 6.1 will give the estimate of $|\nabla \log f_i|^2$ up to ∂E. Therefore, we only need to show that we can still estimate $|\nabla \log f_i|^2$ if its maximum point occurs on ∂E. If $x_0 \in \partial E$ is the maximum point of $\phi |\nabla \log f_i|^2$, hence a maximum point of $|\nabla \log f_i|^2$, then by setting $h = \log f_i$, the strong maximum principle asserts that

$$\frac{\partial |\nabla h|^2}{\partial r}(x_0) < 0.$$

On the other hand, since $h = 0$ on ∂E, we can choose an orthonormal frame $\{e_1, \ldots, e_m\}$ at x_0 such that $e_1 = \nabla h/|\nabla h| = -\partial/\partial r$. Hence we have

$$0 > \frac{\partial |\nabla h|^2}{\partial r}$$
$$= -2h_1 h_{11}$$
$$= -2|\nabla h| h_{11}. \tag{20.2}$$

However, the Laplacian of h is given by

$$-|\nabla h|^2 = \Delta h$$
$$= h_{11} + H h_1,$$

where H is the mean curvature of ∂E with respect to the outward normal vector e_1. Therefore combining the above inequality with (20.2), we have

$$0 > -|\nabla h|(-|\nabla h|^2 - H h_1)$$
$$= |\nabla h|^3 + H |\nabla h|^2.$$

This implies that the maximum point cannot occur on ∂E if $\min_{\partial E} H \geq 0$. If $-H_0 = \min_{\partial E} H$ is negative, then

$$|\nabla h|^2(x_0) \leq H_0^2.$$

Since x_0 is the maximum point for $\phi |\nabla h|^2$, we obtain an estimate of $|\nabla h|^2$ on ∂E. Note that in this argument we assume that ∂E is smooth so that $H_0 < \infty$. If this is not the case, we can always smooth out ∂E and the statement of the lemma is still valid for the smooth boundary. In particular, this proves that f is a barrier function and the minimal property of f follows from (20.1) because it applies to any barrier function g.

The boundary conditions on f_i and the fact that it is harmonic imply that

$$\int_{E_p(\rho_i)} |\nabla f_i|^2 = -\int_{\partial E} \frac{\partial f_i}{\partial r}$$

for each i. Hence for any fixed $\rho < \rho_i$,

$$\int_{E_p(\rho)} |\nabla f_i|^2 \leq -\int_{\partial E} \frac{\partial f_i}{\partial r}.$$

Since the right-hand side is uniformly bounded according to the above gradient estimate, by letting $i \to \infty$ we conclude that

$$\int_{E_p(\rho)} |\nabla f|^2 \leq C$$

for some constant $0 < C < \infty$. In fact, since f is harmonic, we see that

$$\int_{\partial E} \frac{\partial f}{\partial r} = \int_{\partial B_p(\rho) \cap E} \frac{\partial f}{\partial r}$$
$$= \lim_{i \to \infty} \int_{\partial B_p(\rho) \cap E} \frac{\partial f_i}{\partial r}$$
$$= \lim_{i \to \infty} \int_{\partial E} \frac{\partial f_i}{\partial r},$$

hence C can be taken to be $\int_{\partial E} \partial f / \partial r$.

Conversely, if the sequence f_i converges uniformly on compact subsets of E to a barrier function f, then E is nonparabolic by definition. $\square$

If E is parabolic, the sequence f_i will still converge to a harmonic function f because of Lemma 17.1. However, f will be identically 1. In this case, by renormalizing the sequence $\{f_i\}$, we can still construct a harmonic function on E that reflects the parabolicity property of E.

Lemma 20.7 *Let E be a parabolic end with respect to $B_p(\rho_0)$. Let ρ_i be an increasing sequence such that $\rho_0 < \rho_1 < \rho_2 < \cdots < \rho_i \to \infty$. There exists a sequence of constants $C_i \to \infty$ such that the sequence of positive harmonic functions g_i defined on $E_p(\rho_i) = B_p(\rho_i) \cap E$, satisfying*

$$g_i = 0 \quad \text{on} \quad \partial E$$

and

$$g_i = C_i \quad \text{on} \quad \partial B_p(\rho_i) \cap E,$$

has a convergent subsequence that converges uniformly on compact subsets of $E \cup \partial E$ to a positive harmonic function g. Moreover, g will have the properties that

$$g = 0 \quad \text{on} \quad \partial E$$

and

$$\sup_{y \in E} g = \infty.$$

Proof The fact that E is parabolic implies that the sequence f_i in Lemma 17.1 converges to the constant function 1. Let us define

$$g_i = C_i(1 - f_i).$$

20 Parabolicity

Obviously g_i will have the appropriate boundary conditions. If we set

$$C_i = \left(1 - \inf_{\partial B_p(\rho_1) \cap E} f_i\right)^{-1},$$

then $\sup_{\partial B_p(\rho_1) \cap E} g_i = 1$. Since $f_i \to 1$, $C_i \to \infty$. Also Lemma 17.1 implies that, after passing to a subsequence, $g_i \to g$ on compact subset of E where g is a harmonic function. We will also call g a barrier function of E. $\square$

Using this characterization of parabolicity, a theorem of Royden [Ro] follows as a corollary.

Corollary 20.8 (Royden) *Parabolicity is a quasi-isometric invariant.*

Proof It suffices to show that if E is a nonparabolic end with respect to the metric ds^2, then for any other metric ds_1^2 which is uniformly equivalent to ds^2, the end E remains nonparabolic with respect to ds_1^2. Let ∇_1, Δ_1, and ν_1 be respectively the gradient, the Laplacian, and the unit outward normal to ∂E with respect to the metric ds_1^2. According to Lemma 20.6, there exists a sequence of harmonic functions, $\{f_i\}$, with respect to ds^2, converging to a nonconstant harmonic function f. Moreover, they satisfy the property stated in Lemma 20.6. In particular,

$$\int_{\partial E} \frac{\partial f_i}{\partial \nu} \, dA = \int_{E_p(\rho_i)} |\nabla f_i|^2 \, dV. \tag{20.3}$$

On the other hand, let h_i be the harmonic function with respect to ds_1^2 satisfying

$$h_i = 1 \quad \text{on} \quad \partial E$$

and

$$h_i = 0 \quad \text{on} \quad \partial B_p(\rho_i) \cap E,$$

then we also have

$$\int_{\partial E} \frac{\partial h_i}{\partial \nu_1} \, dA_1 = \int_{E_p(\rho_i)} |\nabla_1 h_i|^2 \, dV_1. \tag{20.4}$$

However, the fact that there exists a constant $C > 1$ such that

$$C^{-1} ds^2 \leq ds_1^2 \leq C \, ds^2$$

implies that

$$\int_{E_p(\rho_i)} |\nabla_1 h_i|^2 \, dV_1 \geq C_1 \int_{E_p(\rho_i)} |\nabla h_i|^2 \, dV \tag{20.5}$$

for some constant $C_1 > 0$. On the other hand, the Dirichlet integral minimizing property of harmonic functions implies that

$$\int_{E_p(\rho_i)} |\nabla h_i|^2 \, dV \geq \int_{E_p(\rho_i)} |\nabla f_i|^2 \, dV.$$

Hence combining this inequality with (20.3), (20.4), and (20.5) we conclude that

$$\int_{\partial E} \frac{\partial h_i}{\partial \nu_1} \, dA_1 \geq C_1 \int_{\partial E} \frac{\partial f_i}{\partial \nu} \, dA.$$

Since we know that $f_i \to f$ by passing to a subsequence, and similarly $h_i \to h$, we have

$$\int_{\partial E} \frac{\partial h}{\partial \nu_1} \, dA_1 \geq C_1 \int_{\partial E} \frac{\partial f}{\partial \nu} \, dA.$$

The fact that f is harmonic and nonconstant implies that the right-hand side must be positive. On the other hand, we also know that h is harmonic and the fact that

$$\int_{\partial E} \frac{\partial h}{\partial \nu_1} \, dA_1 > 0$$

implies that h is nonconstant, hence it is a barrier function with respect to the metric ds_1^2. $\square$

The following theorem from [LT4] gives a necessary condition for a manifold to be nonparabolic.

Theorem 20.9 (Li–Tam) *Let M be a complete manifold. If M is nonparabolic, then for any point $p \in M$ we must have*

$$\int_1^\infty \frac{dt}{A_p(t)} < \infty,$$

where $A_p(r)$ denotes the area of $\partial B_p(r)$. Moreover, if $G(p, y)$ is the minimal Green's function, then

$$\int_1^r \frac{dt}{A_p(t)} \leq \sup_{y \in \partial B_p(1)} G(p, y) - \inf_{y \in \partial B_p(r)} G(p, y)$$

for all $r > 1$.

Proof For $p \in M$, let $G(p, y)$ be the minimal positive Green's function with a pole at p. In view of the construction in Chapter 17, we may assume that $G(p, y)$ is the limit of the sequence $\{G_i(p, y)\}$, where G_i is the Dirichlet Green's function defined on $B_p(\rho_i)$.

20 Parabolicity

For any $1 < \rho < \rho_i$, let

$$s_i(1) = \sup_{y \in \partial B_\rho(1)} G_i(p, y)$$

and

$$i_i(\rho) = \inf_{y \in \partial B_\rho(\rho)} G_i(p, y).$$

Let f be the harmonic function defined on $B_p(\rho) \setminus B_p(1)$ satisfying the boundary conditions

$$f(y) = s_i(1) \quad \text{on} \quad \partial B_p(1)$$

and

$$f(y) = G_i(p, y) \quad \text{on} \quad \partial B_p(\rho).$$

The maximum principle implies that

$$f(y) \geq G_i(p, y) \quad \text{on} \quad B_p(\rho) \setminus B_p(1).$$

In particular, we have

$$\frac{\partial f}{\partial r} \leq \frac{\partial G_i}{\partial r} \quad \text{on} \quad \partial B_p(\rho). \tag{20.6}$$

On the other hand, the harmonicity of f and Stokes' theorem imply that

$$0 = \int_{B_p(\rho) \setminus B_p(1)} \Delta f$$

$$= \int_{\partial B_p(\rho)} \frac{\partial f}{\partial r} - \int_{\partial B_p(1)} \frac{\partial f}{\partial r}.$$

Also, we observe that

$$\int_{\partial B_p(\rho)} \frac{\partial G_i}{\partial r} = \int_{B_p(\rho)} \Delta G_i$$

$$= -1$$

and (20.6) implies that

$$\int_{\partial B_p(1)} \frac{\partial f}{\partial r} \leq -1. \tag{20.7}$$

Let us now consider a harmonic function h defined on $B_p(\rho) \setminus B_p(1)$ satisfying the boundary conditions

$$h(y) = s_i(1) \quad \text{on} \quad \partial B_p(1)$$

and
$$h(y) = i_i(\rho) \quad \text{on} \quad \partial B_p(\rho).$$

Again the maximum principle implies that
$$h(y) \leq f(y) \quad \text{on} \quad B_p(\rho) \setminus B_p(1),$$

and
$$\frac{\partial h}{\partial r} \leq \frac{\partial f}{\partial r} \quad \text{on} \quad \partial B_p(1).$$

Hence combining this inequality with (20.7), we obtain
$$\int_{\partial B_p(\rho)} \frac{\partial h}{\partial r} = \int_{\partial B_p(1)} \frac{\partial h}{\partial r}$$
$$\leq -1. \tag{20.8}$$

If we define the function
$$g(r) = (s_i(1) - i_i(\rho)) \left(\int_1^\rho \frac{dt}{A_p(t)}\right)^{-1} \int_r^\rho \frac{dt}{A_p(t)} + i_i(\rho),$$

then $g(r(y))$ has the same boundary conditions as $h(y)$. The Dirichlet integral minimizing property for harmonic functions implies that
$$\int_{B_p(\rho) \setminus B_p(1)} |\nabla h|^2 \leq \int_{B_p(\rho) \setminus B_p(1)} |\nabla g|^2$$
$$= \int_1^\rho \left((s_i(1) - i_i(\rho)) \left(\int_1^\rho \frac{dt}{A_p(t)}\right)^{-1} \frac{1}{A_p(r)}\right)^2 A_p(r)\, dr$$
$$= (s_i(1) - i_i(\rho))^2 \left(\int_1^\rho \frac{dt}{A_p(t)}\right)^{-1}. \tag{20.9}$$

On the other hand, integration by parts and (20.8) yield
$$\int_{B_p(\rho) \setminus B_p(1)} |\nabla h|^2 = i_i(\rho) \int_{\partial B_p(\rho)} \frac{\partial h}{\partial r} - s_i(1) \int_{\partial B_p(1)} \frac{\partial h}{\partial r},$$
$$\geq s_i(1) - i_i(\rho)$$

because $s_i(1) \geq s_i(\rho) \geq i_i(\rho)$ and
$$s_i(r) = \sup_{y \in \partial B_p(r)} G_i(p, y)$$

is a decreasing function of r. Combining this with (20.9) gives the estimate

$$\int_1^\rho \frac{dt}{A_p(t)} \leq s_i(1) - i_i(\rho).$$

Letting $i \to \infty$ and using the fact that G is nonconstant, we arrive with the estimate

$$\int_1^\rho \frac{dt}{A_p(t)} \leq s(1) - i(\rho),$$

where

$$s(1) = \sup_{y \in \partial B_p(1)} G(p, y)$$

and

$$i(\rho) = \inf_{y \in \partial B_p(\rho)} G(p, y).$$

Letting $\rho \to \infty$, the finiteness of

$$\int_1^\infty \frac{dt}{A_p(t)}$$

follows. $\square$

The following lemma gives a criterion for an end to be nonparabolic. It was first proved by Cao, Shen and Zhu in [CSZ] for minimal submanifolds. However, their arugment can be generalized to the following context.

Lemma 20.10 *Let E be an end of a complete Riemannian manifold. Suppose for some $\nu \geq 1$ and $C > 0$, E satisfies a Sobolev type inequality of the form*

$$\left(\int_E |u|^{2\nu} \right)^{1/\nu} \leq C \int_E |\nabla u|^2$$

for all compactly supported functions $u \in H_c^{1,2}(E)$ defined on E, then E must either have finite volume or be nonparabolic.

Proof Let E be an end of M. For ρ sufficiently large, let us consider the set $E(\rho) = E \cap B_p(\rho)$, where $B_p(\rho)$ is the geodesic ball of radius ρ in M centered at some point $p \in M$. Let us denote by r the distance function of M to the point p. Suppose the function f_ρ is the solution to the equation

$$\Delta f_\rho = 0 \quad \text{on} \quad E(\rho),$$

with boundary conditions

$$f_\rho = 1 \quad \text{on} \quad \partial E$$

and

$$f_\rho = 0 \quad \text{on} \quad E \cap \partial B_p(\rho).$$

According to Lemma 20.6, f_ρ converges to a nonconstant harmonic function f if and only if E is nonparabolic.

For a fixed $0 < \rho_0 < \rho$ such that $E(\rho_0) \neq \emptyset$, let ϕ be a nonnegative cutoff function satisfying the properties that

$$\phi = 1 \quad \text{on} \quad E(\rho) \setminus E(\rho_0),$$

$$\phi = 0 \quad \text{on} \quad \partial E,$$

and

$$|\nabla \phi| \leq C_1.$$

Applying the inequality in the assumption and using the fact that f_ρ is harmonic, we obtain

$$\left(\int_{E(\rho)} (\phi f_\rho)^{2\nu} \right)^{1/\nu}$$

$$\leq C \int_{E(\rho)} |\nabla (\phi f_\rho)|^2$$

$$= C \left(\int_{E(\rho)} |\nabla \phi|^2 f_\rho^2 + 2 \int_{E(\rho)} \phi f_\rho \langle \nabla \phi, \nabla f_\rho \rangle + \int_{E(\rho)} \phi^2 |\nabla f_\rho|^2 \right)$$

$$= C \left(\int_{E(\rho)} |\nabla \phi|^2 f_\rho^2 + \tfrac{1}{2} \int_{E(\rho)} \langle \nabla(\phi^2), \nabla (f_\rho^2) \rangle + \int_{E(\rho)} \phi^2 |\nabla f_\rho|^2 \right)$$

$$= C \int_{E(\rho)} |\nabla \phi|^2 f_\rho^2.$$

In particular, for a fixed ρ_1 satisfying $\rho_0 < \rho_1 < \rho$, we have

$$\left(\int_{E(\rho_1) \setminus E(\rho_0)} f_\rho^{2\nu} \right)^{1/\nu} \leq C_2 \int_{E(\rho_0)} f_\rho^2.$$

If E is parabolic, then the limiting function f is identically 1. Letting $\rho \to \infty$, we obtain

$$(V_E(\rho_1) - V_E(\rho_0))^{1/\nu} \leq C \, V_E(\rho_0),$$

where $V_E(\rho)$ denotes the volume of the set $E(\rho)$. Since $\rho_1 > \rho_0$ is arbitrary, this implies that E must have finite volume and the theorem is proved. □

Lemma 20.11 *Let M be a complete Riemannian manifold. Given a geodesic ball $B_p(\rho_0)$, suppose there exist constants $v > 1$ and $C > 0$ such that a Sobolev type inequality of the form*

$$\left(\int_{B_p(\rho_0)} |u|^{2v}\right)^{1/v} \leq C \int_{B_p(\rho_0)} |\nabla u|^2$$

is valid for all compactly supported functions $u \in H_c^{1,2}(B_p(\rho_0))$. Then there exists a constant $C_1 > 0$ depending only on C and v such that

$$V_p(\rho_0) \geq C_1 \rho_0^{2v/(v-1)}.$$

In particular, if an end E of M with respect to the compact set $B_p(\rho_0)$ satisfies the Sobolev inequality

$$\left(\int_E |u|^{2v}\right)^{1/v} \leq C \int_E |\nabla u|^2$$

for all compactly supported functions $u \in H_c^{1,2}(E)$, then the end must have volume growth given by

$$V_E(\rho_0 + \rho) \geq C_1 \rho^{2v/(v-1)}$$

for all $\rho > 0$, where $V_E(\rho_0 + \rho)$ denotes the volume of the set $E_p(\rho_0 + \rho)$.

Proof For a fixed point $p \in M$ and $0 < \rho < \rho' < \rho_0$, let us consider the function

$$u(x) = \begin{cases} 1 & \text{on } B_p(\rho), \\ \dfrac{\rho' - r(x)}{\rho' - \rho} & \text{on } B_p(\rho') \setminus B_p(\rho), \\ 0 & \text{on } M \setminus B_p(\rho'). \end{cases}$$

where $r(x)$ is the distance function to the point p. Plugging this into the Sobolev inequality, we conclude that

$$V_p^{1/v}(\rho) \leq C(\rho' - \rho)^{-2} V_p(\rho'). \tag{20.10}$$

Let us define $\rho_k = \left(1 - \sum_{i=1}^k 2^{-(i+1)}\right) \rho_0$ for $k = 1, 2, \ldots$. Using $\rho' = \rho_{k-1}$ and $\rho = \rho_k$, (20.10) becomes

$$V_p^{\frac{1}{v}}(\rho_k) \leq C 4^{k+1} \rho_0^{-2} V_p(\rho_{k-1}).$$

This can be rewritten as

$$\left(\frac{V_p(\rho_k)}{\rho_k^{2\nu/(\nu-1)}}\right)^{1/\nu} \leq C 4^{k+1} \rho_0^{-2} \frac{\rho_{k-1}^{2\nu/(\nu-1)}}{\rho_k^{2/(\nu-1)}} \frac{V_p(\rho_{k-1})}{\rho_{k-1}^{2\nu/(\nu-1)}}$$

$$= C 4^{k+1} \frac{\left(1 - \sum_{i=1}^{k-1} 2^{-(i+1)}\right)^{2\nu/(\nu-1)}}{\left(1 - \sum_{i=1}^{k} 2^{-(i+1)}\right)^{2/(\nu-1)}} \frac{V_p(\rho_{k-1})}{\rho_{k-1}^{2\nu/(\nu-1)}}. \quad (20.11)$$

Using the estimate

$$\tfrac{1}{2} \leq 1 - \sum_{i=1}^{k} 2^{-(i+1)} \leq 1$$

for all $1 \leq k$, (20.11) yields

$$\left(\frac{V_p(\rho_k)}{\rho_k^{2\nu/(\nu-1)}}\right)^{1/\nu} \leq C_1 4^{k+1} \frac{V_p(\rho_{k-1})}{\rho_{k-1}^{2\nu/(\nu-1)}},$$

where $C_1 > 0$ depends on C and ν. Iterating this inequality k times, we obtain

$$\left(\frac{V_p(\rho_k)}{\rho_k^{2\nu/(\nu-1)}}\right)^{\nu^{-k}} \leq C_1^{\sum_{i=1}^{k} \nu^{-i}} 4^{\sum_{i=1}^{k}(i+1)\nu^{-i}} \frac{V_p(\rho_0)}{\rho_0^{2\nu/(\nu-1)}}. \quad (20.12)$$

Observing that $\rho_k \to \rho_0/2$ and $\nu^{-k} \to 0$ as $k \to \infty$, we conclude that the left-hand side of (20.12) tends to 1. Hence (20.12) becomes

$$C_1^{-\sum_{i=1}^{\infty} \nu^{-i}} 4^{-\sum_{i=1}^{\infty}(i+1)\nu^{-i}} \rho_0^{2\nu/(\nu-1)} \leq V_p(\rho_0).$$

The lower bound of the volume follows by observing that

$$\sum_{i=1}^{\infty} \nu^{-i} < \sum_{i=1}^{\infty} (i+1)\nu^{-i} < \infty,$$

and the constants on the left-hand side are finite because $\nu > 1$.

If the Sobolev inequality is valid on an end E with respect to $\overline{B_p(\rho_0)}$, then we apply the above volume estimate to the ball $B_x(\rho)$ with $x \in \partial B_p(\rho_0 + \rho) \cap E$. Hence we obtain

$$V_E(\rho_0 + 2\rho) \geq V_x(\rho)$$
$$\geq C_1 \rho^{2\nu/(\nu-1)},$$

and the volume growth of E follows. $\square$

Combining Lemma 20.10 and Lemma 20.11, we obtain the following corollary for the Sobolev type inequality with $\nu > 1$. The case when $\nu = 1$ is just the Dirichlet Poincaré inequality and will be addressed separately in Chapter 22. In that case, it is possible to have a finite volume end given by a cusp.

Corollary 20.12 *Let E be an end of a complete Riemannian manifold. Suppose that for some $\nu > 1$ and $C > 0$, E satisfies a Sobolev type inequality of the form*

$$\left(\int_E |u|^{2\nu}\right)^{1/\nu} \leq C \int_E |\nabla u|^2,$$

for all compactly supported functions $u \in H_c^{1,2}(E)$ defined on E, then E must be nonparabolic.

21
Harmonic functions and ends

In this chapter, we will construct harmonic functions defined on M by extending the barrier functions defined at all of the ends of M. This construction was first proved by Tam and Li in [LT1] for manifolds with nonnegative sectional curvature near infinity. They later gave a construction for arbitrary complete manifolds in [LT6]. In [STW], Sung, Tam, and Wang presented the construction in a more systematic manner. It is their version that is given here.

Theorem 21.1 (Li–Tam and Sung–Tam–Wang) *Let M be a complete manifold and let Ω be a smooth compact subdomain in M. Suppose g is a harmonic function defined on $M \setminus \Omega$ which is smooth up to the boundary $\partial \Omega$ of Ω. If M is nonparabolic, then there exists a harmonic function h defined on M and a constant $C > 0$ such that*

$$|g(x) - h(x)| \leq C\, G(p, x),$$

where $G(p, x)$ is the minimal positive Green's function and $p \in \Omega$. Moreover, the function $g - h$ has a finite Dirichlet integral on $M \setminus \Omega$.

Proof Let $p \in \Omega$ be a fixed point and let $\rho_0 > 0$ be such that $\Omega \subset B_p(\rho_0)$. Let ϕ be a smooth nonnegative function satisfying

$$\phi = \begin{cases} 0 & \text{on} \quad B_p(\rho_0), \\ 1 & \text{on} \quad M \setminus B_p(2\rho_0). \end{cases}$$

If M is nonparabolic and $G(x, y)$ is the minimal positive Green's function on M, then we define the function h by

$$h(x) = \phi(x)\, g(x) + \int_M G(x, y)\, \Delta(\phi\, g)(y)\, dy.$$

Since
$$\Delta(\phi g) = \Delta g = 0 \quad \text{on} \quad M \setminus B_p(2\rho_0),$$
the function $\Delta(\phi g)$ has compact support and h is harmonic on M. For $x \notin B_p(4\rho_0)$, we have

$$|h(x) - g(x)| = |h(x) - \phi(x)g(x)|$$

$$= \left| \int_{B_p(2\rho_0)} G(x, y) \Delta(\phi g)(y) \, dy \right|$$

$$\leq \sup_{y \in B_p(2\rho_0)} G(x, y) \int_{B_p(2\rho_0)} |\Delta(\phi g)|(y) \, dy.$$

Since $x \notin B_p(4\rho_0)$, $G(x, y)$ as a function of y is harmonic on $B_p(4\rho_0)$. Hence the local Harnack inequality asserts that

$$G(x, y) \leq C\, G(x, p)$$

for $y \in B_p(2\rho_0)$, where C is a constant depending only on the lower bound of the Ricci curvature of $B_p(4\rho_0)$, m, and ρ_0. This proves that

$$|h(x) - g(x)| \leq C\, G(p, x)$$

for all $x \notin B_p(4\rho_0)$. Since G is positive, compactness implies that the same bound is valid for all $x \in M$ by possibly adjusting the constant C.

Note that from the construction of G in Chapter 17, the bound on the function $|h - g|$ can be written as

$$|h(x) - g(x)| \leq C\, f(x) \quad \text{on} \quad M \setminus B_p(4\rho_0), \tag{21.1}$$

where f is the minimal barrier function of Chapter 20 satisfying the properties that

$$f = 1 \quad \text{on} \quad \partial B_p(4\rho_0),$$

$$\inf_{y \to \infty} f(y) = 0,$$

and

$$\int_{M \setminus B_p(4\rho_0)} |\nabla f|^2 < \infty.$$

For any $\rho > 4\rho_0$, let u_ρ be the harmonic function defined on $B_p(\rho) \setminus B_p(4\rho_0)$ satisfying the boundary conditions

$$u_\rho = h - g \quad \text{on} \quad \partial B_p(4\rho_0) \tag{21.2}$$

and
$$u_\rho = 0 \quad \text{on} \quad \partial B_p(\rho). \tag{21.3}$$

Since (21.1) implies that
$$|u_\rho - (h-g)| \leq Cf \quad \text{on} \quad \partial B_p(4\rho_0) \cup \partial B_p(\rho),$$
the maximum principle asserts that
$$|u_\rho - (h-g)| \leq Cf \quad \text{on} \quad B_p(\rho) \setminus B_p(4\rho_0).$$

Hence there exists a sequence of $\{\rho_i\}$ such that $u_i - (h-g)$, with $u_i = u_{\rho_i}$, converges uniformly on compact subset of $M \setminus B_p(4\rho_0)$ to a harmonic function $u - (h-g)$. Moreover,
$$u - (h-g) = 0 \quad \text{on} \quad \partial B_p(4\rho_0)$$
and
$$|u - (h-g)| \leq Cf.$$

In particular, the harmonic function
$$f - C^{-1}(u - (h-g))$$
is nonnegative and has boundary value 1 on $\partial B_p(4\rho_0)$. Hence by the minimality property of f,
$$f \leq f - C^{-1}(u - (h-g))$$
and
$$u - (h-g) \leq 0.$$

Running through the same argument using the function $(h-g) - u$, we conclude that
$$(h-g) - u \leq 0,$$
hence $h - g = u$. In particular, the sequence of functions $\{u_i\}$ converges to $h - g$. However, since
$$\int_{B_p(\rho_i) \setminus B_p(4\rho_0)} |\nabla u_i|^2 = -\int_{\partial B_p(4\rho_0)} u_i \frac{\partial u_i}{\partial r}$$
$$= -\int_{\partial B_p(4\rho_0)} (h-g) \frac{\partial u_i}{\partial r},$$

we conclude that for any $\rho > 4\rho_0$

$$\int_{B_p(\rho)\setminus B_p(4\rho_0)} |\nabla u_i|^2 \leq -\int_{\partial B_p(4\rho_0)} (h-g)\frac{\partial u_i}{\partial r}.$$

Letting $i \to \infty$ yields

$$\int_{B_p(\rho)\setminus B_p(4\rho_0)} |\nabla(h-g)|^2 \leq -\int_{\partial B_p(4\rho_0)} (h-g)\frac{\partial(h-g)}{\partial r}.$$

Since ρ is arbitrary, this completes the proof of the theorem. $\square$

Theorem 21.2 (Li–Tam and Sung–Tam–Wang) *Let M be a complete manifold and Ω be a smooth compact subdomain in M. Suppose g is a harmonic function defined on $M\setminus\Omega$ which is smooth up to the boundary $\partial\Omega$ of Ω. If M is parabolic, then there exists a harmonic function h defined on M such that $|g-h|$ is bounded on $M\setminus\Omega$ if*

$$\int_{\partial\Omega} \frac{\partial g}{\partial \nu} = 0,$$

where ν is the outward unit normal to $\partial\Omega$. Moreover, $|g-h|$ must have finite Dirichlet integral on $M\setminus\Omega$.

Proof Following an argument similar to that in the proof of Theorem 21.1, we define the harmonic function

$$h(x) = \phi(x)g(x) + \int_M G(x,y)\Delta(\phi g)(y)\,dy,$$

where $G(x,y)$ is a symmetric Green's function constructed in Chapter 17, and ϕ is given by

$$\phi = \begin{cases} 0 & \text{on} \quad B_p(\rho_0), \\ 1 & \text{on} \quad M\setminus B_p(2\rho_0). \end{cases}$$

Again when for $x \notin B_p(4\rho_0)$, we have

$$|h(x) - g(x)| = |h(x) - \phi(x)g(x)|$$

$$= \left|\int_{B_p(2\rho_0)} G(x,y)\Delta(\phi g)(y)\,dy\right|$$

$$\leq \left| \int_{B_p(2\rho_0)} (G(x,y) - G(x,p)) \, \Delta(\phi g)(y) \, dy \right|$$

$$+ \left| G(x,p) \int_{B_p(2\rho_0)} \Delta(\phi g)(y) \, dy \right|. \quad (21.4)$$

However, since

$$\int_{B_p(2\rho_0)} \Delta(\phi g)(y) \, dy = \int_{\partial B_p(2\rho_0)} \frac{\partial g}{\partial r}$$

$$= \int_{\partial \Omega} \frac{\partial g}{\partial \nu}$$

$$= 0$$

and

$$\left| \int_{B_p(2\rho_0)} (G(x,y) - G(x,p)) \, \Delta(\phi g)(y) \, dy \right|$$

$$\leq \sup_{y \in B_p(2\rho_0)} |G(x,y) - G(x,p)| \int_{B_p(2\rho_0)} |\Delta(\phi g)(y)| \, dy,$$

(21.4) becomes

$$|h(x) - g(x)| \leq C \sup_{y \in B_p(2\rho_0)} |G(x,y) - G(x,p)|.$$

On the other hand, from the construction of G, we know that $G(y,x) - G(p,x)$ is a bounded harmonic function on $M \setminus B_p(4\rho_0)$ for $y \in B_p(2\rho_0)$. This implies that $|h - g|$ is bounded on $M \setminus \Omega$.

To see that $h - g$ has a finite Dirichlet integral, we follow a similar argument to that in the proof of Theorem 21.1. By solving the Dirichlet boundary problem (21.2) and (21.3), we obtain a harmonic function u with the properties that

$$u - (h - g) = 0 \quad \text{on} \quad \partial B_p(4\rho_0)$$

and

$$|u - (h - g)| \leq C.$$

Unless $u - (g - h)$ is identically 0, then either

$$\sup_{M \setminus B_p(8\rho_0)} u - (g - h) = \alpha$$

or

$$\inf_{M \setminus B_p(8\rho_0)} u - (g - h) = -\alpha$$

for some $\alpha > 0$. In the first case, the function $f = \alpha - u + (h - g)$ is a non-constant positive harmonic function whose infimum is achieved at infinity. In the second case, the function $f = u - (h - g) - \alpha$ has the same property. This contradicts the assumption that M is parabolic, and we must have $u = h - g$. The rest of the argument is exactly the same as in the proof of Theorem 21.1. □

There is one more property resulting from the above constructions that is important to point out. Recall that when M is nonparabolic the minimal Green's function G is given by the limit of a sequence of Dirichlet Green's function defined on a compact exhaustion. Let us assume that the compact exhaustion is given by $\{B_p(\rho_i)\}$. Similarly, for a parabolic manifold, the constructed Green's function can be obtained by taking the limit of $G_i - a_i$ where a_i is a sequence of constants. In either case, if we define the sequence of harmonic functions

$$h_i(x) = \phi(x) g(x) + \int_{B_p(\rho_i)} G_i(x, y) \Delta(\phi g)(y) \, dy$$

for $\rho_i \geq 4\rho_0$, then h_i will solve the Dirichlet problem

$$\Delta h_i = 0 \quad \text{on} \quad B_p(\rho_i) \tag{21.5}$$

and

$$h_i = g \quad \text{on} \quad \partial B_p(\rho_i). \tag{21.6}$$

In the nonparabolic case, obviously $h_i \to h$, since $G_i \to G$. In the parabolic case, this is also true because

$$\int_{B_p(\rho_i)} a_i \, \Delta(\phi g)(y) \, dy = 0,$$

hence

$$h_i(x) = \phi(x) g(x) + \int_{B_p(\rho_i)} (G_i(x, y) - a_i) \, \Delta(\phi g)(y) \, dy,$$

which converges to h. This implies that the harmonic function constructed in Theorem 21.1 and Theorem 21.2 can be obtained by taking limits of harmonic functions h_i solving the Dirichlet problem (21.5) and (21.6). By the maximum principle, we conclude that

$$\inf_{\partial B_p(\rho_i)} g \leq h_i(x) \leq \sup_{\partial B_p(\rho_i)} g$$

for all $x \in B_p(\rho_i)$. Taking the limit as $i \to \infty$, we obtain

$$\liminf_{y \to \infty} g \leq h(x) \leq \limsup_{y \to \infty} g.$$

We are now ready to construct harmonic functions that reflect the geometry and topology of a complete manifold.

Theorem 21.3 (Li–Tam) *Let M be a complete manifold that is nonparabolic. There exist spaces of harmonic functions $\mathcal{K}^0(M)$ and $\mathcal{K}'(M)$, with (possibly infinite) dimensions given by $k^0(M)$ and $k'(M)$, respectively, such that*

$$k^0(M) = N^0(M)$$

and

$$k'(M) = N'(M).$$

In particular,

$$k^0(M) + k'(M) = N(M).$$

Moreover, $\mathcal{K}^0(M)$ is a subspace of the space of bounded harmonic functions with a finite Dirichlet integral on M, and $\mathcal{K}'(M)$ is spanned by a set of positive harmonic functions.

Proof The assumption that M is nonparabolic implies that $N^0(M) \geq 1$. If $N^0(M) = 1$, then the space $\mathcal{K}^0(M)$ is given by the constant functions. Let us now assume that $N^0(M) \geq 2$ and $E_1, \ldots, E_{N^0_\rho(M)}$ is the set of all nonparabolic ends with respect to the compact set $B_p(\rho)$. It suffices to show that we can construct $N^0_\rho(M)$ many linearly independent bounded harmonic functions with finite Dirichlet integral. Since ρ is arbitrary, this will prove that $\mathcal{K}^0(M)$ exists.

For each $1 \leq i \leq N^0_\rho(M)$, let us define the harmonic function ψ_i on $M \setminus B_p(\rho)$ given by

$$\psi_i = \begin{cases} 1 & \text{on } E_i, \\ 0 & \text{on } M \setminus (B_p(\rho) \cup E_i). \end{cases}$$

By Theorem 21.1, there exists a harmonic function h_i defined on M such that $h_i - \psi_i$ is bounded and has a finite Dirichlet integral on $M \setminus B_p(\rho)$. By the remark following Theorem 21.2, h_i must also be bounded between 0 and 1

21 Harmonic functions and ends

and has a finite Dirichlet integral on M. Moreover, on any nonparabolic end E_k, Theorem 21.1 also asserts that

$$|\psi_i - h_i| \leq C f_k \quad \text{on} \quad E_k,$$

where f_k is the minimal barrier function of E_k. Applying this to E_i, it implies that if $\{x_j^i\}$ is a sequence of points in E_i with $x_j^i \to E_i(\infty)$ as $j \to \infty$, such that $f_i(x_j^i) \to 0$, then $h_i(x_j^i) \to 1$. Similarly, this also shows that for any sequence $\{x_j^\alpha\} \subset E_\alpha$ with $x_j^\alpha \to E_\alpha(\infty)$ for $\alpha \neq i$ as $j \to \infty$ such that $f_\alpha(x_j^\alpha) \to 0$, then $h_i(x_j^\alpha) \to 0$.

Obviously, this construction yields $N_\rho^0(M)$ bounded harmonic functions $\{h_i\}$ with finite Dirichlet integrals and satisfying

$$\lim_{j \to \infty} h_i(x_j^i) = 1$$

and

$$\lim_{j \to \infty} h_i(x_j^\alpha) = 0 \quad \text{for} \quad \alpha \neq i.$$

The linear independence of this set of functions follows and $\mathcal{K}^0$ is constructed.

Now let $E_1', \ldots E_{N_\rho'(M)}'$ be the set of parabolic ends with respect to $B_p(\rho)$. We will construct a set of $N_\rho'(M)$ linearly independent positive harmonic functions. Again, since ρ is arbitrary, this will prove that a space $\mathcal{K}'$ exists with the properties that $\dim \mathcal{K}' = k'(M) = N'(M)$ and $\mathcal{K}'$ is spanned by positive harmonic functions. Therefore $\mathcal{K}(M) = \mathcal{K}^0(M) + \mathcal{K}'(M)$ and the theorem follows.

Modifying the above argument, for each E_i' let g_i be a barrier function for E_i' given by Lemma 20.7. Let us define the function ψ_i by

$$\psi_i = \begin{cases} g_i & \text{on} \quad E_i', \\ 0 & \text{on} \quad M \setminus (B_p(\rho) \cup E_i'). \end{cases}$$

Using Theorem 21.1 again, there exists a harmonic function h_i with the property that

$$|\psi_i - h_i| \leq C \quad \text{on} \quad E_\alpha' \text{ for } \alpha \neq i$$

and

$$|\psi_i - h_i| \leq C f_k \quad \text{on} \quad E_k.$$

In particular, since $g_i(y) \to \infty$ as $y \to E_i'(\infty)$, we conclude that

$$\limsup_{y \to E_i'(\infty)} h_i(y) = \infty.$$

Also h_i is bounded on $M \setminus E_i'$. Note that according to the remark preceding Theorem 21.2, h_i must be positive since it can be viewed as a limit of a sequence of positive harmonic functions satisfying (21.5) and (21.6). Moreover, since there is a nonparabolic end, $\inf_M h_i = 0$. Clearly, the functions in the set $\{h_i\}$ are linearly independent and they span the space $\mathcal{K}'_\rho(M)$. This completes the theorem. $\square$

A similar construction also gives a corresponding theorem for parabolic manifolds.

Theorem 21.4 (Li–Tam) *Let M be a complete manifold that is parabolic. There exists a space of harmonic functions $\mathcal{K}(M)$, with (possibly infinite) dimension given by $k(M)$, such that*

$$k(M) = N(M).$$

Moreover, $\mathcal{K}(M)$ is spanned by a set of harmonic functions which are bounded from either above or below when restricted on each end of M.

Proof Let E_i' for $1 \leq i \leq N_\rho(M)$ be the set of parabolic ends with respect to $B_p(\rho)$. Following the same notation as before, we let g_i be the barrier functions for the ends. Note that for each i, from the construction g_i achieves its minimum value on $\partial E_i'$. Therefore, by the maximum principle, we conclude that

$$\int_{\partial E_i'} \frac{\partial g_i}{\partial r} = \alpha_i$$

for some $\alpha_i > 0$. Let us define the function ψ_i by

$$\psi_i = \begin{cases} g_1 & \text{on} & E_1', \\ -\frac{\alpha_1}{\alpha_i} g_i & \text{on} & E_i', \\ 0 & \text{on} & M \setminus (E_1' \cup E_i'). \end{cases}$$

Clearly,

$$\int_{\partial B_p(\rho)} \frac{\partial \psi_i}{\partial r} = 0,$$

and we can apply Theorem 21.2 to obtain a harmonic function h_i on M with the property that $|\psi_i - h_i|$ is bounded and has finite Dirichlet integral. In particular, the function h_i has the properties that

$$\limsup_{y \to E'_1(\infty)} h_i(y) = \infty,$$

$$\liminf_{y \to E'_i(\infty)} h_i(y) = -\infty,$$

and

$$|h_i(y)| \leq C \quad \text{on} \quad M - \left(B_p(\rho) \cup E'_1 \cup E'_i\right).$$

Moreover, since the barrier functions $\{g_i\}$ are positive, h_i must be bounded from either above or below on each end. Clearly, the set of functions $\{h_i\}$ together with a constant function forms a linear independent set and spans the space $\mathcal{K}_\rho(M)$. This completes the proof of the theorem. □

Combining Theorem 21.4 with Theorem 7.4, we obtain the following finiteness theorem.

Theorem 21.5 (Li–Tam) *Let M be an m-dimensional complete noncompact Riemannian manifold without boundary. Suppose that the Ricci curvature of M is nonnegative on $M \setminus B_p(1)$ for some unit geodesic ball centered at $p \in M$. Let us assume that the lower bound of the Ricci curvature on $B_p(1)$ is given by*

$$\mathcal{R}_{ij} \geq -(m-1)R$$

for some constant $R \geq 0$. Then M must have finitely many ends. Moreover, there exists a constant $C(m, R) > 0$ depending only on m and R such that

$$N(M) \leq C(m, R).$$

Proof If M is nonparabolic, we define the space of harmonic functions $\mathcal{H}'(M) = \mathcal{K}^0(M) \bigoplus \mathcal{K}'(M)$ as constructed in Theorem 21.3. Since functions in $\mathcal{K}^0(M)$ are all bounded harmonic functions and any finite dimensional subspace of $\mathcal{K}'(M)$ is spanned by harmonic functions that are positive and each tends to infinity at each end, their linear combination must be either bounded above or bounded below on each end. Hence we can apply Theorem 7.4 to estimate $\dim \mathcal{H}'(M)$, therefore obtaining the bound

$$N(M) \leq C(m, R)$$

by combining with Theorem 21.3.

If M is parabolic, we simply combine Theorem 21.4 and Theorem 7.4 to obtain the estimate on $N(M)$. □

We would like to point out that when M has nonnegative Ricci curvature everywhere, this argument also recovers the splitting theorem (Theorem 4.4) of Cheeger and Gromoll [CG2]. The interested reader may note that the estimate on the number of ends actually holds on manifolds whose Ricci curvature need not be nonnegative at infinity. In fact, it was proved by Li and Tam [LT1] that if we assume that the Ricci curvature of M is bounded from below by

$$R_{ij}(x) \geq -(m-1)k(r(x)),$$

for some nonincreasing function $k(r) \geq 0$ satisfying the property

$$\int_1^\infty k(r) r^{m-1} \, dr < \infty,$$

then $N(M)$ is finite and can be estimated. However, the proof is more involved and might not substantially add to our overall educational purpose.

22
Manifolds with positive spectrum

In this chapter, we will consider manifolds whose spectrum of the Laplacian, $\mathrm{Spec}(\Delta)$, acting on L^2 functions is bounded from below by a positive number. If we denote

$$\mu_1(M) = \inf \mathrm{Spec}(\Delta)$$

to be the infimum of the spectrum of Δ, it can be characterized by

$$\mu_1(M) = \inf_{f \in H_c^{1,2}(M)} \frac{\int_M |\nabla f|^2}{\int_M f^2},$$

where the infimum is taken over all compactly supported functions in the Sobolev space $H_c^{1,2}(M)$. In particular,

$$\mu_1(M) = \inf_{i \to \infty} \mu_1(\Omega_i)$$

for any compact exhaustion $\{\Omega_i\}$, where $\mu_1(\Omega_i)$ is the Dirichlet first eigenvalue of Δ on Ω_i. In 1975, Cheng and Yau [CgY] gave a necessary condition for a complete manifold to have $\mu_1(M) > 0$ by showing that if M has polynomial volume growth, then $\mu_1(M) = 0$.

Another important quantity of the spectrum is the infimum of the essential spectrum $\mu_e(M)$ of Δ. It has the property that

$$\mu_e(M) \geq \mu_1(M)$$

and given any $\epsilon > 0$ there exists a compact set $\Omega \subset M$ such that

$$\mu_e(M) \leq \mu_1(\Omega') + \epsilon$$

267

for any compact set $\Omega' \subset M \setminus \Omega$. In view of this remark, most of the statements proved in this chapter with the assumption of $\mu_1(E) > 0$ on an end can be stated as $\mu_e(M) > 0$.

Throughout this chapter, we will assume that E is an end of M with respect to the compact set $B_p(\rho_0)$. We also assume that the infimum of the Dirichlet spectrum of Δ on E is positive. In particular,

$$0 < \mu_1(E) \leq \frac{\int_E |\nabla f|^2}{\int_E f^2}$$

for any compactly supported function $f \in H_c^{1,2}(E)$ defined on E. Our goal is to give decay L^2-estimates on the harmonic functions $f \in \mathcal{K}^0(M)$ constructed in Theorem 21.1. This is also equivalent to estimating the barrier functions on a nonparabolic end, and hence the Green's function on a nonparabolic manifold. Note that by taking ρ_0 sufficiently large, the condition $\mu_1(E) > 0$ is guaranteed if $\mu_e(M) > 0$.

In 1981, Brooks [Br] improved Cheng and Yau's theorem and showed that if M has infinite volume and if we denote the volume entropy, $\tau(M)$, by

$$\tau(M) = \limsup_{\rho \to \infty} \frac{\log V_p(\rho)}{\rho},$$

then

$$\mu_e(M) \leq \frac{\tau^2(M)}{4}.$$

In a paper of Li and Wang [LW5], the authors proved a sharp estimate of the barrier functions. As one of the consequences, this implies an estimate on the volume growth (Corollary 22.6) and can be viewed as an improvement of Brooks' result.

Theorem 22.1 (Li–Wang) *Let M be a complete Riemannian manifold. Suppose E is an end of M with respect to $B_p(\rho_0)$ such that $\mu_1(E) > \mu$ for some constant $\mu \geq 0$. Let f be a nonnegative function defined on E satisfying the differential inequality*

$$\Delta f \geq -\mu f.$$

If f satisfies the growth condition

$$\int_{E(\rho)} f^2 e^{-2ar} = o(\rho)$$

as $\rho \to \infty$, with $a = \sqrt{\mu_1(E) - \mu}$, then it must satisfy the decay estimate

$$\int_{E(\rho+1)\setminus E(\rho)} f^2 \leq C(a)(1 + (\rho - \rho_0)^{-1}) e^{-2a\rho} \int_{E(\rho_0+1)\setminus E(\rho_0)} e^{2ar} f^2$$

for some constant $C(a) > 0$ depending on a and for all $\rho \geq 2(\rho_0 + 1)$, where $E(\rho) = B_p(\rho) \cap E$.

Proof We will first prove that for any $0 < \delta < 1$, there exists a constant $0 < C < \infty$ such that

$$\int_E e^{2\delta ar} f^2 \leq C.$$

Indeed, let $\phi(r(x))$ be a nonnegative cutoff function with support on E with $r(x)$ being the geodesic distance to the fixed point p. Then for any function $h(r(x))$, integration by parts yields

$$\int_E |\nabla(\phi e^h f)|^2$$

$$= \int_E |\nabla(\phi e^h)|^2 f^2 + \int_E (\phi e^h)^2 |\nabla f|^2 + 2 \int_E \phi e^h f \langle \nabla(\phi e^h), \nabla f \rangle$$

$$= \int_E |\nabla(\phi e^h)|^2 f^2 + \int_E \phi^2 e^{2h} |\nabla f|^2 + \frac{1}{2} \int_E \langle \nabla(\phi^2 e^{2h}), \nabla(f^2) \rangle$$

$$= \int_E |\nabla(\phi e^h)|^2 f^2 + \int_E \phi^2 e^{2h} |\nabla f|^2 - \frac{1}{2} \int_E \phi^2 e^{2h} \Delta(f^2)$$

$$\leq \int_E |\nabla(\phi e^h)|^2 f^2 + \mu \int_E \phi^2 e^{2h} f^2$$

$$= \int_E |\nabla \phi|^2 e^{2h} f^2 + 2 \int_E \phi e^{2h} \langle \nabla \phi, \nabla h \rangle f^2$$

$$+ \int_E \phi^2 |\nabla h|^2 e^{2h} f^2 + \mu \int_E \phi^2 e^{2h} f^2. \tag{22.1}$$

On the other hand, using the variational principle for $\mu_1(E)$, we have

$$\mu_1(E) \int_E \phi^2 e^{2h} f^2 \leq \int_E |\nabla(\phi e^h f)|^2,$$

hence (22.1) becomes

$$a^2 \int_E \phi^2 e^{2h} f^2$$
$$\leq \int_E |\nabla\phi|^2 e^{2h} f^2 + 2 \int_E \phi e^{2h} \langle \nabla\phi, \nabla h \rangle f^2 + \int_E \phi^2 |\nabla h|^2 e^{2h} f^2. \quad (22.2)$$

Let us now choose

$$\phi(r(x)) = \begin{cases} r(x) - \rho_0 & \text{on} \quad E(\rho_0 + 1) \setminus E(\rho_0), \\ 1 & \text{on} \quad E(\rho) \setminus E(\rho_0 + 1), \\ \rho^{-1}(2\rho - r(x)) & \text{on} \quad E(2\rho) \setminus E(\rho), \\ 0 & \text{on} \quad E \setminus E(2\rho) \end{cases}$$

and

$$h(r) = \begin{cases} \delta a r & \text{for} \quad r \leq A/(1+\delta)a, \\ A - ar & \text{for} \quad r \geq A/(1+\delta)a \end{cases}$$

for some fixed constant $A > (\rho_0 + 1)(1+\delta)a$. When $\rho \geq A/(1+\delta)a$, we see that

$$|\nabla h|^2 = \begin{cases} \delta^2 a^2 & \text{for} \quad r \leq A/(1+\delta)a, \\ a^2 & \text{for} \quad r \geq A/(1+\delta)a \end{cases}$$

and

$$\langle \nabla\phi, \nabla h \rangle = \begin{cases} \delta a & \text{on} \quad E(\rho_0 + 1) \setminus E(\rho_0), \\ \rho^{-1}a & \text{on} \quad E(2\rho) \setminus E(\rho), \\ 0 & \text{otherwise.} \end{cases}$$

Substituting the above inequality into (22.2), we obtain

$$a^2 \int_E \phi^2 e^{2h} f^2 \leq \int_{E(\rho_0+1)\setminus E(\rho_0)} e^{2h} f^2 + \rho^{-2} \int_{E(2\rho)\setminus E(\rho)} e^{2h} f^2$$
$$+ 2\delta a \int_{E(\rho_0+1)\setminus E(\rho_0)} e^{2h} f^2 + 2\rho^{-1} a \int_{E(2\rho)\setminus E(\rho)} e^{2h} f^2$$
$$+ \delta^2 a^2 \int_{E(A((1+\delta)a)^{-1})\setminus E(\rho_0)} \phi^2 e^{2h} f^2 + a^2$$
$$\times \int_{E(2\rho)\setminus E(A((1+\delta)a)^{-1})} \phi^2 e^{2h} f^2.$$

This can be rewritten as

$$a^2 \int_{E(A((1+\delta)a)^{-1})\setminus E(\rho_0+1)} e^{2h} f^2$$

$$\leq a^2 \int_{E(A((1+\delta)a)^{-1})} \phi^2 e^{2h} f^2$$

$$\leq \int_{E(\rho_0+1)\setminus E(\rho_0)} e^{2h} f^2 + \rho^{-2} \int_{E(2\rho)\setminus E(\rho)} e^{2h} f^2$$

$$+ 2\delta a \int_{E(\rho_0+1)\setminus E(\rho_0)} e^{2h} f^2 + 2\rho^{-1}a \int_{E(2\rho)\setminus E(\rho)} e^{2h} f^2$$

$$+ \delta^2 a^2 \int_{E(A((1+\delta)a)^{-1})\setminus E(\rho_0)} \phi^2 e^{2h} f^2,$$

hence

$$(1-\delta^2)a^2 \int_{E(A((1+\delta)a)^{-1})\setminus E(\rho_0+1)} e^{2h} f^2$$

$$\leq (\delta^2 a^2 + 2\delta a + 1) \int_{E(\rho_0+1)\setminus E(\rho_0)} e^{2h} f^2$$

$$+ \rho^{-2} \int_{E(2\rho)\setminus E(\rho)} e^{2h} f^2 + 2\rho^{-1}a \int_{E(2\rho)\setminus E(\rho)} e^{2h} f^2.$$

The definition of h and the assumption on the growth estimate on f imply that the last two terms on the right-hand side tend to 0 as $\rho \to \infty$, and we obtain the estimate

$$(1-\delta^2)a^2 \int_{E(A((1+\delta)a)^{-1})\setminus E(\rho_0+1)} e^{2\delta ar} f^2$$

$$\leq (\delta^2 a^2 + 2\delta a + 1) \int_{E(\rho_0+1)\setminus E(\rho_0)} e^{2\delta ar} f^2.$$

Since the right-hand side is independent of A, by letting $A \to \infty$ we conclude that

$$\int_{E\setminus E(\rho_0+1)} e^{2\delta ar} f^2 \leq C_1, \tag{22.3}$$

with

$$C_1 = \frac{\delta^2 a^2 + 2\delta a + 1}{(1-\delta^2)a^2} \int_{E(\rho_0+1)\setminus E(\rho_0)} e^{2\delta ar} f^2.$$

Our next step is to improve this estimate by setting $h = ar$ in the preceding argument. Note that (22.2) asserts that

$$-2a \int_E \phi \, e^{2ar} \langle \nabla \phi, \nabla r \rangle f^2$$
$$\leq \int_E |\nabla \phi|^2 e^{2ar} f^2.$$

For $\rho_0 < \rho_1 < \rho$, let us now choose ϕ to be

$$\phi(x) = \begin{cases} \dfrac{r(x) - \rho_0}{\rho_1 - \rho_0} & \text{on} \quad E(\rho_1) \setminus E(\rho_0), \\ \dfrac{\rho - r(x)}{\rho - \rho_1} & \text{on} \quad E(\rho) \setminus E(\rho_1). \end{cases}$$

We conclude that

$$\frac{2a}{(\rho - \rho_1)^2} \int_{E(\rho) \setminus E(\rho_1)} (\rho - r) e^{2ar} f^2$$
$$\leq \frac{1}{(\rho_1 - \rho_0)^2} \int_{E(\rho_1) \setminus E(\rho_0)} e^{2ar} f^2 + \frac{1}{(\rho - \rho_1)^2} \int_{E(\rho) \setminus E(\rho_1)} e^{2ar} f^2$$
$$+ \frac{2a}{(\rho_1 - \rho_0)^2} \int_{E(\rho_1) \setminus E(\rho_0)} (r - \rho_0) e^{2ar} f^2.$$

On the other hand, for any $0 < t < \rho - \rho_1$, since

$$\frac{2at}{(\rho - \rho_1)^2} \int_{E(\rho - t) \setminus E(\rho_1)} e^{2ar} f^2$$
$$\leq \frac{2a}{(\rho - \rho_1)^2} \int_{E(\rho) \setminus E(\rho_1)} (\rho - r) e^{2ar} f^2,$$

we deduce that

$$\frac{2at}{(\rho - \rho_1)^2} \int_{E(\rho - t) \setminus E(\rho_1)} e^{2ar} f^2$$
$$\leq \left(\frac{2a}{\rho_1 - \rho_0} + \frac{1}{(\rho_1 - \rho_0)^2} \right) \int_{E(\rho_1) \setminus E(\rho_0)} e^{2ar} f^2$$
$$+ \frac{1}{(\rho - \rho_1)^2} \int_{E(\rho) \setminus E(\rho_1)} e^{2ar} f^2. \tag{22.4}$$

Observe that if we take $\rho_1 = \rho_0 + 1$, $t = a^{-1}$, and set

$$g(\rho) = \int_{E(\rho) \setminus E(\rho_0 + 1)} e^{2ar} f^2,$$

then (22.4) can be written as
$$g(\rho - a^{-1}) \leq C_2 \rho^2 + \tfrac{1}{2} g(\rho),$$
where
$$C_2 = \frac{2a+1}{2} \int_{E(\rho_0+1) \setminus E(\rho_0)} e^{2ar} f^2$$
is independent of ρ. Iterating this inequality, for any positive integer k and $\rho \geq 1$, we obtain
$$g(\rho) \leq C_2 \sum_{i=1}^{k} \frac{(\rho + ia^{-1})^2}{2^{i-1}} + 2^{-k} g(\rho + ka^{-1})$$
$$\leq C_2 \rho^2 \sum_{i=1}^{\infty} \frac{(1 + ia^{-1})^2}{2^{i-1}} + 2^{-k} g(\rho + ka^{-1})$$
$$\leq C_3 \rho^2 + 2^{-k} g(\rho + ka^{-1}),$$
where
$$C_3 = C_2 \sum_{i=1}^{\infty} \frac{(1 + ia^{-1})^2}{2^{i-1}}.$$
However, our previous estimate (22.3) implies that
$$g(\rho + ka^{-1}) = \int_{E(\rho+ka^{-1}) \setminus E(\rho_0+1)} e^{2ar} f^2$$
$$\leq e^{2a(\rho+ka^{-1})(1-\delta)} \int_{E(\rho+ka^{-1}) \setminus E(\rho_0+1)} e^{2\delta ar} f^2$$
$$\leq C e^{2a(\rho+ka^{-1})(1-\delta)}.$$
Hence,
$$2^{-k} g(\rho + ka^{-1}) \to 0$$
as $k \to \infty$ by choosing $2(1 - \delta) < \ln 2$. This proves the estimate
$$\int_{E(\rho) \setminus E(\rho_0+1)} e^{2ar} f^2 \leq C_3 \rho^2 \tag{22.5}$$
for all $\rho \geq \rho_0 + 1$.

Using inequality (22.4) again and choosing $\rho_1 = \rho_0 + 1$ and $t = \rho/2$ this time, we conclude that

$$a\rho \int_{E(\rho/2)\setminus E(\rho_0+1)} e^{2ar} f^2$$

$$\leq (2a+1)(\rho - \rho_0 - 1)^2 \int_{E(\rho_0+1)\setminus E(\rho_0)} e^{2ar} f^2 + \int_{E(\rho)\setminus E(\rho_0+1)} e^{2ar} f^2.$$

Applying the estimate (22.5) to the second term on the right-hand side, we have

$$\int_{E(\rho/2)\setminus E(\rho_0+1)} e^{2ar} f^2 \leq C_5 \rho,$$

where

$$C_5 = (2a+1)a^{-1}\left(\frac{(\rho - \rho_0 - 1)^2}{\rho^2} + \sum_{i=1}^{\infty} \frac{(1+ia^{-1})^2}{2^i}\right)$$

$$\times \int_{E(\rho_0+1)\setminus E(\rho_0)} e^{2ar} f^2.$$

Therefore, for $\rho \geq 2(\rho_0 + 1)$, we have

$$\int_{E(\rho)} e^{2ar} f^2 \leq C(a) \rho \int_{E(\rho_0+1)\setminus E(\rho_0)} e^{2ar} f^2, \qquad (22.6)$$

where $C(a)$ denotes a constant depending only on a.

We are now ready to prove the lemma by using (22.6). Setting $t = 2a^{-1}$ and $\rho_1 = \rho - 4a^{-1}$ in (22.4), we obtain

$$\int_{E(\rho-2a^{-1})\setminus E(\rho-4a^{-1})} e^{2ar} f^2$$

$$\leq \left(\frac{8}{a(\rho - \rho_0 - 4a^{-1})} + \frac{4}{a^2(\rho - \rho_0 - 4a^{-1})^2}\right) \int_{E(\rho-4a^{-1})\setminus E(\rho_0)} e^{2ar} f^2$$

$$+ \frac{1}{4} \int_{E(\rho)\setminus E(\rho-4a^{-1})} e^{2ar} f^2.$$

According to (22.6), the first term of the right-hand side is bounded by

$$C(a)(1 + (\rho - \rho_0 - 4a^{-1})^{-1}) \int_{E(\rho_0+1)\setminus E(\rho_0)} e^{2ar} f^2$$

for $\rho \geq 2(\rho_0 + 1)$. Hence, by renaming ρ, the above inequality can be rewritten as

$$\int_{E(\rho+2a^{-1})\setminus E(\rho)} e^{2ar} f^2 \leq C(a)(1 + (\rho - \rho_0)^{-1})$$

$$\times \int_{E(\rho_0+1)\setminus E(\rho_0)} e^{2ar} f^2 + \frac{1}{3} \int_{E(\rho+4a^{-1})\setminus E(\rho+2a^{-1})} e^{2ar} f^2.$$

Iterating this inequality k times, we arrive at

$$\int_{E(\rho+2a^{-1})\setminus E(\rho)} e^{2ar} f^2 \leq C(a)(1 + (\rho - \rho_0)^{-1}) \sum_{i=0}^{k-1} 3^{-i} + 3^{-k}$$

$$\times \int_{E(\rho+2a^{-1}(k+1))\setminus E(\rho+2a^{-1}k)} e^{2ar} f^2.$$

However, using (22.6) again, we conclude that the second term is bounded by

$$3^{-k} \int_{E(\rho+2(k+1))\setminus E(\rho+2k)} e^{2ar} f^2$$

$$\leq C(a) \, 3^{-k} (\rho + 2(k+1)) \int_{E(\rho_0+1)\setminus E(\rho_0)} e^{2ar} f^2,$$

which tends to 0 as $k \to \infty$. Hence

$$\int_{E(\rho+2a^{-1})\setminus E(\rho)} e^{2ar} f^2 \leq C(a)(1 + (\rho - \rho_0)^{-1})$$

for $\rho + 4a^{-1} \geq 2(\rho_0 + 1)$, and the theorem follows when $a \leq 2$. If $a \geq 2$, we simply sum the above estimate $[a/2]$ times by dividing the interval $[\rho, \rho+1]$ into $[a/2]$ components. □

The following proposition is helpful in finding a lower bound for $\mu_1(M)$.

Proposition 22.2 *Let M be a compact Riemannian manifold with smooth boundary ∂M. If there exists a positive function f defined on M satisfying*

$$\Delta f \leq -\mu f,$$

then the first Dirichlet eigenvalue, $\mu_1(M)$, of M must satisfy

$$\mu_1(M) \geq \mu.$$

In particular, if M is complete, noncompact, and without boundary, and the positive function f described above is defined on M, then $\mu_1(M) \geq \mu$.

Proof Let us first assume that M is compact with smooth boundary ∂M. Let u be the first eigenfunction satisfying

$$\Delta u = -\mu_1(M) u \quad \text{on} \quad M$$

and

$$u = 0 \quad \text{on} \quad \partial M.$$

We may assume that $u \geq 0$ on M, and the regularity of u asserts that $u > 0$ in the interior of M. Integration by parts yields

$$(\mu_1(M) - \mu) \int_M uf \geq \int_M u \, \Delta f - \int_M f \, \Delta u$$

$$= \int_{\partial M} u \frac{\partial f}{\partial \nu} - \int_{\partial M} f \frac{\partial u}{\partial \nu}$$

$$\geq 0,$$

where ν is the outward unit normal of M. Hence $\mu_1(M) \geq \mu$ and the first part of the proposition is proved. When M is complete, noncompact, and without boundary, we apply the previous argument to any compact smooth subdomain $D \subset M$ and obtain

$$\mu_1(D) \geq \mu.$$

The second part of the proposition follows from the fact that

$$\inf_{D \subset M} \mu_1(D) = \mu_1(M). \qquad \square$$

We will now show that the hypothesis of Theorem 22.1 is the best possible. Indeed, if we consider the hyperbolic space form of constant -1 sectional curvature, then the metric in terms of polar coordinates is given by

$$ds^2_{\mathbb{H}^m} = dr^2 + \sinh^2 t \, ds^2_{\mathbb{S}^{m-1}}.$$

The volume growth is given by

$$V_p(r) = \alpha_{m-1} \int_0^r \sinh^{m-1} t \, dt \sim C \, e^{(m-1)r}$$

and

$$\mu_1(\mathbb{H}^m) \leq \frac{(m-1)^2}{4} \tag{22.7}$$

by Cheng's theorem (Corollary 6.4). In fact, if we consider

$$\beta(x) = \lim_{t \to \infty} (t - r(\gamma(t), x))$$

to be the Buseman function with respect to some geodesic ray γ, then following the computation in Chapter 4, we conclude that

$$\Delta \beta(x) = -\lim_{t \to \infty} \Delta r(\gamma(t), x)$$
$$= -(m-1).$$

Defining the positive function

$$h(x) = \exp\left(\frac{m-1}{2} \beta(x)\right),$$

a direct computation implies that

$$\Delta h = -\frac{(m-1)^2}{4} h.$$

Proposition 22.2 implies that $\mu_1(\mathbb{H}^m) \geq (m-1)^2/4$, and hence when combined with (22.7) yields $\mu_1(\mathbb{H}^m) = (m-1)^2/4$.

Let us now consider any nonconstant bounded harmonic function f, then

$$\int_{B_p(\rho)} f^2 e^{-2\sqrt{\mu_1} r} = O(\rho).$$

Of course, if the conclusion of Theorem 22.1 is valid, then f will be in $L^2(\mathbb{H}^m)$, which implies that f is identically constant by Yau's theorem (Lemma 7.1). However, it is known that $\mathbb{H}^m$ has an infinite dimensional space of bounded harmonic functions. This implies that the hypothesis of Theorem 22.1 cannot be relaxed.

Corollary 22.3 *Let M be a complete Riemannian manifold. Suppose E is an end of M such that $\mu_1(E) > 0$. Then for any harmonic function $f \in \mathcal{K}^0(M)$, there exists a constant a such that $f - a$ must be in $L^2(E)$. Moreover, the function $f - a$ must satisfy the decay estimate*

$$\int_{E(\rho+1) \setminus E(\rho)} (f-a)^2 \leq C \exp\left(-2\rho \sqrt{\mu_1(E)}\right)$$

for some constant $C > 0$ depending on f, $\mu_1(E)$, and m, where $E(\rho) = B_p(\rho) \cap E$.

Proof It suffices to prove the corollary for those functions f constructed in Theorem 21.1 because the decay property is preserved under linear combinations. Following the remark preceding Theorem 21.2, for a nonparabolic end E_1, let f_ρ be a sequence of harmonic functions by solving the Dirichlet boundary problem

$$\Delta f_\rho = 0 \quad \text{on} \quad B_p(\rho),$$

$$f_\rho = 1 \quad \text{on} \quad \partial B_p(\rho) \cap E_1$$

and

$$f_\rho = 0 \quad \text{on} \quad \partial B_p(\rho) \setminus E_1.$$

A subsequence of this sequence converges to $f \in \mathcal{K}^0(M)$ uniformly on compact subsets of M. For any fixed end E, since f_ρ has the boundary value either 0 or 1 on $\partial E(\rho)$, by considering either the function f_ρ or $1 - f_\rho$, we may assume that f_ρ has the boundary value 0 on $\partial B_p(\rho) \cap E$. Let us define the function g_ρ by

$$g_\rho(x) = \begin{cases} f_\rho(x) & \text{on} \quad E(\rho), \\ 0 & \text{on} \quad E \setminus E(\rho). \end{cases}$$

Clearly, g_ρ is a nonnegative subharmonic function defined on E. Moreover, since g_ρ has compact support, the hypothesis of Theorem 22.1 is satisfied and the decay estimate holds for g_ρ. The corollary follows by taking $\rho \to \infty$. □

We point out that Corollary 22.3 also holds for any function f with $a = 0$ provided that f is the limit of a sequence of harmonic functions f_ρ on $E(\rho)$ satisfying $f_\rho = 0$ on $\partial E(\rho)$ regardless of their boundary values on ∂E.

In particular, when applying Corollary 22.3 to the Green's function, we obtain the following sharp decay estimate.

Corollary 22.4 *Let M be a complete manifold with $\mu_1(M) > 0$. Then the minimal positive Green's function $G(p, \cdot)$ with a pole at $p \in M$ must satisfy the decay estimate*

$$\int_{B_p(\rho+1) \setminus B_p(\rho)} G^2(p, x) \, dx \leq C \exp\left(-2\rho \sqrt{\mu_1(M)}\right)$$

for $\rho \geq 1$.

In the case when $M = \mathbb{H}^m$, (17.8) asserts that the Green's function is given by

$$G(p, x) = \int_{r(x)}^\infty \frac{dt}{A_p(t)},$$

where $A_p(t) = \alpha_{m-1} \sinh^{(m-1)} t$ is the area of the boundary of the geodesic ball of radius t centered at $p \in \mathbb{H}^m$. One computes readily that

$$\int_{B_p(\rho+1) \setminus B_p(\rho)} G^2(p, x)\, dx \sim C \exp(-(m-1)\rho).$$

Since $\mu_1(\mathbb{H}^m) = (m-1)^2/4$, the quantity $2\sqrt{\mu_1(\mathbb{H}^m)}$ is exactly $(m-1)$, indicating the sharpness of Corollary 22.4.

Applying Corollary 22.3, we obtain volume estimates for those ends with positive spectrum. As pointed out in the introduction, these estimates are sharp. The sharp growth estimate is realized by the hyperbolic space $\mathbb{H}^m$, while the sharp decay estimate is realized by a hyperbolic cusp (see Example 21.1). To state our estimate, let us denote the volume of the set $E(\rho)$ by $V_E(\rho)$ and the volume of the end E by $V(E)$.

Theorem 22.5 (Li–Wang) *Let E be an end of complete manifold M with $\mu_1(E) > 0$.*

(1) If E is a parabolic end, then E must have exponential volume decay given by

$$V_E(\rho + 1) - V_E(\rho) \leq C (1 + (\rho - \rho_0)^{-1}) (V_E(\rho_0 + 1) - V_E(\rho_0))$$
$$\times \exp\left(-2(\rho - \rho_0)\sqrt{\mu_1(E)}\right)$$

for some constant $C > 0$ depending on the $\mu_1(E)$. In particular, we have

$$V(E) - V_E(\rho) \leq C (V_E(\rho_0 + 1) - V_E(\rho_0))$$
$$\times \exp\left(-2(\rho - \rho_0)\sqrt{\mu_1(E)}\right).$$

(2) If E is a nonparabolic end, then E must have exponential volume growth given by

$$V_E(\rho) \geq C \exp\left(2\rho \sqrt{\mu_1(E)}\right)$$

for all $\rho \geq \rho_0 + 1$ and for some constant $C > 0$ depending on the end E.

Proof Let f_ρ be the harmonic function on $E(\rho)$ with $f_\rho = 1$ on ∂E and $f_\rho = 0$ on $\partial E(\rho)$. The assumption that E is parabolic implies that f_ρ converges to $f = 1$ as $\rho \to \infty$. Hence the estimate from Theorem 22.1 yields the first estimate of the theorem. Letting $\rho = \rho + i$ for $i = 0, 1, \ldots$ and summing over i, we conclude that

$$V(E) - V_E(\rho) \leq C \left(V_E(\rho_0 + 1) - V_E(\rho_0)\right) \sum_{i=0}^{\infty} (1 + (\rho + i - \rho_0)^{-1})$$

$$\times \exp\left(-2(\rho + i - \rho_0)\sqrt{\mu_1(E)}\right)$$

$$\leq C \left(V_E(\rho_0 + 1) - V_E(\rho_0)\right) \exp\left(-2(\rho - \rho_0)\sqrt{\mu_1(E)}\right).$$

This proves the volume decay estimate for the case of parabolic ends.

If E is nonparabolic, then f_ρ converges to a nonconstant harmonic function f on E. Thus, we conclude that there exists a positive constant C such that for $r \geq \rho_0$

$$C = \int_{\partial E} \frac{\partial f}{\partial \nu}$$

$$= \int_{\partial B_p(r) \cap E} \frac{\partial f}{\partial \nu}$$

$$\leq \int_{\partial B_p(r) \cap E} |\nabla f|$$

$$\leq A_E^{1/2}(r) \left(\int_{\partial B_p(r) \cap E} |\nabla f|^2\right)^{1/2},$$

or equivalently

$$\frac{C}{A_E(r)} \leq \int_{\partial B_p(r) \cap E} |\nabla f|^2.$$

Integrating the preceding inequality with respect to r from ρ to $\rho + 1$ and using Corollary 22.3 and Lemma 7.1, we obtain

$$\int_{\rho}^{\rho+1} \frac{1}{A_E(r)} \, dr \leq C \int_{E(\rho+1) \setminus E(\rho)} |\nabla f|^2$$

$$\leq C \exp\left(-2\rho \sqrt{\mu_1(E)}\right).$$

Therefore,
$$1 \le \int_\rho^{\rho+1} A_E(r)\, dr \int_\rho^{\rho+1} \frac{1}{A_E(r)} dr$$
$$\le C \exp\left(-2\rho \sqrt{\mu_1(E)}\right) (V_E(\rho+1) - V_E(\rho))$$
$$\le C \exp\left(-2\rho \sqrt{\mu_1(E)}\right) V_E(\rho+1).$$

Since ρ is arbitrary, we conclude that
$$V_E(\rho) \ge C \exp\left(2\rho \sqrt{\mu_1(E)}\right)$$
by adjusting the constant C, and the theorem is proved. $\square$

Corollary 22.6 *Let M be a complete manifold with infinite volume. Suppose the essential spectrum of M has a positive lower bound, i.e.,*
$$\mu_e(M) > 0,$$
then for any $\epsilon > 0$ there exists $C > 0$, depending on ϵ, such that
$$V_p(\rho) \ge C \exp\left(2\rho \sqrt{\mu_e(M) - \epsilon}\right).$$

Proof For $\epsilon > 0$, let ρ_0 be sufficiently large that $\mu_1(M \setminus B_p(\rho_0)) + \epsilon \ge \mu_e(M)$. Theorem 22.5 and the infinite volume assumption on M assert that M must have at least one infinite volume end E with respect to $B_p(\rho_0)$. In fact, the volume growth of this end must satisfy
$$V_E(\rho) \ge C \exp\left(2\rho \sqrt{\mu_e(M) - \epsilon}\right),$$
hence
$$V_p(\rho) \ge C \exp\left(2\rho \sqrt{\mu_e(M) - \epsilon}\right). \quad \square$$

Our last corollary concerns L^q harmonic functions on an end with positive bottom spectrum.

Corollary 22.7 *Let E be an end of M with $\mu_1(E) > 0$. Let $\mathcal{L}^q(E)$ be the space of L^q harmonic functions on E. If $u \in \mathcal{L}^q(E)$ with $q \ge 2$, then u must be bounded and it must satisfy the estimate*
$$\int_{E(\rho+1) \setminus E(\rho)} u^2 \le C \exp\left(-2\rho \sqrt{\mu_1(E)}\right).$$

If $u \in \mathcal{L}^q(E)$ with $1 < q < 2$, then the same conclusion is true provided that the volume growth of E is bounded by

$$V_E(\rho) \leq C \exp\left(\frac{2q}{2-q} \rho \sqrt{\mu_1(E)}\right).$$

Proof Let $u \in \mathcal{L}^q(E)$ be an L^q harmonic function. Define f_ρ to be the harmonic function on $E(\rho)$ satisfying

$$f_\rho = 0 \quad \text{on} \quad \partial B_p(\rho) \cap E$$

and

$$f_\rho = u \quad \text{on} \quad \partial E.$$

Clearly, the maximum principle asserts that a subsequence of f_ρ as $\rho \to \infty$ converges to a function $f \in \mathcal{L}^\infty(E)$ with $u = f$ on ∂E. Moreover, by Corollary 22.3, f satisfies the estimate

$$\int_{E(\rho+1) \setminus E(\rho)} f^2 \leq C \exp\left(-2\rho \sqrt{\mu_1(E)}\right). \tag{22.8}$$

If $q \geq 2$, the boundedness of f and (22.8) imply that $f \in \mathcal{L}^q(E)$. In particular, the function $u - f$ is in $\mathcal{L}^q(E)$ with 0 boundary condition on ∂E. Applying the uniqueness theorem of Yau [Y2] (Lemma 7.1) for L^q harmonic functions, we conclude that $u = f$.

For $1 < q < 2$, the Schwarz inequality, (22.8), and the volume growth bound give

$$\int_{E(\rho+1) \setminus E(\rho)} f^q \leq \left(\int_{E(\rho+1) \setminus E(\rho)} f^2\right)^{q/2} (V_E(\rho+1) - V_E(\rho))^{(2-q)/2}$$

$$\leq C \exp\left(-q \rho \sqrt{\mu_1(E)}\right) \exp\left(q \rho \sqrt{\mu_1(E)}\right)$$

$$\leq C,$$

implying that the L^q-norm of f is at most of linear growth. Again, by applying the argument of Lemma 7.1 to the subharmonic function $g = |f - u|$ defined on E with boundary condition

$$g = 0 \quad \text{on} \quad \partial E,$$

we conclude that $g = 0$. To see this, let ϕ be a cutoff function satisfying

$$\phi = \begin{cases} 1 & \text{on} \quad E(\rho), \\ 0 & \text{on} \quad E \setminus E(2\rho), \end{cases}$$

and
$$|\nabla \phi| \leq C \rho^{-1} \quad \text{on} \quad E(2\rho) \setminus E(\rho).$$
Integration by parts yields
$$0 \leq \int_E \phi^2 g^{q-1} \Delta g$$
$$= -2 \int_E \phi g^{q-1} \langle \nabla \phi, \nabla g \rangle - (q-1) \int_E \phi^2 g^{q-2} |\nabla g|^2.$$
On the other hand, applying the Schwarz inequality
$$-2 \int_E \phi g^{q-1} \langle \nabla \phi, \nabla g \rangle \leq \frac{q-1}{2} \int_E \phi^2 g^{q-2} |\nabla g|^2 + \frac{2}{q-1} \int_E |\nabla \phi|^2 g^q,$$
we conclude that
$$\int_E \phi^2 g^{q-2} |\nabla g|^2 \leq \frac{4}{(q-1)^2} \int_E |\nabla \phi|^2 g^q.$$
Using the definition of ϕ, this implies that
$$\int_{E(\rho)} g^{q-2} |\nabla g|^2 \leq \frac{C}{\rho^2} \int_{E(2\rho) \setminus E(\rho)} g^q.$$

The growth estimate on the L^q-norm of f and the fact that $u \in L^q$ imply that the right-hand side tends to 0 as $\rho \to \infty$. Hence g must be identically constant. The boundary condition of g asserts that it must be identically 0, and $f = u$.

In both cases, since $f = u$, the function u must satisfy (22.7). This concludes the corollary. □

23
Manifolds with Ricci curvature bounded from below

In this chapter, we will assume that M^m has Ricci curvature bounded from below by $-(m-1)R$ for some constant $R > 0$. After normalizing, we may assume that $R = 1$. Note that the Bishop volume comparison theorem asserts that for any $x \in M$, the ratio of the volumes of geodesic balls $B_x(\rho_1)$ and $B_x(\rho_2)$ for $\rho_1 < \rho_2$ must satisfy

$$\frac{V_x(\rho_2)}{V_x(\rho_1)} \leq \frac{\bar{V}(\rho_2)}{\bar{V}(\rho_1)}, \tag{23.1}$$

where $\bar{V}(\rho)$ is the volume of a geodesic ball of radius ρ in the m-dimensional hyperbolic space form $\mathbb{H}^m$ of constant -1 curvature. In particular, by taking $x = p$, $\rho_1 = 0$, and $\rho_2 = \rho$ this implies that

$$V_p(\rho) \leq C_3 \exp((m-1)\rho) \tag{23.2}$$

for sufficiently large ρ. On the other hand, if we let $x \in \partial B_p(\rho)$, $\rho_1 = 1$, and $\rho_2 = \rho + 1$, (23.1) implies

$$V_x(1) \geq C_4 \, V_x(\rho+1) \, \exp(-(m-1)\rho)$$
$$\geq C_4 \, V_p(1) \, \exp(-(m-1)\rho). \tag{23.3}$$

Hence we have the following volume estimate.

Proposition 23.1 *Let M^m be a complete manifold with Ricci curvature bounded from below by*

$$\mathcal{R}_{ij} \geq -(m-1).$$

The volume growth of M must satisfy the upper bound

$$V_p(\rho) \leq \bar{V}(\rho) \leq C_3 \exp((m-1)\rho)$$

for some constant $C_3 > 0$ depending only on m. Also the volume cannot decay faster than exponentially and must satisfy the estimate

$$V_x(1) \geq C_4 \, V_p(1) \, \exp(-(m-1)r(x)),$$

where $r(x)$ denotes the distance from p to x. In particular,

$$V_p(\rho + 1) - V_p(\rho - 1) \geq C_4 \, V_p(1) \, \exp(-(m-1)\rho).$$

Note that, in view of Theorem 22.5, if we also assume that $\mu_1(M) > 0$, then one has both upper and lower control of the volume growth for a parabolic end and also for a nonparabolic end.

Recall that by Cheng's theorem (Corollary 6.4), under the Ricci curvature assumption

$$\mu_1(M) \leq \frac{(m-1)^2}{4}.$$

While $\mathbb{H}^m$ achieves equality, it is not the only manifold with this property. In fact, let us consider the following example.

Example 23.2 Let $M^m = \mathbb{R} \times N^{m-1}$ be the complete manifold with the warped product metric

$$ds_M^2 = dt^2 + \exp(2t) \, ds_N^2.$$

According to the computation in (A.3) and (A.4) (see Appendix A), by setting $f(t) = \exp(t)$, the Ricci curvature on M is given by

$$\mathcal{R}_{1j} = -(m-1)\delta_{1j}$$

and

$$\mathcal{R}_{\alpha\beta} = \exp(-2t) \, \tilde{\mathcal{R}}_{\alpha\beta} - (m-1)\delta_{\alpha\beta},$$

where $\tilde{\mathcal{R}}_{\alpha\beta}$ is the Ricci tensor on N and $e_1 = \partial/\partial t$. In particular, if the Ricci curvature of N is nonnegative, then

$$\mathcal{R}_{ij} \geq -(m-1).$$

Moreover, N is Ricci flat if and only if M is Einstein with

$$\mathcal{R}_{ij} = -(m-1).$$

Let us consider the function $g(t) = \exp(-\alpha t)$ for a constant $(m-1)/2 \leq \alpha \leq m-1$. One computes that

$$\Delta g = \frac{d^2 g}{dt^2} + (m-1)\frac{dg}{dt}$$
$$= \alpha^2 g - (m-1)\alpha g$$
$$= -\alpha(m-1-\alpha)g$$

and

$$|\nabla g|^2 = \alpha^2 g^2.$$

If we let $\alpha = (m-1)/2$, then the function

$$g(t) = \exp\left(-\frac{(m-1)t}{2}\right)$$

satisfies the equation

$$\Delta g = -\frac{(m-1)^2}{4} g.$$

Since g is positive, Proposition 22.2 implies that $\mu_1(M) \geq (m-1)^2/4$. Using the upper bound of Cheng (Corollary 6.4), we conclude that $\mu_1(M) = (m-1)^2/4$.

It turns out that Example 23.2 gives the only class of manifolds with dimension $m \geq 3$ that have more than one end on which Cheng's inequality is achieved.

There is another important example that is relevant to the issues being considered in this chapter.

Example 23.3 For $m \geq 3$, let $M^m = \mathbb{R} \times N^{m-1}$ be the product manifold endowed with the warped product metric

$$ds_M^2 = dt^2 + \cosh^2(t)\, ds_N^2,$$

where N is a compact manifold with Ricci curvature bounded from below by

$$\tilde{\mathcal{R}}_{\alpha\beta} \geq -(m-2).$$

Following the notation and the computation in Appendix A, by setting $f(t) = \cosh(t)$, the Ricci curvature on M is given by

$$\mathcal{R}_{1j} = -(m-1)\delta_{1j}$$

and
$$\mathcal{R}_{\alpha\beta} = \cosh^{-2}(t)\tilde{\mathcal{R}}_{\alpha\beta} - \left(1 + (m-2)\tanh^2(t)\right)\delta_{\alpha\beta}.$$

Hence $\mathcal{R}_{ij} \geq -(m-1)$ because of the assumption that $\tilde{\mathcal{R}}_{\alpha\beta} \geq -(m-2)$. Moreover, M is Einstein with $\mathcal{R}_{ij} = -(m-1)$ if and only if N is Einstein with $\tilde{\mathcal{R}}_{\alpha\beta} = -(m-2)$.

Note that if we define $g(t) = \cosh^{-(m-2)}(t)$, then

$$\Delta g = \left(\frac{\partial^2}{\partial t^2} + (m-1)\frac{\sinh(t)}{\cosh(t)}\frac{\partial}{\partial t}\right)\cosh^{-(m-2)}t$$

$$= -(m-2)\cosh^{-(m-2)}(t)$$

$$= -(m-2)g(t).$$

Proposition 22.2 implies that $\mu_1(M) \geq (m-2)$. If we compute the L^2-norm of g on the set $[-t,t] \times N$, then

$$\int_{[-t,t] \times N} g^2 = \int_{-t}^{t} V(N)\cosh^{m-1}(t)\cosh^{-2(m-2)}(t)\,dt$$

$$= V(N)\int_{-t}^{t} \cosh^{3-m}(t)\,dt.$$

This implies that $g \in L^2(M)$ for $m > 3$, and g is an eigenfunction, hence $\mu_1(M) \leq m-2$. When $m = 3$, the L^2-norm of g over $[-t,t] \times N$ grows linearly in t. Although g is not an eigenfunction, it is still sufficient to show that $\mu_1(M) \leq m-2$, as will be seen in the next theorem. In any event, we have

$$\mu_1(M) = m - 2.$$

The following improved version of the Bochner formula for harmonic functions was first proved by Yau [Y1]. It is by now standard in the literature.

Lemma 23.4: (Yau) *Let M^m be a complete manifold whose Ricci curvature is bounded by*

$$\mathcal{R}_{ij}(x) \geq -(m-1)k(x)$$

for some function $k(x)$. Suppse f is a harmonic function defined on M, then

$$\Delta|\nabla f|^q \geq -q(m-1)k(x)|\nabla f|^q$$

$$+ \frac{1}{q}\left(\frac{m}{m-1} + q - 2\right)|\nabla f|^{-q}|\nabla(|\nabla f|^q)|^2$$

for all $q > 0$. In particular, if $q \geq (m-2)/(m-1)$, then

$$\Delta |\nabla f|^q \geq -q(m-1)k(x)|\nabla f|^q.$$

Proof Let $\{e_1, \ldots, e_m\}$ be an orthonormal frame near a point $x \in M$ so that $|\nabla f| e_1 = \nabla f$ at x. We will follow a similar computation to that in the proof of Lemma 5.6. Since f is harmonic, a direct computation and the commutation formula yield

$$\Delta |\nabla f|^2 \geq 2 f_{ij}^2 - 2(m-1)k(x)|\nabla f|^2.$$

Also, just like (5.13) and (5.14), we have

$$|\nabla |\nabla f|^2|^2 = 4|\nabla f|^2 \sum_{j=1}^m f_{1j}^2$$

and

$$f_{ij}^2 \geq f_{11}^2 + 2\sum_{\alpha=2}^m f_{1\alpha}^2 + \frac{(\Delta f - f_{11})^2}{m-1}$$

$$\geq \frac{m}{m-1} \sum_{j=1}^m f_{1j}^2.$$

Combining the above three inequalities, we obtain

$$\Delta |\nabla f|^2 \geq \frac{m}{2(m-1)} |\nabla f|^{-2} |\nabla |\nabla f|^2|^2 - 2(m-1)k(x)|\nabla f|^2$$

$$= \frac{2m}{(m-1)} |\nabla |\nabla f||^2 - 2(m-1)k(x)|\nabla f|^2.$$

The lemma follows directly from this inequality. $\square$

Theorem 23.5 (Li–Wang) *Let M^m be a complete Riemannian manifold of dimension $m \geq 3$. Suppose*

$$\mathcal{R}_{ij} \geq -(m-1)$$

and

$$\mu_1(M) \geq m-2,$$

then either:

(1) M has only one end with infinite volume; or
(2) M is the warped product given by Example 23.3.

23 Manifolds with Ricci curvature bounded from below

Proof Assume that M has more than one infinite volume end. According to Theorem 21.3, there exists a nonconstant function $f \in \mathcal{K}^0(M)$ given by the construction. Let g be the function defined by

$$g = |\nabla f|^{(m-2)/(m-1)}.$$

Lemma 23.4 asserts that

$$\Delta g \geq -(m-2)g.$$

We now claim that the function g must satisfy the integral condition

$$\int_{B_p(2\rho) \setminus B_p(\rho)} g^2 \leq C\rho.$$

To see this, let us apply Hölder's inequality and get

$$\int_{B_p(2\rho) \setminus B_p(\rho)} g^2$$

$$\leq \left(\int_{B_p(2\rho) \setminus B_p(\rho)} \exp\left(2r \sqrt{\mu_1(M)}\right) |\nabla f|^2 \right)^{(m-2)/(m-1)}$$

$$\times \left(\int_{B_p(2\rho) \setminus B_p(\rho)} \exp\left(-2(m-2)r \sqrt{\mu_1(M)}\right) \right)^{1/(m-1)}. \quad (23.4)$$

Using the lower bound on the Ricci curvature, the volume comparison theorem asserts that

$$A_p(r) \leq C \exp((m-1)r),$$

for some constant $C > 0$ depending only on m alone. The Fubini theorem now yields

$$\int_{B_p(2\rho) \setminus B_p(\rho)} \exp\left(-2(m-2)r \sqrt{\mu_1(M)}\right)$$

$$\leq C \int_\rho^{2\rho} \exp\left(-2(m-2)r \sqrt{\mu_1(M)}\right) \exp((m-1)r) \, dr$$

$$= C \int_\rho^{2\rho} \exp\left((m-1)r - 2(m-2)r \sqrt{\mu_1(M)}\right) dr. \quad (23.5)$$

Using the lower bound of $\mu_1(M)$, we conclude that the right-hand side of (23.5) is at most linear in ρ when $m = 3$, and exponentially decays to 0 when $m \geq 4$.

On the other hand, recall that the decay estimate in Theorem 22.1 implies that

$$\int_{E(\rho+1)\setminus E(\rho)} (f-a)^2 \leq C \, \exp\left(-2\rho \sqrt{\mu_1(M)}\right),$$

where a is the asymptotic value of f at infinity of an end E given by either 1 or 0. Combining this inequality with Lemma 7.1 by choosing $\alpha = 1$, $\rho_1 = \rho$, $\rho_2 = \rho + 1$, $\rho_3 = \rho + 2$, and $\rho_4 = \rho + 3$, we have

$$\int_{E(\rho+2)\setminus E(\rho+1)} |\nabla f|^2 \leq 8C \, \exp\left(-2\rho \sqrt{\mu_1(M)}\right).$$

In particular, there exists a constant $C_1 > 0$ independent of ρ such that

$$\int_{E(\rho+2)\setminus E(\rho+1)} \exp\left(2r \sqrt{\mu_1(M)}\right) |\nabla f|^2 \leq C_1$$

and

$$\int_{B_p(2\rho)\setminus B_p(\rho)} \exp\left(2r \sqrt{\mu_1(M)}\right) |\nabla f|^2 \leq C_1 \, \rho.$$

Applying this and (23.5) to (23.4), we conclude that

$$\int_{B_p(2\rho)\setminus B_p(\rho)} g^2 \leq C_2 \, \rho$$

for the case when $n = 3$, and

$$\int_{B_p(2\rho)\setminus B_p(\rho)} g^2 \to 0$$

when $n \geq 4$. This proves our claim on the L^2 estimate of g.

To complete our proof of the theorem, we consider ϕ to be a nonnegative compactly supported function on M, then

$$\int_M |\nabla(\phi g)|^2 = \int_M |\nabla \phi|^2 g^2 + 2 \int_M \phi g \, \langle \nabla \phi, \nabla g \rangle + \int_M \phi^2 \, |\nabla g|^2. \quad (23.6)$$

23 Manifolds with Ricci curvature bounded from below

The second term on the right-hand side can be written as

$$2\int_M \phi g \langle \nabla\phi, \nabla g\rangle = \frac{1}{2}\int_M \langle \nabla(\phi^2), \nabla(g^2)\rangle$$

$$= -\int_M \phi^2 g \Delta g - \int_M \phi^2 |\nabla g|^2$$

$$= (m-2)\int_M \phi^2 g^2 - \int_M \phi^2 |\nabla g|^2$$

$$-\int_M \phi^2 g (\Delta g + (m-2)g). \tag{23.7}$$

Combining (23.6) with (23.7) and the variational charaterization of $\mu_1(M) \geq m-2$, this implies that

$$(m-2)\int_M \phi^2 g^2 \leq \int_M |\nabla(\phi g)|^2$$

$$= (m-2)\int_M \phi^2 g^2 + \int_M |\nabla\phi|^2 g^2$$

$$-\int_M \phi^2 g (\Delta g + (m-2)g),$$

hence we have

$$\int_M \phi^2 g (\Delta g + (m-2)g) \leq \int_M |\nabla\phi|^2 g^2. \tag{23.8}$$

For $\rho > 0$, let us choose ϕ to satisfy the properties that

$$\phi = \begin{cases} 1 & \text{on} \quad B_p(\rho), \\ 0 & \text{on} \quad M\setminus B_p(2\rho), \end{cases}$$

and

$$|\nabla\phi| \leq C\rho^{-1} \quad \text{on} \quad B_p(2\rho)\setminus B_p(\rho)$$

for some constant $C > 0$. Then the right-hand side of (23.8) can be estimated by

$$\int_M |\nabla\phi|^2 g^2 \leq C\rho^{-2}\int_{B_p(2\rho)\setminus B_p(\rho)} g^2.$$

By the L^2 estimate of g, this tends to 0 as $\rho \to \infty$. Therefore taking the above inequality together with (23.8), we conclude that either g must be identically 0 or it must satisfy

$$\Delta g = -(m-2)g.$$

If M has more than one infinite volume end, then by Theorem 21.3 there must exist a nonconstant f, hence $g \neq 0$. So all the inequalities used in deriving Lemma 23.4 become equalities.

We can now argue to conclude that $M = \mathbb{R} \times N$ with the warped product metric $ds^2 = dt^2 + \cosh^2(t) \, ds_N^2$ for some compact manifold N with Ricci curvature satisfying $\tilde{R} \geq -(m-2)$.

Indeed, since $\Delta f = 0$, the Hessian of f must be of the form

$$(f_{ij}) = \begin{pmatrix} -(m-1)\mu & 0 & 0 & \cdots & 0 \\ 0 & \mu & 0 & \cdots & 0 \\ 0 & 0 & \mu & \cdots & 0 \\ \vdots & \vdots & \vdots & \ddots & \\ 0 & 0 & 0 & \cdots & \mu \end{pmatrix}.$$

The fact that $f_{1\alpha} = 0$ for all $\alpha \neq 1$ implies that $|\nabla f|$ is identically constant along the level set of f. In particular, the level sets of $|\nabla f|$ and f coincide. Moreover,

$$\mu \, \delta_{\alpha\beta} = f_{\alpha\beta}$$
$$= h_{\alpha\beta} \, f_1$$

with $(h_{\alpha\beta})$ being the second fundamental form of the level set of f with respect to e_1. Hence

$$f_{11} = -H \, f_1, \tag{23.9}$$

where H is the mean curvature of the level set of f with respect to e_1. Applying the same computation to the function g, we obtain

$$-(m-2) \, g = \Delta g$$
$$= g_{11} + H \, g_1. \tag{23.10}$$

On the other hand, since $g = |\nabla f|^{(m-2)/(m-1)}$, we have

$$g_1 = \left(|\nabla f|^{(m-2)/(m-1)} \right)_1$$
$$= \frac{m-2}{m-1} |\nabla f|^{-m/(m-1)} f_i \, f_{i1}$$
$$= \frac{m-2}{m-1} f_1^{-1/(m-1)} f_{11}.$$

Hence, combining this identity with (23.9), we conclude that

$$H = -f_1^{-1} f_{11}$$
$$= -\frac{m-1}{m-2} g_1 g^{-1}. \qquad (23.11)$$

Substituting this into (23.10) yields

$$g_{11} - \frac{m-1}{m-2}(g_1)^2 g^{-1} + (m-2)g = 0.$$

Setting $u = g^{-1/(m-2)} = |\nabla f|^{-1/(m-1)}$, this differential equation becomes

$$u_{11} - u = 0.$$

Viewing this as an ordinary differential equation along the integral curve generated by the vector field e_1, one concludes that

$$u(t) = A \exp(t) + B \exp(-t).$$

Since u must be nonnegative, A and B must be nonnegative. Moreover, $\nabla f \neq 0$.

Note that M is assumed to have at least two infinite volume ends. We claim that any fixed level set N of $|\nabla f|$ must be compact. Indeed, by the facts that f has no critical points and that the level set of f coincides with the level set of $|\nabla f|$, M must be topologically the product $\mathbb{R} \times N$. If N is noncompact, then M has only one end, hence N must be compact.

If both A and B are nonzero, then the function u must have its minimum along N; hence by reparameterizing, we may assume N is given by $t = 0$. Moreover, by scaling f, we may also assume that $N = \{|\nabla f| = 1\}$. Therefore,

$$0 = u'(0) = A - B$$

and

$$1 = u(0) = A + B.$$

This implies that

$$u(t) = \cosh(t)$$

and

$$g(t) = \cosh^{-(m-2)}(t).$$

Using (23.11), we conclude that
$$H(t) = (m-1)\tanh(t)$$
and
$$\vec{II}_{\alpha\beta}(t) = \delta_{\alpha\beta}\tanh(t).$$
This implies that the metric on $M = \mathbb{R} \times N$ must be of the form
$$ds_M = dt^2 + \cosh^2(t)\, ds_N^2$$
as claimed.

If either A or B is 0, say $B = 0$, then
$$u(t) = A\exp(t).$$
In this case, after normalizing $A = 1$, we have
$$g(t) = \exp(-(m-2)t).$$
Using (23.11), we conclude that
$$H(t) = (m-1)$$
and
$$h_{\alpha\beta}(t) = \delta_{\alpha\beta}.$$
This implies that the metric on $M = \mathbb{R} \times N$ must be of the form
$$ds_M^2 = dt^2 + \exp(2t)\, ds_N^2.$$
However, this implies that M has only one infinite volume end which contradicts our assumption. This completes the proof of the theorem. The curvature restriction on N follows by direct computation and from the curvature assumption on M. $\square$

Theorem 23.6 (Li–Wang) *Let M^m be a complete m-dimensional manifold with $m \geq 3$. Suppose that*
$$\mathcal{R}_{ij} \geq -(m-1)$$
and
$$\mu_1(M) \geq \frac{(m-1)^2}{4}.$$

23 Manifolds with Ricci curvature bounded from below

Then:

(1) M has only one end;
(2) $M = \mathbb{R} \times N$ with the warped product metric
$$ds_M^2 = dt^2 + \exp(2t)\, ds_N^2,$$
where N is a compact manifold with nonnegative Ricci curvature; or
(3) M is of dimension 3 and $M^3 = \mathbb{R} \times N^2$ with the warped product metric
$$ds_M^2 = dt^2 + \cosh^2(t)\, ds_N^2,$$
where N is a compact surface with Gaussian curvature bounded from below by -1.

Proof Assuming that M has more than one end, then Theorem 23.5 implies that there must only be one nonparabolic end unless $m = 3$, in which case $m - 2 = (m-1)^2/4$, accounting for case (3) in our theorem. We may now assume that M has only one nonparabolic end E_1 and at least one parabolic end E_2. Theorem 21.3 implies that there exists a nonconstant positive harmonic function $f \in \mathcal{K}'(M)$ defined on M with the property that $\limsup_{x \to E_2(\infty)} f(x) = \infty$. The gradient estimate of Theorem 6.1 asserts that
$$|\nabla f|^2 \le (m-1)^2 f^2.$$
Combining this estimate with the fact that f is harmonic we obtain
$$\Delta f^{1/2} = -\frac{1}{4} f^{-3/2} |\nabla f|^2$$
$$\le -\frac{(m-1)^2}{4} f^{1/2}. \tag{23.12}$$
If we write $h = f^{1/2}$, then for any nonnegative cutoff function ϕ we have
$$\int_M |\nabla(\phi h)|^2 = \int_M |\nabla \phi|^2 h^2 - \int_M \phi^2 h\, \Delta h$$
$$\le \int_M |\nabla \phi|^2 h^2 + \frac{(m-1)^2}{4} \int_M \phi^2 h^2$$
$$- \int_M \phi^2 h \left(\frac{(m-1)^2}{4} h + \Delta h \right).$$
On the other hand, the assumption $\mu_1(M) \ge (m-1)^2/4$ implies that
$$\frac{(m-1)^2}{4} \int_M \phi^2 h^2 \le \int_M |\nabla(\phi h)|^2.$$

Hence, we conclude that

$$\int_M \phi^2 h \left(\frac{(m-1)^2}{4} h + \Delta h \right) \leq \int_M |\nabla \phi|^2 h^2. \qquad (23.13)$$

Integrating the gradient estimate of Theorem 6.1 along geodesics, we deduce that f must satisfy the growth estimate

$$f(x) \leq C \exp((m-1)r(x)),$$

where $r(x)$ is the geodesic distance from x to a fixed point $p \in M$. In particular, the above inequality when restricted on the parabolic end E_2 together with the volume estimate of Theorem 22.5 imply

$$\int_{E_2(\rho)} f \leq C\rho. \qquad (23.14)$$

On the other hand, Corollary 22.3 asserts that on E_1, the function f must satisfy the decay estimate

$$\int_{E_1(\rho+1) \setminus E_1(\rho)} f^2 \leq C \exp(-(m-1)\rho)$$

for ρ sufficiently large. In particular, Schwarz inequality implies that

$$\int_{E_1(\rho+1) \setminus E_1(\rho)} f \leq C \exp\left(-\frac{(m-1)}{2}\rho\right) V_{E_1}^{1/2}(\rho+1),$$

where $V_{E_1}(r)$ denotes the volume of $E_1(r)$. Combining this with the volume estimate of Proposition 23.1, we conclude that

$$\int_{E_1(\rho+1) \setminus E_1(\rho)} f \leq C$$

for some constant C independent of ρ. In particular,

$$\int_{E_1(\rho)} f \leq C\rho,$$

and taking this together with (23.14) we obtain the growth estimate

$$\int_{B_p(\rho)} f \leq C\rho. \qquad (23.15)$$

Choosing ϕ to be

$$\phi = \begin{cases} 1 & \text{on} \quad B_p(\rho), \\ \dfrac{2\rho - r}{\rho} & \text{on} \quad B_p(2\rho) \setminus B_p(\rho), \\ 0 & \text{on} \quad M \setminus B_p(2\rho) \end{cases}$$

23 Manifolds with Ricci curvature bounded from below

in (23.13), we conclude that the right-hand side is given by

$$\int_M |\nabla \phi|^2 h^2 = \rho^{-2} \int_{B_p(2\rho) \setminus B_p(\rho)} h^2.$$

Inequality (23.15) implies that

$$\int_M |\nabla \phi|^2 h^2 \to 0$$

as $\rho \to \infty$. Hence we obtain

$$\Delta h = -\frac{(m-1)^2}{4} h$$

and the inequalities in deriving the estimate

$$\Delta h \geq -\frac{(m-1)^2}{4} h$$

all become equalities. In particular,

$$|\nabla f| = (m-1) f$$

and

$$|\nabla (\log f)|^2 = (m-1)^2, \tag{23.16}$$

hence the inequalities used to prove Theorem 6.1 must all become equalities. More specifically,

$$(\log f)_{1\alpha} = 0$$

for all $2 \leq \alpha \leq m$ and

$$(\log f)_{\alpha\beta} = -\frac{\delta_{\alpha\beta}}{m-1} |\nabla (\log f)|^2$$

$$= -(m-1)\delta_{\alpha\beta}$$

for all $2 \leq \alpha, \beta \leq m$. On the other hand, since e_1 is the unit normal to the level set of $\log f$, the second fundamental form $(h_{\alpha\beta})$ of the level set is given by

$$(\log f)_{\alpha\beta} = (\log f)_1 h_{\alpha\beta}$$

$$= (m-1) h_{\alpha\beta},$$

hence implying that $h_{\alpha\beta} = -\delta_{\alpha\beta}$.

Moreover, (23.16) also implies that if we set $t = \log(f)/(m-1)$, then t must be the distance function between the level sets of f, and hence also for the level set of log f. The fact that $h_{\alpha\beta} = -\delta_{\alpha\beta}$ implies that the metric on M can be written as

$$ds_M^2 = dt^2 + \exp(-2t)\, ds_N^2.$$

Since M has two ends, N must be compact. A direct computation shows that the condition $\mathcal{R}_{ij} \geq -(m-1)$ implies that $\tilde{\mathcal{R}}_{ij} \geq 0$. This proves the theorem. □

24
Manifolds with finite volume

In this chapter, we assume that M^m is a complete manifold with finite volume. We assume that the Ricci curvature is bounded from below by

$$\mathcal{R}_{ij} \geq -(m-1).$$

Since the constant functions are L^2 harmonic functions, this implies that 0 is an eigenvalue for the L^2-spectrum of the Laplacian. We define the quantity $\lambda_1(M)$ by the Rayleigh quotient

$$\lambda_1(M) = \inf_{\phi \in H^{1,2}(M), \int_M \phi = 0} \frac{\int_M |\nabla \phi|^2}{\int_M \phi^2},$$

where the infimum is taken over all functions ϕ in the Sobolev space $H^{1,2}(M)$ satisfying $\int_M \phi = 0$. This plays the role of a generalized first nonzero Neumann eigenvalue, although $\lambda_1(M)$ might not necessarily be an eigenvalue. Note that

$$\lambda_1(M) \leq \max\{\mu_1(\Omega_1), \mu_1(\Omega_2)\},$$

for any two disjoint domains Ω_1 and Ω_2 of M, where $\mu_1(\Omega_1)$ and $\mu_1(\Omega_2)$ are their first Dirichlet eigenvalues, respectively. In particular, we have $\lambda_1(M) \leq \mu_e(M)$, the greatest lower bound for the essential spectrum of M. Therefore, according to Cheng's theorem [Cg1] (Corollary 6.4), one always has

$$\lambda_1(M) \leq \frac{(m-1)^2}{4}.$$

The main purpose of this chapter is to prove the following theorem.

Theorem 24.1 *Let M^m be a complete Riemannian manifold with Ricci curvature bounded from below by*

$$\mathcal{R}_{ij} \geq -(m-1).$$

Assume that M has a finite volume given by $V(M)$ and

$$\lambda_1(M) \geq \frac{(m-1)^2}{4}.$$

Then there exists a constant $C(m) > 0$, depending only on m, such that

$$N(M) \leq C(m) \left(\frac{V(M)}{V_p(1)}\right)^2 \ln\left(\frac{V(M)}{V_p(1)} + 1\right),$$

where $V_p(1)$ denotes the volume of the unit ball centered at any point $p \in M$.

The assumption $\lambda_1(M) \geq (m-1)^2/4$ implies that 0 is the only eigenvalue below $(m-1)^2/4$, so the spectrum of the M satisfies $\mathrm{Spec}(M) \subset \{0\} \cup [(m-1)^2/4, \infty)$.

In the special case when $m = 2$, our estimate is less effective than using the Cohn–Vossen–Hartman formula (see [LT4]). Indeed, the assumption that $K \geq -1$ implies that the negative part of the Gaussian curvature defined by

$$K_- = \begin{cases} 0 & \text{if } K > 0, \\ -K & \text{if } K \leq 0 \end{cases}$$

is at most 1. In particular,

$$\int_M K_- \leq V(M)$$

and M has finite total curvature. Hartman's theorem then implies that M must be conformally equivalent to a compact Riemannian surface of genus g with $N(M)$ punctures. Moreover, since M has finite volume, the Cohn–Vossen–Hartman formula (see [LT4]) asserts that

$$-V(M) \leq \int_M K$$
$$= 2\pi \chi(M).$$

In particular, using the fact that the Euler characteristic is given by

$$\chi(M) = 2 - 2g - N(M),$$

we conclude that

$$N(M) \leq 2 - 2g + (2\pi)^{-1} V(M).$$

This indicates that the dependency on $V(M)$ is better than the one provided by Theorem 24.1, and the value of the theorem lies in the case when $m \geq 3$.

Note that when $M = \mathbb{H}^m / \Gamma$ is a hyperbolic manifold, with its universal covering given by the hyperbolic m-space $\mathbb{H}^m$, the L^2-spectrum of the Laplacian on $\mathbb{H}^m$ is the interval $[(m-1)^2/4, \infty)$. In this special case, the assumption that $\lambda_1(M) \geq (m-1)^2/4$ can be expressed as $\lambda_1(M) = \mu_1(\mathbb{H}^m)$, where $\mu_1(\mathbb{H}^m)$ is the greatest lower bound of the spectrum of Δ on $\mathbb{H}^m$. With this point of view, it is also possible to prove a theorem analogous to Theorem 24.1 for locally symmetric spaces of finite volume. However, stronger results are available by using Margulis' thick–thin decomposition. For any finite volume, locally symmetric space $M = X/\Gamma$, where X is a symmetric space and Γ a discrete subgroup of the isometry group of X, the number of ends of M is always bounded by $N(M) \leq C(X) V(M)$, where the constant C depends on X and $V(M)$ is the volume of M. The interested reader should refer to [Ge] for more details.

Theorem 24.2 *Let M^m be a complete Riemannian manifold with Ricci curvature bounded from below by*

$$\mathcal{R}_{ij} \geq -(m-1).$$

Assume that M has a finite volume given by $V(M)$, and

$$\mu_1(M \setminus B_p(\rho_0)) \geq \frac{(m-1)^2}{4}.$$

Then there exists a constant $C(m) > 0$ depending only on m, such that the number of ends of M satisfies

$$N(M) \leq C(m) V(M) V_p^{-1}(1) \exp((m-1)\rho_0),$$

where $V_p(1)$ denotes the volume of the unit ball centered at point $p \in M$.

Proof According to Theorem 22.5, if we let $V_p(\rho)$ be the volume of the geodesic ball $B_p(\rho)$, then for all $\rho \geq 2(\rho_0 + 1)$ we have

$$V_p(\rho + 2) - V_p(\rho) \leq C \left(1 + (\rho - \rho_0)^{-1}\right) \exp(-(m-1)(\rho - \rho_0))$$
$$\times (V_p(\rho_0 + 1) - V_p(\rho_0)). \tag{24.1}$$

On the other hand, Proposition 23.1 asserts that if $y \in \partial B_p(\rho + 1)$, then

$$V_y(1) \geq C_1^{-1} \exp(-(m-1)\rho) V_p(1). \tag{24.2}$$

Obviously, if $N_\rho(M)$ denotes the number of ends with respect to $B_p(\rho)$, then there exists $N_\rho(M)$ number of points $\{y_i \in \partial B_p(\rho)\}$ such that $B_{y_i}(1) \cap$

$B_{y_j}(1) = \emptyset$ for $i \neq j$. In particular applying (24.2) to each of the y_i and combining the result with (24.1), we conclude that

$$N_p(M) C_1^{-1} \exp(-(m-1)\rho) V_p(1) \leq \sum_{i=1}^{N_p(M)} V_{y_i}(1)$$

$$\leq V_p(\rho+2) - V_p(\rho)$$

$$\leq C (1 + (\rho - \rho_0)^{-1}) \exp(-(m-1)$$

$$\times (\rho - \rho_0))(V_p(\rho_0 + 1) - V_p(\rho_0)).$$

This implies that

$$N_p(M) \leq C\, C_1 (1 + (\rho - \rho_0)^{-1})$$

$$\times \exp((m-1)\rho_0) (V_p(\rho_0 + 1) - V_p(\rho_0)) V_p^{-1}(1).$$

Letting $\rho \to \infty$, we conclude that the number of ends $N(M)$ of M is bounded by

$$N(M) \leq C\, C_1 \exp((m-1)\rho_0) (V_p(\rho_0 + 1) - V_p(\rho_0)) V_p^{-1}(1) \qquad (24.3)$$

and the result follows. $\square$

We remark that if M has finitely many eigenvalues $0 < \mu_1 \leq \mu_2 \leq \cdots \leq \mu_k$ below $(m-1)^2/4$, then it is easy to see there exists $\rho_0 > 0$ such that

$$\mu_1(M \setminus B_p(\rho_0)) \geq \frac{(m-1)^2}{4}.$$

In particular, Theorem 24.1 implies that M must have finitely many ends. However, the estimate of the number of ends is not effective as it is unclear to us at this moment how to control the size of ρ_0 in terms of the eigenvalues below $(m-1)^2/4$. On the other hand, we will demonstrate below that ρ_0 can be effectively controlled if $k = 1$.

The following lemma allows us to estimate $\mu_1(B_p(\rho))$ of a geodesic ball centered at p with radius ρ in terms of the volume of the ball. Note that we do not need to impose any curvature assumptions on M.

Lemma 24.3 *Let M be a complete Riemannian manifold. Then for any $0 < \delta < 1$, $\rho > 2$, and $p \in M$, we have*

$$\mu_1(B_p(\rho)) \leq \frac{1}{4\delta^2 (\rho-1)^2} \left(\ln \left(\frac{V_p(\rho)}{V_p(1)} \right) + \ln \left(\frac{12}{1-\delta} \right) \right)^2.$$

24 Manifolds with finite volume

Proof For the ease of notation, we will use μ_1 to denote $\mu_1(B_p(\rho))$. We may assume

$$\mu_1 \geq 4\rho^{-2}$$

as otherwise the conclusion automatically holds true. Let $r(x)$ denote the distance function to a fixed point $p \in M$. The variational characterization of $\mu_1(B_p(\rho))$ implies that

$$\mu_1 \int_M \phi^2 \exp\left(-2\delta\sqrt{\mu_1}\,r\right)$$
$$\leq \int_M |\nabla(\phi \exp\left(-\delta\sqrt{\mu_1}\,r\right))|^2$$
$$= \int_M |\nabla\phi|^2 \exp\left(-2\delta\sqrt{\mu_1}\,r\right) - 2\delta\sqrt{\mu_1} \int_M \phi \exp\left(-2\delta\sqrt{\mu_1}\,r\right) \langle \nabla\phi, \nabla r \rangle$$
$$+ \delta^2 \mu_1 \int_M \phi^2 \exp\left(-2\delta\sqrt{\mu_1}\,r\right)$$

for any nonnegative Lipschitz function ϕ with support in $B_p(\rho)$. In particular, for $\rho > 2$, if we choose

$$\phi = \begin{cases} 1 & \text{on } B_p\left(\rho - \mu_1^{-\frac{1}{2}}\right), \\ \sqrt{\mu_1}\,(\rho - r) & \text{on } B_p(\rho) \setminus B_p\left(\rho - \mu_1^{-\frac{1}{2}}\right), \\ 0 & \text{on } M \setminus B_p(\rho), \end{cases}$$

then we have

$$(1 - \delta^2)\,\mu_1 \exp\left(-2\delta\sqrt{\mu_1}\right) V_p(1)$$
$$\leq (1 - \delta^2)\,\mu_1 \int_M \phi^2 \exp\left(-2\delta\sqrt{\mu_1}\,r\right)$$
$$= \int_M |\nabla\phi|^2 \exp\left(-2\delta\sqrt{\mu_1}\,r\right) - 2\delta\sqrt{\mu_1} \int_M \phi \exp\left(-2\delta\sqrt{\mu_1}\,r\right) \langle \nabla\phi, \nabla r \rangle$$
$$\leq (1 + 2\delta)\,\mu_1 \exp\left(-2\delta\left(\rho\sqrt{\mu_1} - 1\right)\right) V_p(\rho).$$

Therefore,

$$\exp\left(2\delta\left(\sqrt{\mu_1}\,(\rho - 1)\right)\right) \leq \frac{3e^2}{2(1 - \delta)} \frac{V_p(\rho)}{V_p(1)}.$$

and

$$2\delta \sqrt{\mu_1}(\rho - 1) \leq \ln\left(\frac{12}{1-\delta}\right) + \ln\left(\frac{V_p(\rho)}{V_p(1)}\right).$$

The lemma follows by rewriting this inequality. □

As a corollary to this lemma, one recovers Brook's theorem asserting that the greatest lower bound of the spectrum of any complete manifold M can be estimated in terms of its volume entropy. This result was proved in Corollary 22.6 with a different argument.

Corollary 24.4 *Let M be a complete manifold and $\mu_1(M)$ be the bottom of the spectrum. Then*

$$\mu_1(M) \leq \frac{1}{4}\left(\liminf_{\rho \to \infty} \frac{\ln V_p(\rho)}{\rho}\right)^2.$$

Proof Note that $\mu_1(M) = \lim_{\rho \to \infty} \mu_1(B_p(\rho))$. Now the result follows by first letting ρ go to infinity and then letting δ go to 1 in the estimate of Lemma 24.3. □

We are now ready to prove Theorem 24.1.

Proof Proof of Theorem 24.1 Let $p \in M$ be a fixed point. For any $0 < \delta < 1$, let

$$\rho_0 = \frac{1}{(m-1)\delta}\left(\ln\left(\frac{12}{1-\delta}\right) + \ln\left(\frac{V(M)}{V_p(1)}\right)\right) + 1.$$

Then according to Lemma 24.3,

$$\mu_1(B_p(\rho_0)) \leq \frac{(m-1)^2}{4}. \tag{24.4}$$

We now observe that

$$\mu_1(M \setminus B_p(\rho_0)) \geq \frac{(m-1)^2}{4}.$$

Indeed, if (24.4) is valid and also $\mu_1(M \setminus B_p(\rho_0)) < (m-1)^2/4$, then the variational principle implies that $\lambda_1(M) < (m-1)^2/4$, contradicting our assumption.

By Theorem 24.2, we have

$$N(M) \leq C(m)\, V(M)\, V_p^{-1}(1)\, \exp((m-1)\rho_0).$$

24 Manifolds with finite volume

If $V(M) > eV_p(1)$, then the claimed estimate follows by plugging in the value of ρ_0 and setting

$$\delta = 1 - \frac{1}{\ln(V(M) V_p^{-1}(1))}.$$

On the other hand, if $V(M) \leq eV_p(1)$, then ρ_0 is bounded from above and hence

$$N(M) \leq C(m)$$

implying the estimate again. □

As pointed out earlier, the key ingredients in the proof of Theorem 24.1 rely on the decay estimate of the volume (24.2) given by the upper bound of the greatest lower bound of the spectrum of the model manifold hyperbolic space $\mathbb{H}^m$. Similar theorems for Kähler manifolds and quaternionic Kähler manifolds also follow by using the corresponding comparison results from [LW8] and [KLZ], respectively.

25
Stability of minimal hypersurfaces in a 3-manifold

In this and the next chapter, we will apply the theory of harmonic functions to the study of complete minimal hypersurfaces in a complete manifold with nonnegative curvature.

Let N^{m+1} be a complete manifold with nonnegative Ricci curvature. Suppose M^m is a complete minimal hypersurface in N. If $|\vec{II}|^2$ denotes the square of the length of the second fundamental form of M and $\mathcal{R}^N_{\nu\nu}$ is the Ricci curvature of N in the direction of the unit normal ν to M, then M being stable in N is characterized by the stability inequality (1.4)

$$\int_M \psi^2 |\vec{II}|^2 + \int_M \psi^2 \mathcal{R}^N_{\nu\nu} \leq \int_M |\nabla \psi|^2 \qquad (25.1)$$

for any compactly supported function $\psi \in H^{1,2}_c(M)$. Geometrically, the stability inequality is derived from the second variation formula for the volume functional under normal variations. Hence a stable minimal hypersurface is not only a critical point of the volume functional but its second derivative is nonnegative with respect to any normal variations. The elliptic operator associated with the stability inequality is given by

$$L = \Delta + |\vec{II}|^2 + \mathcal{R}^N_{\nu\nu}.$$

The stability of M is equivalent to the fact that the operator $-L$ is nonnegative. We say that M has finite index when the operator $-L$ has only finitely many negative eigenvalues. This has the geometric interpretation that there is only a finite dimensional space of normal variations violating the stability inequality.

The study of stable minimal hypersurfaces can be viewed as an effort to prove a generalized Bernstein's theorem. Bernstein first established that an entire minimal graph in $\mathbb{R}^3$ must be a plane. Recall that a minimal graph is a minimal hypersurface which is given by a graph of a function defined on $\mathbb{R}^2$.

The validity of Bernstein's theorem in higher dimensions was established for the entire minimal graph in $\mathbb{R}^{m+1}$ for $m \leq 7$ by Simons [S], and many other authors, such as Fleming [Fl], Almgren [A], and De Giorgi [De], for the lower dimensional cases. Counterexamples for $m \geq 8$ were found by Bombieri, De Giorgi, and Guisti [BDG]. Since entire minimal graphs are area minimizing, a natural question to ask is if a Bernstein type theorem is valid for stable minimal hypersurfaces in $\mathbb{R}^{m+1}$.

In 1979, do Carmo and Peng [dCP] proved that a complete, stable, minimally immersed hypersurface M in $\mathbb{R}^3$ must be planar. At the same time, Fischer-Colbrie and Schoen [FCS] independently showed that a complete, stable, minimally immersed hypersurface M in a complete three-dimensional manifold N with nonnegative scalar curvature must be either conformally a plane $\mathbb{R}^2$ or conformally a cylinder $\mathbb{R} \times \mathbb{S}^1$. For the special case when N is $\mathbb{R}^3$, they also proved that M must be planar.

In 1984, Gulliver [Gu1] studied a yet larger class of submanifolds in $\mathbb{R}^3$. He proved that a complete, oriented, minimally immersed hypersurface with finite index in $\mathbb{R}^3$ must have finite total curvature. In particular, applying Huber's theorem, one concludes that the hypersurface must be conformally equivalent to a compact Riemann surface with finitely many punctures. The same result was also independently proved by Fischer-Colbrie in [FC]. In addition, she also proved that a complete, oriented, minimally immersed hypersurface with finite index in a complete three-dimensional manifold with nonnegative scalar curvature must be conformally equivalent to a compact Riemann surface with finite punctures. Shortly after, Gulliver [Gu2] improved the result of Fischer-Colbrie and proved that if the ambient manifold has nonnegative scalar curvature, then a minimal hypersurface with finite index must have quadratic area growth and finite topological type, and the square of the length of the second fundamental form must be integrable. Indeed, a complete surface with quadratic area growth and finite topological type must be conformally equivalent to a compact Riemann surface with finitely many punctures.

In 1997, Cao, Shen, and Zhu [CSZ] considered the high dimensional cases of the theorem of do Carmo–Peng and Fischer-Colbrie–Schoen. They proved that a complete, oriented, stable, minimally immersed hypersurface M^n in $\mathbb{R}^{n+1}$ must have only one end. This theorem was generalized by Li and Wang [LW4] when they showed that a complete, oriented, minimally immersed hypersurface M^n in $\mathbb{R}^{n+1}$ with finite index must have finitely many ends. In a later paper [LW7], Li and Wang also generalized their theorem to minimal hypersurfaces with finite index in a complete manifold with nonnegative sectional curvature.

We will present the relationship between harmonic functions and the stability of minimal hypersurfaces in this chapter. Some two-dimensional results will be proved using this technique, while the higher dimensional results will be presented in the next chapter.

Note that when the ambient manifold is Euclidean space and $n \geq 3$, then M is nonparabolic by applying Corollary 24.2. The key issue is the validity of the Sobolev inequality proved by Michael and Simon [MS] in the form

$$\left(\int_M |u|^{2m/(m-2)}\right)^{(m-2)/m} \leq C \int_M |\nabla u|^2$$

on any minimal submanifolds of $\mathbb{R}^n$. However, when N is only assumed to have nonnegative Ricci (or sectional) curvature, then M can be parabolic as in the case of the cylinder $M = \mathbb{R} \times P$ in $N = \mathbb{R}^2 \times P$. The next theorem states that the case in which M is parabolic is a very special situation.

Theorem 25.1 *Let M^m be a complete, minimally immersed, stable, hypersurface in a manifold, N^{m+1}, with nonnegative Ricci curvature. If M is parabolic, then M must be totally geodesic in N. Moreover, the Ricci curvature $\mathcal{R}_{\nu\nu}$ of N in the normal direction also vanishes, and M must have nonnegative scalar curvature.*

Proof According to Lemma 24.3 and the discussion thereafter, the assumption that M is parabolic can be characterized by the following construction. For a fixed $p \in M$, a given $\rho > 0$, and a sequence $\rho_i > \rho$ with $\rho_i \to \infty$, let f_i be a sequence of harmonic functions satisfying

$$\Delta f_i = 0 \quad \text{on} \quad B_p(\rho_i) \setminus B_p(\rho),$$

with boundary conditions

$$f_i = 1 \quad \text{on} \quad \partial B_p(\rho)$$

and

$$f_i = 0 \quad \text{on} \quad \partial B_p(\rho_i).$$

Then the manifold M is parabolic if and only if f_i converges uniformly on compact subsets of $M \setminus B_p(\rho)$ to the constant function 1. If this is the case, since

$$\int_{B_p(\rho_i) \setminus B_p(\rho)} |\nabla f_i|^2 = \int_{\partial B_p(\rho_i)} f_i \frac{\partial f_i}{\partial r} - \int_{\partial B_p(\rho)} f_i \frac{\partial f_i}{\partial r}$$

$$= -\int_{\partial B_p(\rho)} \frac{\partial f_i}{\partial r},$$

the fact that $f_i \to 1$ implies that the right-hand side must tend to 0 as $\rho_i \to \infty$.

25 Stability of minimal hypersurfaces in a 3-manifold

To prove the proposition, we consider the compactly supported function

$$\psi = \begin{cases} 1 & \text{on} & B_p(\rho), \\ f_i & \text{on} & B_p(\rho_i) \setminus B_p(\rho), \\ 0 & \text{on} & M \setminus B_p(\rho_i). \end{cases}$$

Using this as a test function in the stability inequality (25.1), we conclude that

$$\int_{B_p(\rho)} |\vec{II}|^2 + \int_{B_p(\rho)} \mathcal{R}^N_{\nu\nu} \leq \int_{B_p(\rho_i) \setminus B_p(\rho)} |\nabla f_i|^2.$$

Since the right-hand side vanishes as $i \to \infty$, we have

$$\int_{B_p(\rho)} |\vec{II}|^2 + \int_{B_p(\rho)} \mathcal{R}^N_{\nu\nu} = 0,$$

hence $|\vec{II}|^2$ and $\mathcal{R}^N_{\nu\nu}$ vanish identically on $B_p(\rho)$. Due to the fact that ρ is arbitrary, this implies the vanishing of $\vec{II}$ and $\mathcal{R}^N_{\nu\nu}$ on M. The nonnegativity of the scalar curvature of M follows by applying the Gauss curvature equation and using the assumption that N has nonnegative Ricci curvature. $\square$

Theorem 25.1 reduces our study of stable minimal hypersurfaces to the nonparabolic case. In this case, we will recall the lemma of Schoen and Yau [SY1].

Lemma 25.2 (Schoen–Yau) *Let M^m be a complete, minimally immersed, stable, hypersurface in N^{m+1}. Let $\mathcal{R}^N$ and K^N be the Ricci curvature and the sectional curvature of N, respectively. Suppose u is a harmonic function defined on M. Then the inequality*

$$\int_M |\nabla \phi|^2 |\nabla u|^2 \geq \frac{1}{m} \int_M \phi^2 |\vec{II}|^2 |\nabla u|^2 + \int_M \phi^2 \mathcal{R}^N_{\nu\nu} |\nabla u|^2$$
$$+ \sum_{\alpha=1}^{m} \int_M \phi^2 K^N(e_1, e_\alpha) |\nabla u|^2 + \frac{1}{m-1} \int_M \phi^2 |\nabla|\nabla u||^2$$

holds for any compactly supported, nonnegative function $\phi \in H^{1,2}_c(M)$.

Proof The Bochner formula of Lemma 23.4 asserts that

$$\Delta |\nabla u|^2 \geq 2\mathcal{R}_{ij} u_i u_j + \frac{m}{2(m-1)} |\nabla u|^{-2} |\nabla |\nabla u|^2|^2. \quad (25.2)$$

If $\{e_1, e_2, \ldots, e_m\}$ is an orthonormal frame of M and (h_{ij}) is the second fundamental form of M, then the Gauss curvature equation asserts that

$$\mathcal{R}_{11} = \sum_{\alpha=2}^{m} K^N(e_1, e_\alpha) + \sum_{\alpha=2}^{m} h_{11} h_{\alpha\alpha} - \sum_{\alpha=2}^{m} h_{1\alpha}^2$$

$$= \sum_{\alpha=2}^{m} K^N(e_1, e_\alpha) - h_{11}^2 - \sum_{\alpha=2}^{m} h_{1\alpha}^2,$$

where we have used the assumption that M is minimal and $K^N(e_1, e_\alpha)$ is the sectional curvature for the 2-plane section spanned by e_1 and e_α. On the other hand, using the inequality

$$|\vec{II}|^2 = \sum_{i,j=1}^{m} h_{ij}^2$$

$$\geq h_{11}^2 + \sum_{\alpha=2}^{m} h_{\alpha\alpha}^2 + 2 \sum_{\alpha=2}^{m} h_{1\alpha}^2$$

$$\geq h_{11}^2 + \frac{\left(\sum_{\alpha=2}^{m} h_{\alpha\alpha}\right)^2}{m-1} + 2 \sum_{\alpha=2}^{m} h_{1\alpha}^2$$

$$\geq \frac{m}{m-1} \left(h_{11}^2 + \sum_{\alpha=2}^{m} h_{1\alpha}^2 \right),$$

we conclude that

$$\mathcal{R}_{11} \geq \sum_{\alpha=2}^{m} K^N(e_1, e_\alpha) - \frac{m-1}{m} |\vec{II}|^2.$$

Choosing the orthonormal frame so that $\nabla u = |\nabla u| e_1$, this implies that

$$\mathcal{R}_{ij} u_i u_j \geq \sum_{\alpha=2}^{m} K^N(e_1, e_\alpha) |\nabla u|^2 - \frac{m-1}{m} |\vec{II}|^2 |\nabla u|^2.$$

Substituting this into (25.2) yields

$$\Delta |\nabla u| \geq \sum_{\alpha=2}^{m} K^N(e_1, e_\alpha) |\nabla u| - \frac{m-1}{m} |\vec{II}|^2 |\nabla u| + \frac{|\nabla |\nabla u||^2}{(m-1)|\nabla u|}. \quad (25.3)$$

By setting $\psi = \phi|\nabla u|$, where ϕ is a nonnegative compactly supported function on M, the stability inequality (25.1) implies that

$$\int_M \phi^2 |\vec{II}|^2 |\nabla u|^2 + \int_M \phi^2 \mathcal{R}^N_{\nu\nu} |\nabla u|^2$$
$$\le \int_M |\nabla \phi|^2 |\nabla u|^2 + 2 \int_M \phi |\nabla u| \langle \nabla \phi, \nabla|\nabla u|\rangle$$
$$+ \int_M \phi^2 |\nabla|\nabla u||^2$$
$$= \int_M |\nabla \phi|^2 |\nabla u|^2 - \int_M \phi^2 |\nabla u| \Delta|\nabla u|.$$

Combining this inequality with (25.3), we conclude that

$$\frac{1}{m}\int_M \phi^2 |\vec{II}|^2 |\nabla u|^2 + \int_M \phi^2 \mathcal{R}^N_{\nu\nu} |\nabla u|^2$$
$$+ \sum_{\alpha=1}^m \int_M \phi^2 K^N(e_1, e_\alpha) |\nabla u|^2 + \frac{1}{m-1}\int_M \phi^2 |\nabla|\nabla u||^2$$
$$\le \int_M |\nabla \phi|^2 |\nabla u|^2.$$
$\square$

As a corollary of Theorem 25.1 and Lemma 25.2, we readily recover the theorem of Fischer-Colbrie and Schoen [FCS]. This theorem was also independently proved by do Carmo and Peng [dCP] for the special case when $N = \mathbb{R}^3$.

In the case of minimal surface in a three-dimensional manifold, if we let S^N be the scalar curvature of N and K be the Gaussian curvature of M, then using the Gauss curvature equation the stability inequality can be written as

$$\frac{1}{2}\int_M \psi^2 |\vec{II}|^2 + \frac{1}{2}\int_M \psi^2 S^N - \int_M \psi^2 K \le \int_M |\nabla \psi|^2 \qquad (25.4)$$

for any nonnegative, compactly supported function $\psi \in H^{1,2}_c(M)$.

Theorem 25.3 (Fischer-Colbrie–Schoen) *Let M^2 be an oriented, complete, stable, minimal hypersurface in a complete manifold N^3 with nonnegative scalar curvature. Then M must be conformally equivalent to either the complex plane $\mathbb{C}$ or the cylinder $\mathbb{R} \times \mathbb{S}^1$. If M is conformally equivalent to the cylinder and has finite total curvature, then it must be totally geodesic and the scalar curvature of N along M must be identically 0.*

Proof Let $\tilde{M}$ be the universal covering of M. By the uniformization theorem, $\tilde{M}$ must be conformally equivalent to either the unit disk $\mathbb{D}^2$ or the complex plane $\mathbb{C}$. We claim that $\tilde{M}$ cannot be conformally equivalent to $\mathbb{D}^2$.

To see this, we first observe that the stability inequality still holds on $\tilde{M}$ by lifting the functions $|\vec{II}|^2$ and S^N to $\tilde{M}$. In fact, if ψ is a nonnegative, compactly supported function on $\tilde{M}$ and $\pi : \tilde{M} \to M$ is the covering map, then the function $\bar{\psi}$ defined by

$$\bar{\psi}^2(x) = \sum_{\tilde{x} \in \pi^{-1}(x)} \psi^2(\tilde{x})$$

is a nonnegative, compactly supported function on M. The stability inequality on M asserts that

$$\tfrac{1}{2}\int_{\tilde{M}} \psi^2 |\vec{II}|^2 + \tfrac{1}{2}\int_{\tilde{M}} \psi^2 S^N - \int_{\tilde{M}} \psi^2 K$$
$$= \tfrac{1}{2}\int_M \bar{\psi}^2 |\vec{II}|^2 + \tfrac{1}{2}\int_M \bar{\psi}^2 S^N - \int_M \bar{\psi}^2 K$$
$$\leq \int_M |\nabla \bar{\psi}|^2.$$

On the other hand, the Schwarz inequality implies that

$$|\bar{\psi}\nabla\bar{\psi}|^2 = \left| \sum_{\tilde{x}\in\pi^{-1}(x)} \psi \nabla \psi \right|^2$$
$$\leq \sum_{\tilde{x}\in\pi^{-1}(x)} \psi^2 \sum_{\tilde{x}\in\pi^{-1}(x)} |\nabla\psi|^2$$
$$= \bar{\psi}^2 \sum_{\tilde{x}\in\pi^{-1}(x)} |\nabla\psi|^2.$$

Therefore we conclude that

$$\int_M |\nabla\bar{\psi}|^2 \leq \int_{\tilde{M}} |\nabla\psi|^2$$

and the stability inequality is valid on $\tilde{M}$. If $\tilde{M}$ is conformally equivalent to $\mathbb{D}^2$, the invariance of the Laplace operator in dimension 2 asserts that there exist nonconstant, bounded harmonic functions with a finite Dirichlet integral on $\tilde{M}$. Moreover, if u is such a harmonic function, then Lemma 7.3 asserts that

$$\Delta |\nabla u| \geq K |\nabla u| + \frac{|\nabla|\nabla u||}{|\nabla u|}$$

and the proof of Lemma 25.2 applied to the stability inequality (25.4) on $\tilde{M}$ implies that

$$\tfrac{1}{2}\int_{\tilde{M}} \phi^2\,|\vec{II}|^2\,|\nabla u|^2 + \tfrac{1}{2}\int_{\tilde{M}} \phi^2\, S^N\,|\nabla u|^2$$
$$+ \int_{\tilde{M}} \phi^2\,|\nabla|\nabla u||^2 \leq \int_{\tilde{M}} |\nabla \phi|^2\,|\nabla u|^2 \qquad (25.5)$$

for any nonnegative, compactly supported function $\phi \in H_c^{1,2}(\tilde{M})$. However, choosing

$$\phi = \begin{cases} 1 & \text{on} \quad B_p(\rho), \\ \dfrac{2\rho - r}{\rho} & \text{on} \quad B_p(2\rho), \\ 0 & \text{on} \quad \tilde{M}\setminus B_p(2\rho), \end{cases}$$

the right-hand side becomes

$$\int_{\tilde{M}} |\nabla \phi|^2\,|\nabla u|^2 = \frac{1}{\rho^2}\int_{B_p(2\rho)\setminus B_p(\rho)} |\nabla u|^2.$$

The fact that u has a finite Dirichlet integral implies that this tends to 0 as $\rho \to \infty$. Therefore, (25.5) asserts the vanishing of $\vec{II}$ and $|\nabla|\nabla u||$. In particular, $|\nabla u|$ must be identically constant and $\tilde{M}$ has a finite volume because u has a finite Dirichlet integral. This contradicts the assumption and $\tilde{M}$ must be conformally equivalent to $\mathbb{C}$.

Using the uniformization theorem again, we conclude that M must be conformally equivalent to either the complex plane $\mathbb{C}$ or the cylinder $\mathbb{R}\times\mathbb{S}^1$. This proves the first part of the theorem.

If M is conformally equivalent to $\mathbb{R}\times\mathbb{S}^1$ and has finite total curvature, then applying the proof of Theorem 25.1 to the stability inequality (25.4) we conclude that

$$\tfrac{1}{2}\int_M |\vec{II}|^2 + \tfrac{1}{2}\int_M S^N - \int_M K \leq 0. \qquad (25.6)$$

In particular, the assumption that N has nonnegative scalar curvature implies that

$$\int_M K \geq 0.$$

The Cohn-Vossen inequality [CV] then asserts that

$$2\pi\,\chi(M) \geq \int_M K \geq 0.$$

Since M is the cylinder, we conclude that $\int_M K = 0$ and (25.6) asserts that $|\vec{II}|^2$ and S^N are both identically 0 on M as claimed. □

Combining this theorem with Theorem 25.1, we obtained the following corollary, which was also proved in [FCS].

Corollary 25.4 (Fischer-Colbrie–Schoen) *Let M^2 be an oriented, complete, stable, minimal hypersurface in a complete manifold N^3 with nonnegative Ricci curvature. Then M must be totally geodesic in N and the Ricci curvature of N in the normal direction to M must be identically zero along M. Moreover, M is either*

(1) *conformally equivalent to the complex plane $\mathbb{C}$; or*
(2) *isometrically the cylinder $\mathbb{R} \times \mathbb{S}^1$.*

Moreover, if $N = \mathbb{R}^3$, then M must be planar in $\mathbb{R}^3$.

26
Stability of minimal hypersurfaces in a higher dimensional manifold

Let us now assume that N^{m+1} is a complete manifold with nonnegative sectional curvature. For a fixed point $p \in N$, let $\gamma : [0, \infty) \to N$ be a normal geodesic ray emanating from p. Recall that the Buseman function, β_γ, with respect to γ is defined by

$$\beta_\gamma(x) = \lim_{t \to \infty} (t - d(x, \gamma(t))).$$

We define the Buseman function, β, with respect to the point p by

$$\beta(x) = \sup_\gamma \beta_\gamma(x).$$

It was proved in [CG1] that $\beta(x)$ is a convex exhaustion function on N. Suppose M^m is a minimally immersed hypersurface in N^{m+1}, then a direct computation implies that the restriction of β on M is subharmonic with respect to the induced metric.

The following lemma is useful for the construction of harmonic functions on a manifold possessing subharmonic functions with certain properties. Since the manifold is only assumed to be complete, it is likely that there are other applications of this lemma.

Lemma 26.1 (Li–Wang) *Let M^m be a complete manifold, and E be an end of M with respect to $B_p(1)$. Suppose g is a subharmonic function defined on E with the property that its maximum is not achieved on ∂E. Let us define $E(\rho) = B_p(\rho) \cap E$ and $s(\rho) = \sup_{\partial B_p(\rho) \cap E} g$. For any sequence $\{\rho_i\}$ with $\rho_i \to \infty$, there exists a subsequence, which we also denote by $\{\rho_i\}$, and a sequence of* positive constants $\{A_i\}$ such that the sequence of solutions $\{u_i\}$ to the boundary value problem

$$\Delta u_i = 0 \quad \text{on} \quad E(\rho_i),$$
$$u_i = 0 \quad \text{on} \quad \partial E,$$

and

$$u_i = A_i \quad \text{on} \quad \partial E(\rho_i) \setminus \partial E$$

converges on compact subsets of E to a positive harmonic function u with boundary value

$$u = 0 \quad \text{on} \quad \partial E.$$

Moreover, the sequence $\{A_i\}$ satisfies the bound

$$0 < A_i \leq C\, s(\rho_i)$$

for some constant $0 < C < \infty$, and

$$\int_{E(\rho_i)} |\nabla u_i|^2 = A_i.$$

Proof Lemma 20.7 asserts that there exists a sequence $\rho_i \to \infty$, a sequence of constants $C_i \to \infty$, and a sequence of positive functions f_i satisfying

$$\Delta f_i = 0 \quad \text{on} \quad E(\rho_i),$$

$$f_i = 0 \quad \text{on} \quad \partial E,$$

and

$$f_i = C_i \quad \text{on} \quad \partial B_p(\rho_i) \cap E,$$

which converges to a positive harmonic function f. Moreover,

$$\int_{E(\rho_i)} |\nabla f_i|^2 = \int_{\partial B_p(\rho_i) \cap E} f_i \frac{\partial f_i}{\partial r} - \int_{\partial E} f_i \frac{\partial f_i}{\partial r}$$
$$= C_i \int_{\partial B_p(\rho_i) \cap E} \frac{\partial f_i}{\partial r}$$
$$= C_i \int_{\partial E} \frac{\partial f_i}{\partial r}.$$

26 Minimal hypersurface stability in a higher dimensional manifold

Let us define
$$u_i = \left(\int_{\partial E} \frac{\partial f_i}{\partial r}\right)^{-1} f_i,$$
then
$$u_i = C_i \left(\int_{\partial E} \frac{\partial f_i}{\partial r}\right)^{-1} \quad \text{on} \quad \partial B_p(\rho_i) \cap E.$$

The sequence u_i will also converge to the harmonic function
$$u = \left(\int_{\partial E} \frac{\partial f}{\partial r}\right)^{-1} f.$$

We now claim that there exists a constant $0 < C < \infty$ such that the sequence $\{A_i = C_i \left(\int_{\partial E} \partial f_i/\partial r\right)^{-1}\}$ satisfies
$$\limsup_{i \to \infty} \frac{A_i}{s(\rho_i)} = C.$$

Indeed, the maximum principle asserts that
$$\frac{A_i}{s(\rho_i)}(g(x) - s(1)) \leq u_i(x) \tag{26.1}$$

on $E(\rho_i)$ since this inequality is satisfied on $\partial E(\rho_i)$. The assumption that g does not achieve its maximum on ∂E and the maximum principle imply that $s(\rho) > s(1)$ for some $\rho > 1$. For this particular ρ, let $x \in \partial E(\rho) \setminus \partial E$ such that $g(x) = s(\rho)$. Applying (26.1) and letting $i \to \infty$, we conclude that
$$\limsup_{i \to \infty} \frac{A_i}{s(\rho_i)} \leq \frac{u(x)}{s(\rho) - s(1)},$$
$$< \infty,$$

confirming the existence of C.

Integrating by parts yields
$$\int_{E(\rho_i)} |\nabla u_i|^2 = \int_{\partial E(\rho_i) \setminus \partial E} u_i \frac{\partial u_i}{\partial r}$$
$$= A_i \int_{\partial E(\rho_i) \setminus \partial E} \frac{\partial u_i}{\partial r}.$$

On the other hand, since

$$0 = \int_{E(\rho_i)} \Delta u_i$$
$$= \int_{\partial E(\rho_i) \setminus \partial E} \frac{\partial u_i}{\partial r} - \int_{\partial E} \frac{\partial u_i}{\partial r},$$

we conclude that

$$\int_{E(\rho_i)} |\nabla u_i|^2 = A_i \int_{\partial E} \frac{\partial u_i}{\partial r}$$
$$= A_i.$$

□

Theorem 26.2 (Li–Wang) *Let M^m be a complete, stable, minimally immersed hypersurface in N^{m+1}. Suppose N is a complete manifold with nonnegative sectional curvature. If M is parabolic, then it must be totally geodesic and have nonnegative sectional curvature. In particular, M either has only one end, or $M = \mathbb{R} \times P$ with the product metric, where P is compact with nonnegative sectional curvature. If M is nonparabolic, then it must only have one nonparabolic end. In this case, any parabolic end of M must be contained in a bounded subset of N.*

Proof If M is parabolic, Theorem 25.1 implies that M must be totally geodesic and the Ricci curvature of N in the normal direction must vanish along M. In particular, the Gauss curvature equation asserts that M has nonnegative sectional curvature. Moreover, if M has more than one end, then Theorem 4.4 implies that M must be isometrically $\mathbb{R} \times P$ for some compact manifold P with nonnegative sectional curvature.

From this point on, we may assume that M is nonparabolic and has at least two ends. Suppose E and F are two ends of M with respect to the compact set $B_p(\rho_0)$. The nonparabolicity of M implies that at least one of the ends is nonparabolic, hence we may assume E is a nonparabolic end.

Let us assume either that F is nonparabolic, or that the Buseman function β when restricted to F is unbounded, hence satisfying the hypothesis of Lemma 26.1. Note that if M is properly immersed in N, then β is unbounded on each end of M. Using the Buseman function β in the role of g in Lemma 26.1, we observe that because

$$|\nabla \beta| \leq 1,$$

$s(\rho)$ has at most linear growth. In particular, there exists a sequence of harmonic functions $\{u_i\}$ defined on $F(\rho_i)$ that converges to a positive harmonic function u defined on F. Moreover, they satisfy

$$u_i = 0 \quad \text{on} \quad \partial F$$

and

$$\int_{\partial F} \frac{\partial u_i}{\partial r} = 1.$$

Similarly, because E is assumed to be nonparabolic, by passing to a subsequence if necessary, there exists a sequence of harmonic functions $\{v_i\}$ defined on $E(\rho_i)$ that converges to a bounded harmonic function v defined on E. The sequence also satisfies

$$v_i = 0 \quad \text{on} \quad \partial E$$

and

$$\int_{\partial E} \frac{\partial v_i}{\partial r} = \int_{E(\rho_i)} |\nabla v_i|^2 = 1.$$

Let us define a sequence of functions k_i on $B_p(\rho_i)$ by

$$k_i(x) = \begin{cases} v_i(x) & \text{on} \quad E(\rho_i), \\ -u_i(x) & \text{on} \quad F(\rho_i), \\ 0 & \text{on} \quad B_p(\rho_i) \setminus (E(\rho_i) \cup F(\rho_i)). \end{cases}$$

Obviously, k_i is harmonic on $E(\rho_i) \cup F(\rho_i)$ and

$$\int_{B_p(1)} \frac{\partial k_i}{\partial r} = \int_{\partial E} \frac{\partial v_i}{\partial r} - \int_{\partial F} \frac{\partial u_i}{\partial r}$$

$$= 0.$$

Let w_i be the solution to the boundary value problem

$$\Delta w_i = 0 \quad \text{on} \quad B_p(\rho_i)$$

and

$$w_i = k_i \quad \text{on} \quad \partial B_p(\rho_i).$$

The fact that w_i minimizes the Dirichlet integral implies that

$$\int_{B_p(\rho_i)} |\nabla w_i|^2 \le \int_{B_p(\rho_i)} |\nabla k_i|^2$$

$$= \int_{E_i} |\nabla v_i|^2 + \int_{F_i} |\nabla u_i|^2$$

$$\le 2C \rho_i.$$

On the other hand, according to Theorem 21.2, since k_i converges to the function

$$k(x) = \begin{cases} v(x) & \text{on} \quad E, \\ -u(x) & \text{on} \quad F, \\ 0 & \text{on} \quad M\setminus(E\cup F), \end{cases}$$

there exists a constant $C_1 > 0$ independent of i such that a subsequence of $\{w_i\}$ converges to a harmonic function w on M satisfying

$$|w - k| \leq C_1.$$

Applying Lemma 25.2, we have

$$(m-1)\int_M \phi^2 |\nabla w_i|^2 |\vec{II}|^2 + m\int_M \phi^2 |\nabla|\nabla w_i||^2$$
$$\leq m(m-1)\int_M |\nabla \phi|^2 |\nabla w_i|^2 \qquad (26.2)$$

for any nonnegative function $\phi \in H_c^{1,2}(M)$ supported on $B_p(\rho_i)$. In particular, for any fixed $\rho_0 < \rho < \rho_i$, if we set

$$\phi(x) = \begin{cases} 1 & \text{on} \quad B_p(\rho), \\ \dfrac{\rho_i - r(x)}{\rho_i - \rho} & \text{on} \quad B_p(\rho_i)\setminus B_p(\rho), \end{cases}$$

then we have

$$(m-1)\int_{B_p(\rho)} |\nabla w_i|^2 |\vec{II}|^2 + m\int_{B_p(\rho)} |\nabla|\nabla w_i||^2$$
$$\leq (m-1)\int_M \phi^2 |\nabla w_i|^2 |\vec{II}|^2 + m\int_M \phi^2 |\nabla|\nabla w_i||^2$$
$$\leq m(m-1)\int_M |\nabla \phi|^2 |\nabla w_i|^2$$
$$\leq \frac{m(m-1)}{(\rho_i - \rho)^2}\int_{B_p(\rho_i)} |\nabla w_i|^2$$
$$\leq \frac{C\rho_i}{(\rho_i - \rho)^2}. \qquad (26.3)$$

Letting $i \to \infty$, this implies that

$$\int_{B_p(\rho)} |\nabla w|^2 |\vec{II}|^2 = 0$$

and

$$\int_{B_p(\rho)} |\nabla|\nabla w||^2 = 0.$$

Since ρ is arbitrary, we conclude that $|\nabla w|$ is constant. Because w is a nonconstant harmonic function, this means that $|\nabla w| \neq 0$, hence $|\vec{II}|^2$ must be identically 0 and M is a totally geodesic submanifold. The Gauss curvature equation then asserts that M must have nonnegative sectional curvature also. The assumption that M has at least two ends and Theorem 4.4 together imply that M must split isometrically into $\mathbb{R} \times P$, where P must be a compact manifold with nonnegative sectional curvature. However, this contradicts the assumption that M is nonparabolic, hence M must have only one end. □

Corollary 26.3 *Let M^m be a complete, properly immersed, stable, minimal hypersurface in N^{m+1}. Suppose N is a complete manifold with nonnegative sectional curvature. Then either*

(1) *M has only one end; or*
(2) *$M = \mathbb{R} \times P$ with the product metric, where P is compact with nonnegative sectional curvature, and M is totally geodesic in N.*

We will now consider the case when the minimal hypersurface has finite index. The following lemma proved by Fischer-Colbrie [FC] gives an effective way to use the finite index assumption.

Lemma 26.4 (Fischer-Colbrie) *Let M be a complete manifold and V be a locally bounded function defined on M. Suppose the spectrum, $\mathrm{Spec}(-L)$, of the elliptic operator*

$$L = \Delta + V(x)$$

on the negative axis is given by only finitely many negative eigenvalues, then there exists a compact set D such that $-L$ is nonnegative on $M \setminus D$.

Proof Let $\Omega \subset M$ be any compact subset of M. The variational principle asserts that the lowest eigenvalue of $-L$ is given by

$$\mu_1(L, \Omega) = \inf \frac{\int_\Omega |\nabla \phi|^2 - \int_\Omega V \phi^2}{\int_\Omega \phi^2},$$

where the infimum is taken over all compactly supported $\phi \in H_c^{1,2}(\Omega)$. In particular, since

$$\int_\Omega |\nabla \phi|^2 - \int_\Omega V \phi^2 \geq \mu_1(\Omega) \int_\Omega \phi^2 - \sup_\Omega V \int_\Omega \phi^2,$$

where $\mu_1(\Omega)$ is the first Dirichlet eigenvalue of the Laplacian on Ω, we conclude that

$$\mu_1(L, \Omega) \geq \mu_1(\Omega) - \sup_\Omega V. \tag{26.4}$$

For a fixed $p \in M$, let us consider the geodesic ball $B_p(\rho)$. Inequality (26.4) asserts that $\mu_1(L, B_p(\rho)) \geq 0$ for ρ sufficiently small because $\mu_1(B_p(\rho)) \to \infty$ as $\rho \to \infty$. Let us define

$$\rho_1 = 2 \sup\{\rho \mid \mu_1(L, B_p(\rho)) \geq 0\}.$$

If $\rho_1 = \infty$, then $-L$ is nonnegative on M and the lemma is proved. Hence we may assume that $\rho_1 < \infty$ and define

$$\rho_2 = 2 \sup\{\rho \mid \mu_1(L, B_p(\rho) \setminus B_p(\rho_1)) \geq 0\}.$$

Again, if $\rho_2 = \infty$, then $-L$ is nonnegative on $M \setminus B_p(\rho_1)$. Inductively we can define

$$\rho_i = 2 \sup\{\rho \mid \mu_1(L, B_p(\rho) \setminus B_p(\rho_{i-1})) \geq 0\}.$$

We claim that some ρ_i must be infinite, hence $-L$ is nonnegative on $M \setminus B_p(\rho_{i-1})$. If not, we get a infinite sequence of $\rho_i \to \infty$. Moreover the definition of ρ_i and the monotonicity of μ_1 imply that $-L$ is negative on all the annuli $B_p(\rho_i) \setminus B_p(\rho_{i-1})$. Let f_i be the eigenfunction on $B_p(\rho_i) \setminus B_p(\rho_{i-1})$ such that

$$L f_i = -\mu_1(L, B_p(\rho_i) \setminus B_p(\rho_{i-1})) f_i,$$

where $\mu_1(L, B_p(\rho_i) \setminus B_p(\rho_{i-1})) < 0$. Since the set of $\{f_i\}$ has disjoint support, they span an infinite-dimensional subspace of $H_c^{1,2}(M)$ such that $-L$ is negative. This contradicts the assumption that $-L$ has only finitely many negative eigenvalues and the lemma is proved. □

Theorem 26.5 (Li–Wang) *Let M^m be a complete, minimally immersed hypersurface with finite index in N^{m+1}. Suppose N is a complete manifold with nonnegative sectional curvature. Let $N(M)$ be the number of ends of M that are either nonparabolic, or parabolic but not contained in any bounded subset* of N, then there exists a constant $C > 0$ depending on a compact set of the M such that $N(M) \leq C$.

Proof Lemma 26.4 asserts that the assumption that M has finite index implies that there exists a compact subset D such that $M \setminus D$ is stable. In particular, the stability inequality (25.1) holds for any compactly supported function ψ defined on $M \setminus D$. Following the argument in the proof of Theorem 26.2, for each pair of such ends E and F, the sequence of harmonic functions $\{w_i\}$ converges to a harmonic function w on M satisfying (26.2) with ϕ being a compact support function on $M \setminus D$. Moreover, Theorem 23.4 asserts that the space of harmonic functions $\mathcal{K} = \{w\}$ constructed above will have dimension equal to the number of ends. Hence, it suffices to estimate the dimension of $\mathcal{K}$.

Note that for any $w \in \mathcal{K}$, we may assume that it can be approximated by a sequence of harmonic functions w_i whose Dirichlet integral has at most linear growth, as indicated by Lemma 26.1.

Let ρ_0 be sufficiently large such that $D \subset B_p(\rho_0)$ and set ϕ to be

$$\phi(x) = \begin{cases} 0 & \text{on} \quad B_p(\rho_0), \\ r(x) - \rho_0 & \text{on} \quad B_p(\rho_0 + 1) \setminus B_p(\rho_0), \\ 1 & \text{on} \quad B_p(\rho_0 + 2) \setminus B_p(\rho_0 + 1), \\ \dfrac{\rho_i - r(x)}{\rho_i - \rho_0 - 2} & \text{on} \quad B_p(\rho_i) \setminus B_p(\rho_0 + 2) \end{cases}$$

for $\rho_0 + 2 < \rho_i$. Following the argument for (26.3), we conclude that

$$(m-1) \int_M \phi^2 |\nabla w_i|^2 |\vec{II}|^2 + m \int_M \phi^2 |\nabla |\nabla w_i||^2$$

$$\leq m(m-1) \int_M |\nabla \phi|^2 |\nabla w_i|^2$$

$$\leq m(m-1) \int_{B_p(\rho_0+1) \setminus B_p(\rho_0)} |\nabla w_i|^2$$

$$+ \frac{m(m-1)}{(\rho_i - \rho_0 - 2)^2} \int_{B_p(\rho_i) \setminus B_p(\rho)} |\nabla w_i|^2$$

$$\leq m(m-1) \int_{B_p(\rho_0+1) \setminus B_p(\rho_0)} |\nabla w_i|^2 + \frac{C \rho_i}{(\rho_i - \rho_0 - 2)^2}.$$

Letting $i \to \infty$, we obtain the estimate

$$(m-1) \int_M \phi^2 |\nabla w|^2 |\vec{II}|^2 + m \int_M \phi^2 |\nabla |\nabla w||^2$$
$$\leq m(m-1) \int_{B_p(\rho_0+1) \setminus B_p(\rho_0)} |\nabla w|^2. \tag{26.5}$$

On the other hand, the Poincaré inequality for mixed boundary conditions asserts that there is a constant $C > 0$ depending on the set $B_p(\rho_0+2) \setminus B_p(\rho_0)$ such that

$$C \int_{B_p(\rho_0+2) \setminus B_p(\rho_0)} f^2 \leq \int_{B_p(\rho_0+2) \setminus B_p(\rho_0)} |\nabla f|^2$$

for any $f \in H^{1,2}(B_p(\rho_0+2) \setminus B_p(\rho_0))$ with the boundary condition

$$f = 0 \quad \text{on} \quad \partial B_p(\rho_0).$$

Applying this to the function $\phi |\nabla w|$, we conclude that

$$C \int_{B_p(\rho_0+2) \setminus B_p(\rho_0)} \phi^2 |\nabla w|^2$$
$$\leq \int_{B_p(\rho_0+2) \setminus B_p(\rho_0)} |\nabla(\phi |\nabla w|)|^2$$
$$\leq 2 \int_{B_p(\rho_0+1) \setminus B_p(\rho_0)} |\nabla \phi|^2 |\nabla w|^2 + 2 \int_{B_p(\rho_0+2) \setminus B_p(\rho_0)} \phi^2 |\nabla |\nabla w||^2$$
$$\leq 2 \int_{B_p(\rho_0+1) \setminus B_p(\rho_0)} |\nabla w|^2 + 2 \int_{B_p(\rho_0+2) \setminus B_p(\rho_0)} \phi^2 |\nabla |\nabla w||^2.$$

Combining this inequality with (26.5), we have

$$C \int_{B_p(\rho_0+2) \setminus B_p(\rho_0+1)} |\nabla w|^2$$
$$\leq C_1 \int_{B_p(\rho_0+2) \setminus B_p(\rho_0)} \phi^2 |\nabla w|^2$$
$$\leq \int_{B_p(\rho_0+1)} |\nabla w|^2. \tag{26.6}$$

Since the function $|\nabla w|$ satisfies the differential inequality

$$\Delta |\nabla w| \geq -|\vec{II}|^2 |\nabla w|,$$

the local mean value inequality (Corollary 14.8) asserts that there exists a constant $C > 0$ depending on the set $B_p(\rho_0 + 2)$, such that if $x \in B_p(\rho_0 + 1)$, then

$$|\nabla w|^2(x) \leq C \int_{B_x(1)} |\nabla w|^2.$$

Together with (26.5), this implies

$$\sup_{B_p(\rho_0+1)} |\nabla w|^2 \leq C \int_{B_p(\rho_0+1)} |\nabla w|^2.$$

The dimension estimate on $\mathcal{K}$ then follows from Lemma 7.3 and the same argument as in the proof of Theorem 7.4. □

We would like to remark that the properness assumption of M in N in Corollary 26.3 is only used to ensure that the Buseman functions satisfy the hypothesis of Lemma 26.1 on a parabolic end. In the event that the end is not properly immersed, it is still possible that one can find a Buseman function satisfying the property that its maximum is not achieved on ∂E. For example, if the end E is not contained in a compact set of N, then the Buseman function is not be bounded on E and the hypothesis of Lemma 26.1 is met. If this is the case, the proof of Corollary 26.3 is still valid for that end. When $N = \mathbb{R}^{m+1}$, we can also recover the theorems in [CSZ] and [LW4].

Corollary 26.6 (Cao–Shen–Zhu) *Let M^m be a minimal hypersurface in $\mathbb{R}^{m+1}$ with $m \geq 3$. If M is stable, then M must have only one end.*

Proof According to Corollary 26.3 and the above remark, we only need to rule out the case when M has a parabolic end. Indeed, by a Sobolev inequality of Michael and Simon [MS] and Corollary 20.12, any end E must be nonparablic, and the corollary follows. □

The above argument also proved the finite index case.

Corollary 26.7 (Li–Wang) *Let M^m be a minimally immersed hypersurface in $\mathbb{R}^{m+1}$ with $m \geq 3$. If M has finite index, then there exists a constant $C > 0$ depending on M such that*

$$N(M) \leq C.$$

27
Linear growth harmonic functions

In this and the next two chapters, we will consider polynomial growth harmonic functions on a complete manifold. Recall that any polynomial growth harmonic function in $\mathbb{R}^m$ is necessarily a polynomial with respect to the variables in rectangular coordinates. Hence the space of polynomial growth harmonic functions of at most order d is given by the space of harmonic polynomials of degree at most d. In particular, these spaces are all finite dimensional (see Appendix B).

Definition 27.1 Let $\mathcal{H}^d(M)$ be the vector space of all polynomial growth harmonic functions defined on M of order at most d, i.e.,

$$\mathcal{H}^d(M) = \{f \mid \Delta f = 0, \text{ and } |f(x)| \leq C\, r^d(x) \text{ for some constant } C > 0\},$$

where $r(x)$ is the distance from x to a fixed point $p \in M$. We also denote the dimension of the vector space $\mathcal{H}^d(M)$ by $h^d(M)$.

Using this notation, one computes (see Appendix B) that

$$h^d(\mathbb{R}^m) = \binom{m+d-1}{d} + \binom{m+d-2}{d-1}.$$

On the other hand, Cheng's theorem (Corollary 6.6) asserts that if M has nonnegative Ricci curvature, then

$$h^d(M) = 1$$

for all $d < 1$. In view of this, Yau conjectured that $h^d(M)$ must be finite dimensional if M has nonnegative Ricci curvature. In 1989, Li and Tam proved that

$$h^1(M) \leq m + 1$$

27 Linear growth harmonic functions

if M has nonnegative Ricci curvature. Note that this estimate is sharp and it is achieved by $\mathbb{R}^m$. In fact, Li [L7] showed that if M is Kähler and has nonnegative holomorphic bisectional curvature, then equality holds if and only if $M = \mathbb{C}^n$ with $2n = m$. The real case of this theorem was later proved by Cheeger, Colding, and Minicozzi [CCM]. They showed that if M has nonnegative Ricci curvature and $h^1(M) = m + 1$, then $M = \mathbb{R}^m$.

We will first present the theorem of Li and Tam [LT3], which is a finer version of the statement mentioned above. The proof relies on Lemma 16.4, which can be viewed as a sharp mean value inequality at infinity

Theorem 27.2 (Li–Tam) *Let M be a complete manifold with nonnegative Ricci curvature. Suppose the volume growth of M satisfies*

$$\limsup_{\rho \to \infty} \rho^{-n} V_p(\rho) < \infty$$

for some $n \le m$, then

$$h^1(M) \le n + 1 \le m + 1.$$

Proof For any $f \in \mathcal{H}^1(M)$, the curvature assumption, Lemma 3.4, and Corollary 6.6 imply that $|\nabla f|^2$ is a bounded subharmonic function on M. In particular, Lemma 16.4 asserts that

$$\lim_{\rho \to \infty} V_p^{-1}(\rho) \int_{B_p(\rho)} |\nabla f|^2 = \sup_M |\nabla f|^2. \tag{27.1}$$

For a fixed point $p \in M$, let us define the subspace of $\mathcal{H}^1(M)$ given by $\mathcal{H}' = \{f \mid f(p) = 0, \, f \in \mathcal{H}^1(M)\}$, and a bilinear form $\bar{D}$ on $\mathcal{H}'$ by

$$\bar{D}(f, g) = \lim_{\rho \to \infty} V_p^{-1}(\rho) \int_{B_p(\rho)} \langle \nabla f, \nabla g \rangle.$$

Observe that the fact that $f - f(p) \in \mathcal{H}'$ for any $f \in \mathcal{H}^1(M)$ implies that the codimension of $\mathcal{H}' \subset \mathcal{H}^1(M)$ is 1. Note that $\bar{D}$ is an inner product on $\mathcal{H}'$ because of (27.1). Given any finite dimensional subspace $\mathcal{H}''$ of $\mathcal{H}'$ with $\dim \mathcal{H}'' = k$, let $\{f_1, \ldots, f_k\}$ be an orthonormal basis of $\mathcal{H}''$ with respect to this inner product. Note that if we define the function $F(x)$ by

$$F^2(x) = \sum_{i=1}^{k} f_i^2(x),$$

then it is independent of orthogonal transformation on the set $\{f_i\}$. Similarly to the argument in Lemma 7.3, for each $x \in M$, we may assume that $f_i(x) = 0$ for all $i \neq 1$. At this point, we see that

$$F^2(x) = f_1^2(x)$$

and

$$F(x) \nabla F(x) = f_1(x) \nabla f_1(x).$$

Hence

$$|\nabla F|(x) \leq |\nabla f_1|(x) \leq 1 \qquad (27.2)$$

because $\sup_M |\nabla f_1|^2 = 1$ due to (27.1). Also, integrating along a geodesic and using the fact that $F(p) = 0$, we have

$$F(x) \leq r(x). \qquad (27.3)$$

Applying (27.2) to the integral

$$2 \sum_{i=1}^{k} \int_{B_p(\rho)} |\nabla f_i|^2 = \int_{B_p(\rho)} \Delta(F^2)$$

$$= 2 \int_{\partial B_p(\rho)} F \frac{\partial F}{\partial r}$$

$$\leq 2 \int_{\partial B_p(\rho)} F \qquad (27.4)$$

and combining this with (27.3), we obtain

$$V_p^{-1}(\rho) \sum_{i=1}^{k} \int_{B_p(\rho)} |\nabla f_i|^2 \leq \frac{\rho A_p(\rho)}{V_p(\rho)}. \qquad (27.5)$$

The fact that $\{f_i\}$ are orthonormal with respect to the inner product $\bar{D}$ implies that for any $\epsilon > 0$, there exists ρ sufficiently large such that

$$k - \epsilon \leq V_p^{-1}(\rho) \sum_{i=1}^{k} \int_{B_p(\rho)} |\nabla f_i|^2.$$

Hence the above inequality together with (27.5) yield

$$\frac{k - \epsilon}{\rho} \leq \frac{A_p(\rho)}{V_p(\rho)}$$

27 Linear growth harmonic functions

for $\rho \geq \rho_0$ with ρ_0 sufficiently large. Integrating this inequality from ρ_0 to ρ gives

$$\left(\frac{\rho}{\rho_0}\right)^{k-\epsilon} \leq \frac{V_p(\rho)}{V_p(\rho_0)}.$$

This implies that $k - \epsilon \leq n$, and the theorem follows since ϵ is arbitrary. □

We are now ready to consider the equality case when $h^1(M) = m + 1$ for the above theorem. First we need the following lemmas.

Lemma 27.3 *Let M^m be a complete manifold with nonnegative Ricci curvature. If*

$$h^1(M) = m + 1,$$

then the function

$$F(x) = \left(\sum_{i=1}^{m} f_i^2(x)\right)^{1/2},$$

defined in the proof of Theorem 27.2, must satisfy

$$\lim_{x \to \infty} \frac{F(x)}{r(x)} = 1.$$

Proof Inequality (27.7) asserts that

$$V_p^{-1}(\rho) \int_{B_p(\rho)} \sum_{i=1}^{m} |\nabla f_i|^2 \leq \frac{\rho}{V_p(\rho)} \int_{\partial B_p(\rho)} \frac{F}{\rho}. \tag{27.6}$$

On the other hand, since $\mathcal{R}_{ij} \geq 0$, the Laplacian comparison theorem asserts that $\Delta r^2 \leq 2m$, hence

$$2m V_p(\rho) \geq \int_{B_p(\rho)} \Delta r^2$$

$$= 2\rho A_p(\rho).$$

Combining this estimate with (27.6), we obtain

$$V_p^{-1}(\rho) \int_{B_p(\rho)} \sum_{i=1}^{m} |\nabla f_i|^2 \leq m A_p^{-1}(\rho) \int_{\partial B_p(\rho)} \frac{F}{\rho}.$$

Letting $\rho \to \infty$ and using (27.3), we conclude that

$$\lim_{\rho \to \infty} A_p^{-1}(\rho) \int_{\partial B_p(\rho)} \frac{F}{\rho} = 1. \tag{27.7}$$

For any given $\delta > 0$, let $m_\delta(\rho)$ be the $(m-1)$-dimensional measure of the set

$$\left\{ x \in \partial B_p(\rho) \left| \frac{F(x)}{\rho} < 1 - \delta \right. \right\}.$$

The limit (27.7) asserts that

$$\lim_{\rho \to \infty} m_\delta(\rho) A_p^{-1}(\rho) = 0. \tag{27.8}$$

We will prove the lemma by assuming the contrary that there exists $\alpha > 0$ and a sequence of $\{x_i\}$ with $x_i \to \infty$ such that

$$F(x_i) r^{-1}(x_i) \leq 1 - \alpha.$$

For any $y \in B_{x_i}(\alpha r(x_i)/4)$, inequality (27.2) asserts that

$$F(y) r^{-1}(y) \leq F(x_i) r^{-1}(y) + r(x_i, y) r^{-1}(y)$$
$$\leq (1 - \alpha) r(x_i) r^{-1}(y) + \frac{\alpha}{4} r(x_i) r^{-1}(y),$$

where $r(x_i, y)$ is the distance between x_i and y. However, since $r(y) \geq (1 - \alpha/4) r(x_i)$, this implies that

$$F(y) r^{-1}(y) \leq 1 - \frac{2\alpha}{4 - \alpha}.$$

Setting $\delta = 2\alpha/(4 - \alpha)$ and using (27.8), we see that for any $\epsilon > 0$, by taking i sufficiently large, we have

$$V_{x_i}(\alpha r(x_i)/4) \leq \int_{r(x_i) - \alpha r(x_i)/4}^{r(x_i) + \alpha r(x_i)/4} m_\delta(r)\, dr$$
$$\leq \epsilon \int_{r(x_i) - \alpha r(x_i)/4}^{r(x_i) + \alpha r(x_i)/4} A_p(r)\, dr$$
$$\leq \epsilon V_p\left(r(x_i) + \frac{\alpha r(x_i)}{4}\right). \tag{27.9}$$

On the other hand, the volume comparison theorem asserts that

$$V_{x_i}\left(\frac{\alpha r(x_i)}{4}\right) \geq \left(\frac{\alpha}{12}\right)^n V_{x_i}(3r(x_i))$$
$$\geq \left(\frac{\alpha}{12}\right)^n V_p(2r(x_i)).$$

27 Linear growth harmonic functions

Combining the volume estimate with (27.9), we conclude that

$$\left(\frac{\alpha}{12}\right)^n \leq \epsilon.$$

Since ϵ is arbitrary, this give a contradiction, and the lemma is proved. $\square$

Using the orthonormal basis $\{f_1, \ldots, f_m\}$ of $\mathcal{H}' = \{f \in \mathcal{H}^1(M) \mid f(p) = 0\}$, we define the map

$$\Phi : M \to \mathbb{R}^m$$

by

$$\Phi(x) = (f_1, \ldots, f_m).$$

Lemma 27.4 *Suppose $X : B_p(\rho) \to \mathbb{R}^m$ is a function defined on $B_p(\rho)$ with $|X(x)| = 1$ for all $x \in B_p(\rho)$. For any subset $A \subset B_p(\rho)$ and any $\theta > 0$, let us define the set*

$$C_{\theta,A} = \{v \in S(A) \mid |d\Phi(v) - X(\pi(v))| < \theta\},$$

where $S(A) = \{v \in T_q(M) \mid q \in A, |v| = 1\}$ is the unit sphere bundle over A. Then for any $0 < \epsilon \leq \theta/4$, we can take ρ sufficiently large, such that

$$\frac{V(C_{\theta,A})}{V(S(B_p(\rho)))} \geq \frac{\bar{V}(\theta - 2\epsilon)}{V(\mathbb{S}^{m-1})} \left(\frac{V(A)}{V_p(\rho)} - \epsilon\right),$$

where $\bar{V}(\theta)$ denotes the $(m-1)$-dimensional measure of the set given by

$$\{v \in \mathbb{S}^{m-1} \mid |v - v_0| \leq \theta\}$$

for some fixed vector $v_0 \in \mathbb{S}^{m-1}$.

Proof For any $\epsilon > 0$, let us define the set

$$D = \{y \in B_p(\rho) \mid |\langle v, w\rangle - \langle d\Phi(v), d\Phi(w)\rangle| < \epsilon, \text{ for all } v, w \in S_y(M)\}.$$

The fact that $\{f_i\}$ form an orthonormal basis with respect to the inner product $\bar{D}_\rho$ implies that

$$\lim_{\rho \to \infty} V_p^{-1}(\rho) \int_{B_p(\rho)} |\langle \nabla f_i, \nabla f_j\rangle - \delta_{ij}| = 0$$

for all $1 \leq i, j \leq m$, hence

$$\frac{V(D)}{V_p(\rho)} \geq 1 - \epsilon$$

for ρ sufficiently large. This implies that

$$V(A \cap D) \geq V(A) - \epsilon V_p(\rho).$$

For any fixed $y \in A \cap D$, and an orthonormal basis $\{e_i\}$ of $T_y(M)$, the set $\{d\Phi(e_i)\}$ forms a basis for $d\Phi(T_y(M))$ if ϵ is sufficiently small. This means that there exists a set of numbers $\{a_i\}_{i=1}^m$ with $\sum_{i=1}^m a_i^2 = 1$ such that

$$\left| X(y) - \sum_{i=1}^m a_i d\Phi(e_i) \right| < \epsilon.$$

Clearly, the set defined by

$$P = \left\{ v \in S_y(M) \, \middle| \, \left| v - \sum_{i=1}^m a_i e_i \right| < \theta - 2\epsilon, \text{ with } y \in D \cap A \right\}$$

satisfies

$$P \subset C_{\theta,A}.$$

Therefore,

$$V(C_{\theta,A}) \geq V(P)$$
$$= \bar{V}(\theta - 2\epsilon) V(D \cap A)$$
$$\geq \bar{V}(\theta - 2\epsilon)(V(A) - \epsilon V_p(\rho)),$$

and this implies the lemma. $\square$

We are now ready to consider the equality case in Theorem 27.2.

Theorem 27.5 (Cheeger–Colding–Minicozzi) *Let M^m be a complete manifold with nonnegative Ricci curvature. If*

$$h^1(M) = m + 1,$$

then $M = \mathbb{R}^m$.

Proof According to the Bishop volume comparison theorem, to show that $M = \mathbb{R}^m$ it suffices to show that for any $\epsilon > 0$ the volume of geodesic balls satisfies

$$V_p(\rho) \geq (1 - \epsilon) \omega_m \rho^m,$$

27 Linear growth harmonic functions

for all ρ sufficiently large. Indeed, this implies that

$$\lim_{\rho \to \infty} \rho^{-m} V_p(\rho) = \omega_m$$

and the equality case of the volume comparison implies that $M = \mathbb{R}^m$.

For any $\epsilon > 0$, let us define A to be a subset of $\mathbb{R}^m$ by

$$A = B_0((1-\epsilon)\rho) \setminus \overline{\Phi(B_p(\rho))}.$$

Let us choose a set of open balls $\mathcal{B}_1 = \{B_{x_i}(\rho_i)\}$ using the following procedure. For any $x \in A$, let ρ_x be the supremum of those values r such that $B_x(r) \subset A$. In particular, $\partial B_x(\rho_x) \cap \partial A \neq \emptyset$. If $\partial B_x(\rho_x) \cap \overline{\Phi(B_p(\rho))} \neq \emptyset$, then we say that $B_x(\rho_x) \in \mathcal{B}_1$. If $\partial B_x(\rho_x) \cap \overline{\Phi(B_p(\rho))} = \emptyset$, then there must be a point $y \in \partial B_x(\rho_x) \cap B_0((1-\epsilon)\rho)$ and x must be contained in the line segment $\gamma(t)$ joining y to 0. Suppose $x = \gamma(t_0)$ for some $0 < t_0 < (1-\epsilon)\rho$. Let us consider the balls $B_{\gamma(t)}(t)$ centered at $\gamma(t)$ of radius t for $t \geq t_0$. It is clear that $y \in \partial B_{\gamma(t)}(t)$ and $x \in B_{\gamma(t)}(t)$. Let t_1 be the first $t > t_0$ such that $B_{\gamma(t_1)}(t_1) \cap \overline{\Phi(B_p(\rho))} \neq \emptyset$. Note that $t_1 < (1-\epsilon)\rho/2$ must exist since $0 = \Phi(p)$ and $B_{\gamma(t_1)}(t_1) \subset B_0((1-\epsilon)\rho)$. In particular, $B_{\gamma(t_1)}(t_1) \subset A$, and we include $B_{\gamma(t_1)}(t_1)$ in our collection $\mathcal{B}_1$. Clearly, $\mathcal{B}_1$ covers A and $\rho_x \leq (1-\epsilon)\rho/2$ for all $B_x(\rho_x) \in \mathcal{B}_1$.

We will choose the set $\{B_{x_i}(\rho_i)\}$ among the collection $\mathcal{B}_1$ inductively. Let $B_{x_1}(\rho_1) = B_{x_1}(\rho_{x_1})$ be chosen such that $\rho_{x_1} \geq \frac{1}{2} \sup_{x \in A} \rho_x$. For a fixed $\alpha > 1$, we define

$$\mathcal{B}_2 = \{B_x(\rho_x) \in \mathcal{B}_1 \mid B_x(\alpha\,\rho_x) \cap B_{x_1}(\alpha\,\rho_1) = \emptyset\}.$$

Let us choose

$$B_{x_2}(\rho_2) = B_{x_2}(\rho_{x_2})$$

such that $\rho_{x_2} \geq \frac{1}{2} \sup_{B_x(\rho_x) \in \mathcal{B}_2} \rho_x$. Inductively, we define

$$\mathcal{B}_{i+1} = \left\{ B_x(\rho_x) \in \mathcal{B}_1 \ \middle|\ B_x(\alpha\,\rho_x) \cap \left(\sum_{j=1}^{i} B_{x_j}(\alpha\,\rho_j)\right) = \emptyset \right\}.$$

If $\mathcal{B}_{i+1} = \emptyset$, then $\{B_{x_j}(\rho_j)\}_{j=1}^{i}$ is our collection. If $\mathcal{B}_{i+1} \neq \emptyset$, then we choose

$$B_{x_{i+1}}(\rho_{i+1}) = B_{x_{i+1}}(\rho_{x_{i+1}})$$

such that $\rho_{x_{i+1}} \geq \frac{1}{2} \sup_{B_x(\rho_x) \in \mathcal{B}_{i+1}} \rho_x$. In any case, we will get a collection $\mathcal{B} = \{B_{x_j}(\rho_j)\}$ whether it is finite or not. This collection will have the properties that

$$B_{x_i}(R_i) \cap \overline{\Phi(B_p(\rho))} \neq \emptyset, \tag{27.10}$$

and
$$B_{x_i}(\alpha \rho_i) \cap B_{x_j}(\alpha \rho_j) = \emptyset \quad \text{for } i \neq j. \tag{27.11}$$

We now claim that
$$A \subset \cup_{\mathcal{B}} B_x(5\alpha \rho_x). \tag{27.12}$$

To see this, suppose $y \in A \setminus \cup_{\mathcal{B}} B_x(5\alpha \rho_x)$, then there exists $B_x(\rho_x) \in \mathcal{B}_1$ such that $y \in B_x(\rho_x)$ and
$$B_x(\alpha \rho_x) \cap (\cup_{\mathcal{B}} B_x(\alpha \rho_x)) \neq \emptyset,$$

otherwise the inductive procedure will continue. Let i be the smallest such number so that
$$B_x(\alpha \rho_x) \cap B_{x_i}(\alpha \rho_i) \neq \emptyset.$$

This implies that
$$B_x(\rho_x) \in \mathcal{B}_i$$

and hence
$$\rho_i \geq \tfrac{1}{2}\rho_x.$$

Therefore, we conclude that
$$y \in B_{x_i}(5\alpha \rho_i),$$

which is a contradiction, and hence (27.12) is valid. Let D be any relatively compact subdomain $D \subset A$ such that $2V(D) \geq V(A)$. Since $\overline{D}$ is compact, there exists a finite subcover $\{B_{x_i}(\rho_i)\}$ of the cover $\mathcal{B}$ such that
$$D \subset \cup_i B_{x_i}(5\alpha \rho_i).$$

In particular, this implies that
$$V(D) \leq \sum_i V_{x_i}(5\alpha \rho_i)$$
$$= (5\alpha)^m \sum_i V_{x_i}(\rho_i).$$

Hence the finite collection $\{B_{x_i}(\rho_i)\}$ not only satisfies (27.10) and (27.11), it also satisfies
$$V(A) \leq 2(5\alpha)^m \sum_i V_{x_i}(\rho_i). \tag{27.13}$$

27 Linear growth harmonic functions

Let $y_i \in \partial B_{x_i}(\rho_i) \cap \overline{\Phi(B_p(\rho))}$ and $p_i \in \overline{B_p(\rho)}$ such that
$$\Phi(p_i) = y_i.$$
Since $|\nabla f_i|^2 \leq 1$, we have
$$|d\Phi(v)| \leq \sqrt{m}|v|$$
for any tangent vector v. Hence, we have
$$\Phi\left(B_{p_i}\left((2\sqrt{m})^{-1}\rho_i\right)\right) \subset B_{y_i}(2^{-1}\rho_i)$$
$$\subset B_{x_i}((2^{-1}+1)\rho_i). \qquad (27.14)$$
For any $\theta > 0$, let us define
$$C_{\theta,i} = \left\{v \in S\left(B_{p_i}\left(\frac{\rho_i}{2\sqrt{m}}\right)\right) \;\middle|\; |d\Phi(v) + \nabla r_{x_i}(\Phi(\pi(v)))| < \theta\right\},$$
where r_{x_i} is the distance function to x_i, and
$$C_\theta = \cup_i C_{\theta,i}.$$
Note that by taking $\alpha = \frac{3}{2}$, (27.14) and (27.11) imply that
$$\Phi(B_{p_i}((2\sqrt{m})^{-1}\rho_i) \cap \Phi(B_{p_j}((2\sqrt{m})^{-1}\rho_j) = \emptyset$$
and
$$C_{\theta,i} \cap C_{\theta,j} = \emptyset$$
for $i \neq j$. Also $B_{p_i}(\rho_i/2\sqrt{m}) \subset B_p((1+(4\sqrt{m})^{-1})\rho)$ since $\rho_i \leq \rho/2$, and after applying Lemma 27.4, we conclude that
$$\frac{V(C_\theta)}{V(S(B_p((1+(4\sqrt{m})^{-1})\rho)))}$$
$$\geq \frac{\bar{V}(\theta/2)}{V(\mathbb{S}^{m-1})}\left(\sum_i V_{p_i}\left(\frac{\rho_i}{2\sqrt{m}}\right)V_p^{-1}((1+(4\sqrt{m})^{-1})\rho) - \epsilon\right). \qquad (27.15)$$
On the other hand, the volume comparison theorem asserts that
$$V_{p_i}\left(\frac{\rho_i}{2\sqrt{m}}\right)V_p^{-1}((1+(4\sqrt{m})^{-1})\rho) \geq V_{p_i}\left(\frac{\rho_i}{2\sqrt{m}}\right)V_{p_i}^{-1}((2+(4\sqrt{m})^{-1})\rho)$$
$$\geq \left(\frac{\rho_i}{(4\sqrt{m}+2^{-1})\rho}\right)^m.$$

Therefore combining the above inequality with (27.15), we have

$$\frac{V(C_\theta)}{V(S(B_p((1+(4\sqrt{m})^{-1})\rho))} \geq \frac{\bar{V}(\theta/2)}{V(\mathbb{S}^{m-1})} \left(\frac{\sum_i \rho_i^m}{((4\sqrt{m}+2^{-1})\rho)^m} - \epsilon \right).$$

However, (27.13) asserts that

$$2\left(\frac{15}{2}\right)^m \omega_m \sum_i \rho_i^m \geq V(A),$$

hence

$$\frac{V(C_\theta)}{V(S(B_p((1+(4\sqrt{m})^{-1})\rho)))} \geq \frac{\bar{V}(\theta/2)}{V(\mathbb{S}^{m-1})} \left(C_1 \frac{V(A)}{\rho^m} - \epsilon \right) \quad (27.16)$$

for some constant $C_1 > 0$ depending only on m.

Let $v \in S(B_{p_i}(\rho_i/2\sqrt{m}))$ be a unit tangent vector. We let γ_v be the normal geodesic emanating in the direction of v, i.e., $\gamma_v'(0) = v$. If $\gamma_v(\rho_i) \in B_p(\rho)$, then $\Phi(\gamma_v(\rho_i)) \notin A$. If $\gamma_v(\rho_i) \notin B_p(\rho)$, then because of Lemma 27.3, for ρ sufficiently large,

$$\rho(\gamma_v(\rho_i)) > (1-\epsilon)r(\gamma_v(\rho_i)),$$

hence

$$\Phi(\gamma_v(\rho_i)) \notin B_0((1-\epsilon)\rho)$$

and

$$\Phi(\gamma_v(\rho_i)) \notin A$$

also. Note that since $\gamma_v(0) \in B_{p_i}(\rho_i/2\sqrt{m})$ and (27.14) imply that $\Phi(\gamma_v(0)) \in B_{x_i}(3\rho_i/2) \setminus B_{x_i}(\rho_i/2)$, we conclude that

$$\Phi(\gamma_v(0)) - \rho_i \nabla r_{x_i}(\Phi(\gamma_v(0))) \in B_{x_i}\left(\frac{\rho_i}{2}\right).$$

Taking this together with the fact that $\Phi(\gamma_v(\rho_i)) \notin A$, we have

$$\tfrac{1}{2} < \left| -\nabla r_{x_i}(\Phi(\pi(v))) - \rho_i^{-1}(\Phi(\gamma_v(\rho_i)) - \Phi(\gamma_v(0))) \right|$$

for all $v \in S(B_{p_i}(\rho_i/2\sqrt{m}))$. Thus for $\theta \leq \tfrac{1}{4}$ and $v \in C_\theta \cap S(B_{p_i}(\rho_i/2\sqrt{m}))$, we obtain

$$\tfrac{1}{4} < |d\Phi(v) - \rho_i^{-1}(\Phi(\gamma_v(\rho_i)) - \Phi(\gamma_v(0)))|.$$

27 Linear growth harmonic functions

Hence we have

$$V(C_\theta) \leq 4 \sum_i \int_{C_\theta \cap S(B_{p_i}(\rho_i/2\sqrt{m}))} |d\Phi(v) - \rho_i^{-1}(\Phi(\gamma_v(\rho_i)) - \Phi(\gamma_v(0)))|.$$

Combining this with (27.16), we obtain the estimate

$$\frac{\bar{V}(\theta/2)}{V(\mathbb{S}^{m-1})} \left(C_1 \frac{V(A)}{\rho^m} - \epsilon \right) \leq 4 V^{-1}(S(B_p((1+2m^{-\frac{1}{2}})\rho)))$$

$$\times \sum_i \int_{C_\theta \cap S(B_{p_i}(\rho_i/2\sqrt{m}))}$$

$$\times |d\Phi(v) - \rho_i^{-1}(\Phi(\gamma_v(\rho_i)) - \Phi(\gamma_v(0)))|. \tag{27.17}$$

We now claim that for any $\epsilon > 0$, we can choose ρ sufficiently large such that the right-hand side of (27.17) is at most ϵ. To see this, we consider

$$|d\Phi(v) - \rho_i^{-1}(\Phi(\gamma_v(\rho_i)) - \Phi(\gamma_v(0)))|$$

$$= \left| d\Phi(v) - \rho_i^{-1} \int_0^{\rho_i} \frac{d}{dt}(\Phi(\gamma_v(t))) \, dt \right|$$

$$= \left| \rho_i^{-1} \int_0^{\rho_i} \left(d\Phi(v) - d\Phi\left(\gamma_v'(t)\right) \right) dt \right|$$

$$= \rho_i^{-1} \left| \int_0^{\rho_i} \int_0^t \frac{d}{ds}\left(d\Phi\left(\gamma_v'(s)\right) \right) ds \, dt \right|$$

$$\leq \rho_i^{-1} \int_0^{\rho_i} \int_0^{\rho_i} |\text{Hess}(\Phi)\left(\gamma_v'(s), \gamma_v'(s)\right)| \, ds \, dt$$

$$\leq \int_0^{\rho_i} |\text{Hess}(\Phi)\left(\gamma_v'(s), \gamma_v'(s)\right)| \, ds.$$

Therefore, we have

$$V^{-1}(S(B_p((1+2m^{-\frac{1}{2}})\rho))) \sum_i \int_{C_\theta \cap S(B_{p_i}(\rho_i))}$$

$$\times |d\Phi(v) - \rho_i^{-1}(\Phi(\gamma_v(\rho_i)) - \Phi(\gamma_v(0)))|$$

$$\leq 2^{-1} \rho \, V^{-1}(B_p((1+2m^{-\frac{1}{2}})\rho)) \int_{B_p((1+2m^{-\frac{1}{2}})\rho)} |\text{Hess}(\Phi)|. \tag{27.18}$$

On the other hand, the Bochner formula implies that

$$\Delta|\nabla f|^2 \geq 2 f_{ij}^2$$

for any harmonic function on M. Let $\phi(t)$ be a cutoff function satisfying the properties

$$\phi(t) = \begin{cases} 1 & \text{for } t \leq \rho, \\ 0 & \text{for } t \geq 2\rho, \end{cases}$$

with $0 \leq \phi(t) \leq 1$, $-C\rho^{-1} \leq \phi'(t) \leq 0$, and $|\phi''(t)| \leq C\rho^{-2}$. Taking the composition with the function $F(x) = \left(\sum_i f_i^2\right)^{\frac{1}{2}}$ and integrating by parts, we obtain

$$2\int_{B_p((1-\epsilon)\rho)} f_{ij}^2 \leq \int_M \phi(F) f_{ij}^2$$

$$\leq \int_M \phi(F) \Delta(|\nabla f|^2 - 1)$$

$$\leq \int_M \Delta\phi(F)(|\nabla f|^2 - 1)$$

$$\leq \int_M (\phi'(F)\Delta F + \phi''(F)|\nabla F|^2)(|\nabla f|^2 - 1). \quad (27.19)$$

The fact that $\Delta F^2 = 2\sum_i |\nabla f_i|^2 \geq 2|\nabla F|^2$ implies that

$$\Delta F \geq 0.$$

Also

$$2F\Delta F + 2|\nabla F|^2 \leq \Delta F^2$$
$$\leq 2m$$

implies that

$$\Delta F \leq mF^{-1}.$$

By taking $f = f_i$ for $i = 1, \ldots, m$ and using the fact that $|\nabla f_i| \leq 1$, (27.18) can be estimated by

$$2\int_{B_p((1-\epsilon)\rho)} f_{jk}^2 \leq C\rho^{-1}\int_{F \leq 2\rho} F^{-1}(1-|\nabla f|^2) + C\rho^{-2}\int_{F \leq 2\rho}(1-|\nabla f|^2)$$

$$\leq C\rho^{-2}\int_{B_p((2+\epsilon)\rho)}(1-|\nabla f|^2)$$

$$\leq \epsilon C\rho^{-2} V_p((2+\epsilon)\rho)$$

$$\leq \epsilon C\rho^2 V_p((1-\epsilon)\rho)\left(\frac{2+\epsilon}{1-\epsilon}\right)^m.$$

This implies that

$$\int_{B_p(\rho)} |\text{Hess}(\Phi)| \leq \left(\int_{B_p(\rho)} |\text{Hess}(\Phi)|^2\right)^{1/2} V_p^{1/2}(\rho)$$

$$\leq \epsilon\, C\, \rho^{-1} V_p(\rho).$$

Combining this estimate with (27.17) and (27.18), we conclude that

$$V(A) \leq \epsilon\, C\, \rho^m,$$

hence implying

$$V(\Phi(B_p(\rho))) \geq V_0((1-\epsilon)\rho) - V(A)$$
$$\geq (1-\epsilon)^m V_0(\rho) - \epsilon\, C\rho^m. \qquad (27.20)$$

On the other hand, since $|d\Phi| \leq 1$, we have

$$V_p(\rho) \geq V(\Phi(B_p(\rho))).$$

Therefore combining this inequality with (27.19), we conclude that

$$V_p(\rho) \geq (1-\epsilon) V_0(\rho)$$

for sufficiently large ρ. This proves the theorem. $\square$

28
Polynomial growth harmonic functions

Yau's conjecture was first proved by Colding and Minicozzi in 1996 [CM1]. Later that year they announced [CM2] the dimension estimate of the form

$$h^d(M) \leq C_1 \, d^{\log C_{\mathcal{VD}}/\log 2},$$

with C_1 depending on the Neumann Poincaré inequality (equivalent to $C_{\mathcal{P}}(M) > 0$ in Definition 18.2) and the volume doubling constant (see Definition 18.1). In particular, it gives a sharp order estimate of the form

$$h^d(M) \leq C_1 \, d^{m-1}$$

for manifolds with nonnegative Ricci curvature. The complete proofs of these announcements were published in [CM3, CM4] in 1997. Meanwhile, in 1997, Li [L8] gave a much simplified proof of this estimate which holds for a larger class of manifolds. In this paper, Li only required the manifold to satisfy a volume comparison condition ($\mathcal{V}_\nu$) (see Definition 28.1) and a mean value inequality ($\mathcal{M}$) (see Definition 28.2). Note that the volume comparison condition and the volume doubling condition can easily be seen to be equivalent. However, the mean value inequality is weaker than the Neumann Poincaré inequality. Moreover, the author's argument can be applied to sections of vector bundles (Theorem 28.4). When Theorem 28.4 is applied to manifolds with nonnegative Ricci curvature (Corollary 28.5), it recovers the sharp order estimate of Colding–Minicozzi. In 1998, Colding and Minicozzi [CM5, CM6] also published a proof for the dimension estimate using the mean value inequality and volume doubling much in the spirit of [L8]. Finally, it is important to note that in [L8] the author proved that finite dimensionality and estimates of $h^d(M)$ actually hold for an even more general class of manifolds (Theorem 28.7), namely those satisfying a weak mean value inequality and

28 Polynomial growth harmonic functions

with polynomial volume growth. However, in this case, the estimate is not as sharp as Theorem 28.4.

Definition 28.1 A manifold is said to satisfy a volume comparison condition $(\mathcal{V}_\mu)$ for some $\mu > 1$, if there exists a constant $C_\mathcal{V} > 0$, such that for any point $x \in M$ and any real numbers $0 < \rho_1 \leq \rho_2 < \infty$ the volume of the geodesic balls centered at x satisfies the inequality

$$V_x(\rho_2) \leq C_\mathcal{V} V_x(\rho_1) \left(\frac{\rho_2}{\rho_1}\right)^\mu.$$

Definition 28.2 A manifold is said to satisfy a mean value inequality $(\mathcal{M})$ if there exists a constant $C_\mathcal{M} > 0$ such that for any $x \in M$, $\rho > 0$ and any nonnegative subharmonic function f defined on M it must satisfy

$$f^2(x) \leq C_\mathcal{M} V_x^{-1}(\rho) \int_{B_x(\rho)} f^2(y) \, dy.$$

Note that if M has nonnegative Ricci curvature, then M satisfies condition $(\mathcal{V}_m)$ with $C_\mathcal{V} = 1$, by the Bishop volume comparison theorem. Moreover, condition $(\mathcal{M})$ is also valid on such a manifold because of Theorem 7.2.

Lemma 28.3 (Li) *Let K be a k-dimensional linear space of sections of a vector bundle E over M. Assume that M has polynomial volume growth at most of order μ, i.e.,*

$$V_p(\rho) \leq C \rho^\mu$$

for $p \in M$ and $\rho \to \infty$. Suppose each section $u \in K$ has polynomial growth of at most degree d such that

$$|u|(x) \leq C r^d(x),$$

where $r(x)$ is the geodesic distance to the fixed point $p \in M$. For any $\beta > 1$, $\delta > 0$, and $\rho_0 > 0$, there exists $\rho > \rho_0$ such that if $\{u_i\}_{i=1}^k$ is an orthonormal basis of K with respect to the inner product $A_{\beta\rho}(u, v) = \int_{B_p(\beta\rho)} \langle u, v \rangle$, then

$$\sum_{i=1}^k \int_{B_p(\rho)} |u_i|^2 \geq k \beta^{-(2d+\mu+\delta)}.$$

Proof For each $\rho > 0$, let A_ρ be the nonnegative bilinear form defined on K given by $A_\rho(u, v) = \int_{B_p(\rho)} \langle u, v \rangle$. Let us denote the trace of the bilinear form A_ρ with respect to $A_{\rho'}$ by $\text{tr}_{\rho'} A_\rho$. Similarly, let $\det_{\rho'} A_\rho$ be the determinant of

A_ρ with respect to $A_{\rho'}$. Assuming that the lemma is false, then for all $\rho > \rho_0$, we have

$$\text{tr}_{\beta\rho} A_\rho < k \beta^{-(2d+\mu+\delta)}.$$

On the other hand, the arithmetic-geometric mean asserts that

$$\left(\det_{\beta\rho} A_\rho\right)^{\frac{1}{k}} \leq k^{-1} \text{tr}_{\beta\rho} A_\rho.$$

This implies that

$$\det_{\beta\rho} A_\rho \leq \beta^{-k(2d+\mu+\delta)}$$

for all $\rho > \rho_0$. Setting $\rho = \beta^i \rho$ for $i = 0, 1, \ldots, j-1$ and iterating this inequality j times yields

$$\det_{\beta^j \rho} A_\rho \leq \beta^{-jk(2d+\mu+\delta)}. \tag{28.1}$$

However, for a fixed A_ρ-orthonormal basis $\{u_i\}_{i=1}^k$ of K, the assumptions on K and on the volume growth imply that there exists a constant $C > 0$ depending on K, such that

$$\int_{B_p(\rho_1)} |u_i|^2 \leq C \left(1 + \rho_1^{2d+\mu}\right)$$

for all $1 \leq i \leq k$ and for all $\rho_1 \geq \rho$. In particular, this implies that

$$\det_\rho A_{\beta^j \rho} \leq k! \, C \, \beta^{jk(2d+\mu)} \rho^{k(2d+\mu)}.$$

This contradicts (28.1) as $j \to \infty$, and the lemma is proved. $\square$

Theorem 28.4 (Li) *Let M^m be a complete manifold satisfying conditions $(\mathcal{V}_\mu)$ and $(\mathcal{M})$. Suppose E is a rank-n vector bundle over M. Let $\mathcal{S}_d(M, E) \subset \Gamma(E)$ be a linear subspace of sections of E such that all $u \in \mathcal{S}_d(M, E)$ satisfy the properties:*

(a) $\Delta |u| \geq 0$; and
(b) $|u|(x) \leq O(r^d(x))$ as $x \to \infty$.

Then the dimension of $\mathcal{S}_d(M, E)$ is finite. Moreover, for all $d \geq 1$, there exists a constant $C > 0$ depending only on μ such that

$$\dim \mathcal{S}_d(M, E) \leq n \, C \, C_\mathcal{M} \, d^{\mu-1}.$$

Proof Let K be a finite dimensional linear subspace of $\mathcal{S}_d(M, E)$ with $\dim K = k$ and $\{u_i\}_{i=1}^k$ be any basis of K. Then for $p \in M$, $\rho > 0$, and any

28 Polynomial growth harmonic functions 343

$0 < \epsilon < \frac{1}{2}$, we claim that they must satisfy the estimate

$$\sum_{i=1}^{k} \int_{B_p(\rho)} |u_i|^2 \leq n\, C\, C_{\mathcal{M}}\, \epsilon^{-(\mu-1)} \sup_{u \in \{\langle A, U \rangle\}} \int_{B_p((1+\epsilon)\rho)} |u|^2, \qquad (28.2)$$

where the supremum is taken over all $u \in K$ of the form $u = \langle A, U \rangle$ for some unit vector $A = (a_1, \ldots, a_k) \in \mathbb{R}^k$ with $U = (u_1, \ldots, u_k)$.

To see this, we will estimate $\sum \int_{B_p(\rho)} |u_i|^2$ by using an argument similar to Lemma 7.3. Observe that for any $x \in B_p(\rho)$, there exists a subspace $K_x \subset K$ which is of at most codimension n such that $u(x) = 0$ for all $u \in K_x$. In particular, by an orthonormal change of basis, we may assume that $u_i \in K_x$ for $n+1 \leq i \leq k$ and $\sum_{i=1}^{k} |u_i|^2(x) = \sum_{i=1}^{n} |u_i|^2(x)$. Since $\Delta |u_i| \geq 0$, if we denote the distance from p to x by $r(x)$, then the mean value inequality $(\mathcal{M})$ implies that

$$\sum_{i=1}^{k} |u_i|^2(x) = \sum_{i=1}^{n} |u_i|^2(x)$$

$$\leq C_{\mathcal{M}}\, V_x^{-1}((1+\epsilon)\rho - r(x)) \sum_{i=1}^{n} \int_{B_x((1+\epsilon)\rho - r(x))} |u_i|^2$$

$$\leq C_{\mathcal{M}}\, V_x^{-1}((1+\epsilon)\rho - r(x))\, n \sup_{u \in \{\langle A, U \rangle\}} \int_{B_p((1+\epsilon)\rho)} |u|^2. \qquad (28.3)$$

However, condition $(\mathcal{V}_\mu)$ and the fact that $r(x) \leq \rho$ imply that

$$C_\mathcal{V}\, V_x((1+\epsilon)\rho - r(x)) \geq \left(\frac{(1+\epsilon)\rho - r(x)}{2\rho} \right)^{\mu} V_x(2\rho)$$

$$\geq \left(\frac{(1+\epsilon)\rho - r(x)}{2\rho} \right)^{\mu} V_p(\rho). \qquad (28.4)$$

Therefore substituting this estimate into (28.3) and integrating over $B_p(\rho)$, we have

$$\sum_{i=1}^{k} \int_{B_p(\rho)} |u_i|^2 \leq \frac{n C_\mathcal{V}\, C_{\mathcal{M}}\, 2^\mu}{V_p(\rho)} \sup_{u \in \{\langle A, U \rangle\}} \int_{B_p((1+\epsilon)\rho)} u^2$$

$$\times \int_{B_p(\rho)} ((1+\epsilon) - \rho^{-1} r(x))^{-\mu}\, dx. \qquad (28.5)$$

On the other hand, if we define
$$f(r) = ((1+\epsilon) - \rho^{-1} r)^{-\mu},$$
then
$$f'(r) = \mu \rho^{-1} ((1+\epsilon) - \rho^{-1} r)^{-(\mu+1)} \geq 0$$
and
$$\int_{B_p(\rho)} ((1+\epsilon) - \rho^{-1} r(x))^{-\mu} dx = \int_0^\rho A_p(t) f(t) dt.$$

To estimate this, we integrate by parts and obtain
$$\int_0^\rho f(t) A_p(t) dt = [f(t) V_p(t)]_0^\rho - \int_0^\rho f'(t) V_p(t) dt. \qquad (28.6)$$

Using $f'(t) \geq 0$ and applying condition $(\mathcal{V}_\mu)$, we have
$$\int_0^\rho f'(t) V_p(t) dt \geq \rho^{-\mu} V_p(\rho) \int_0^\rho f'(t) t^\mu dt$$
$$\geq \rho^{-\mu} V_p(\rho) \left([f(t) t^\mu]_0^\rho - \mu \int_0^\rho f(t) t^{\mu-1} dt \right).$$

Substituting this into (28.6) yields
$$\int_0^\rho f(t) A_p(t) dt = \rho^{-\mu} V_p(\rho) \mu \int_0^\rho f(t) t^{\mu-1} dt$$
$$\leq \rho^{-1} V_p(\rho) \mu \int_0^\rho ((1+\epsilon) - t\rho^{-1})^{-\mu} dt$$
$$= \frac{\mu}{\mu - 1} V_p(\rho) (\epsilon^{-\mu+1} - (1+\epsilon)^{-\mu+1})$$
$$\leq \frac{\mu}{\mu - 1} V_p(\rho) \epsilon^{-(\mu-1)}.$$

The claim follows by combining this with (28.5).

To complete the proof of the theorem, let $\{u_i\}_{i=1}^k$ be an $A_{\beta\rho}$-orthonormal basis of any finite dimensional subspace $K \subset \mathcal{S}_d(M, E)$. Clearly, it suffices to prove the estimate for $k = \dim K$. Note that condition $(\mathcal{V}_\mu)$ implies that the volume growth of M is at most of order ρ^μ, hence Lemma 28.3 implies that there is a $\rho > 0$ such that
$$\sum_{i=1}^k \int_{B_p(\rho)} |u_i|^2 \geq k \beta^{-(2d+\mu+\delta)}.$$

28 Polynomial growth harmonic functions

On the other hand, by setting $\beta = 1 + \epsilon$, (28.2) implies that

$$\sum_{i=1}^{k} \int_{B_p(\rho)} |u_i|^2 \leq n\, C\, C_{\mathcal{M}}\, \epsilon^{-(\mu-1)}$$

because $\int_{B_p((1+\epsilon)\rho)} |u|^2 = 1$ for all $u \in \{\langle A, U \rangle\}$. For $d \geq 1$, the estimate on k follows by setting $\epsilon = (2d)^{-1}$ and observing that $(1 + (2d)^{-1})^{-(2d+\mu+\delta)}$ is bounded from below. □

Note that extra care is used to obtain the order $\mu - 1$ in Theorem 28.4 because this is sharp on Euclidean space (B.3).

Corollary 28.5 (Colding–Minicozzi) *Let M^m be a complete Riemannian manifold with nonnegative Ricci curvature. There exists a constant $C > 0$ depending only on m such that the dimension of the $\mathcal{H}^d(M)$ is bounded by*

$$h^d(M) \leq C\, d^{m-1}$$

for all $d \geq 1$.

Finite dimensionality of $\mathcal{H}^d(M)$ can be obtained by relaxing both the volume comparison condition ($\mathcal{V}_\mu$) and the mean value inequality condition ($\mathcal{M}$). However, in the general case, the order of dependency in d will not be sharp.

Definition 28.6 A complete manifold M is said to satisfy a weak mean value inequality ($\mathcal{WM}$) if there exist constants $C_{\mathcal{WM}} > 0$ and $b > 1$ such that for any nonnegative subharmonic function f defined on M it must satisfy

$$f(x) \leq C_{\mathcal{WM}}\, V_x^{-1}(\rho) \int_{B_x(b\rho)} f(y)\, dy.$$

for all $x \in M$ and $\rho > 0$.

Theorem 28.7 *Let M be a complete manifold satisfying the weak mean value property ($\mathcal{WM}$). Suppose that the volume growth of M satisfies $V_p(\rho) = O(\rho^\mu)$ as $\rho \to \infty$ for some point $p \in M$. Then $\mathcal{H}^d(M)$ is finite dimensional for all $d \geq 0$ and $\dim \mathcal{H}^d(M) \leq C_{\mathcal{WM}} (2b+1)^{(2d+\mu)}$.*

Proof Let us set $\beta = 2b + 1$, where b is the constant in ($\mathcal{WM}$). By Lemma 28.3, for any $\delta > 0$, there exists $\rho > 0$ such that for $u_1, \ldots, u_k$, an orthonormal basis of $\mathcal{H}^d(M)$ with respect to the inner product $A_{\beta\rho}(u, v)$, we have

$$\sum_{i=1}^{k} \int_{B_p(\rho)} u_i^2 \geq k\, \beta^{-(2d+\mu+\delta)}. \tag{28.7}$$

On the other hand, since the function $\sum_{i=1}^k u_i^2(x)$ is subharmonic, the maximum principle implies that there exists a point $q \in \partial B_p(\rho)$ such that

$$\sum_{i=1}^k u_i^2(x) \leq \sum_{i=1}^k u_i^2(q) \tag{28.8}$$

for all $x \in B_p(\rho)$. Again, by an orthonormal change of basis, we may assume that $u_j(q) = 0$ for $2 \leq j \leq k$. Applying the weak mean value inequality ($\mathcal{WM}$) to u_1^2 and noting that $B_p(\rho) \subset B_q(2\rho) \subset B_p((2b+1)\rho)$, we get

$$V_p(\rho)\, u_1^2(q) \leq V_q(2\rho)\, u_1^2(q)$$

$$\leq C_{\mathcal{WM}} \int_{B_q(2b\rho)} u_1^2$$

$$\leq C_{\mathcal{WM}} \int_{B_p((2b+1)\rho)} u_1^2$$

$$= C_{\mathcal{WM}}.$$

Thus integrating (28.8) over the ball $B_p(\rho)$ gives

$$\sum_{i=1}^k \int_{B_p(\rho)} u_i^2 \leq V_p(\rho) \sum_{i=1}^k u_i^2(q)$$

$$\leq V_p(\rho)\, u_1^2(q)$$

$$\leq C_{\mathcal{WM}}. \tag{28.9}$$

From (28.7) and (28.9) we conclude that

$$k\, \beta^{-(2d+\mu+\delta)} \leq C_{\mathcal{WM}},$$

hence $k \leq C_{\mathcal{WM}}\, \beta^{(2d+\mu)}$ as $\delta > 0$ is arbitrary. This completes the proof. $\square$

Corollary 28.8 *Let M be a complete manifold. Suppose there exist constants $C_1, C_2 > 0$ and $\mu > 2$ such that for all $p \in M$, $\rho > 0$, and for all $f \in H_c^{1,2}(M)$ we have*

$$\left(\int_{B_p(\rho)} |f|^{2\mu/(\mu-2)} \right)^{(\mu-2)/\mu} \leq C_1\, V_p(\rho)^{-2/\mu} \rho^2$$

$$\times \int_{B_p(\rho)} |\nabla f|^2 + C_2\, V_p(\rho)^{-2/\mu} \int_{B_p(\rho)} f^2.$$

Then
$$\dim \mathcal{H}^d(M) < C\beta^d$$

for all $d \geq 0$, for some constants $C > 0$ and $\beta > 1$ depending only on C_1, C_2, μ, and b.

Proof Applying the Moser iteration process given by Lemma 19.1 the Sobolev type inequality in the hypothesis implies the weak mean value inequality (WM). Also, by choosing f to be a cutoff function satisfying

$$f(x) = \begin{cases} 1 & \text{if } x \in B_p(1), \\ 0 & \text{if } x \in M \setminus B_p(2), \end{cases}$$

and

$$|\nabla f| \leq 1,$$

and applying to the Sobolev inequality in the hypothesis we conclude that

$$V_p^{(\mu-2)/\mu}(1) \leq C_1 V_p^{-2/\mu}(\rho)(C_1 \rho^2 + C_2) V_p(2).$$

This implies that $V_p(\rho) = O(\rho^\mu)$ as $\rho \to \infty$. The corollary follows from Theorem 28.7. □

In the case when M^m has nonnegative sectional curvature, it was shown by Li and Wang in [LW3] that if $h^d(M) = \dim \mathcal{H}^d(M)$, then $\sum_{i=1}^d h^i(M)$ has an upper bound which is asymptotically sharp as $d \to \infty$. In particular, they showed that if $0 \leq \alpha_0 \leq \omega_m$ is a constant given by

$$\liminf_{\rho \to \infty} \rho^{-m} V_p(\rho) = \alpha_0,$$

then

$$\liminf_{d \to \infty} d^{-(m-1)} h^d \leq \frac{2\alpha_0}{(m-1)!\,\omega_m}.$$

Moreover,

$$\liminf_{d \to \infty} d^{-(m-1)} h^d = \frac{2}{(m-1)!}$$

if and only if $M = \mathbb{R}^m$. The proof of this will not be presented here, however, this motivates the following question.

Question *If M is a complete manifold with nonnegative sectional curvature, then is it true that*

$$h^d(M) \leq h^d(\mathbb{R}^m)$$
$$= \binom{m+d-1}{d} + \binom{m+d-2}{d-1}?$$

29
L^q harmonic functions

Recall that Yau's theorem (Lemma 7.1) asserts that a complete manifold does not admit any nonconstant L^q harmonic functions for $q \in (1, \infty)$. In this chapter, we will discuss the validity of this Liouville property for L^q harmonic functions when $q \in (0, 1]$. The condition to ensure the Liouville property for the case when $q \in (0, 1)$ is quite different from the condition for the case $q = 1$. Both of these cases were first considered by Li and Schoen [LS], when they obtained the sharp curvature condition for $q \in (0, 1)$. In the same paper, they also obtained an almost sharp condition for $q = 1$, while the sharp version was later proved by Li in [L4] using the heat equation method. Theorem 29.1 presented below is the argument from [LS], while Theorem 29.3 is from [L4]. Counterexamples for both theorems when the curvature conditions have been violated were given in [LS] indicating the sharpness of these conditions.

Theorem 29.1 (Li–Schoen) *Let M^m be a complete manifold. There exists a constant $\delta(m) > 0$ depending only on m such that if the Ricci curvature of M satisfies the lower bound*

$$\mathcal{R}_{ij}(x) \geq -\delta(m)\, r^{-2}(x) \quad \text{for} \quad r(x) \to \infty,$$

where $r(x)$ is the distant to a fixed point $p \in M$, then any nonnegative, L^q subharmonic function must be identically zero for $q \in (0, 1)$.

Proof Let f be a nonnegative L^q subharmonic function defined on M. According to the mean value inequality (14.23) of Corollary 14.8, for nonnegative subharmonic functions, we have the estimate

$$f^q(x) \leq C_1 \bar{V}(2\rho)\, \rho^{-m}\, \exp\left(C_2 \rho \sqrt{R}\right) V_x^{-1}(\rho) \int_{B_x(\rho)} f^q,$$

if
$$\mathcal{R}_{ij} \geq -(m-1)R$$

on $B_x(2\rho)$ with constants $C_1, C_2 > 0$ depends only on m and q, and $\bar{V}(2\rho)$ is the volume of the geodesic ball of radius 2ρ in the model space of constant $-R$ curvature. Let us now apply the above mean value inequality to any point x sufficiently far from p with $5\rho \leq r(x)$. The curvature assumption asserts that the quantity

$$\bar{V}(2\rho)\, \rho^{-m} \exp\left(C_2 \rho \sqrt{R}\right)$$

is bounded, hence

$$f^q(x) \leq C_3 V_x^{-1}(\rho), \tag{29.1}$$

where C_3 also depends on the L^q-norm of f. We will now estimate $V_x(\rho)$ using the volume comparison theorem.

Let γ be a minimal normal geodesic joining p to x with $\gamma(0) = p$ and $\gamma(T) = x$. Define the values $\{t_i\}$ with $t_0 = 0$, $t_1 = 1 + \beta$, and $t_i = 2\sum_{j=0}^{i} \beta^j - 1 - \beta^i$, for some fixed

$$\beta > \frac{2}{2^{1/m} - 1} > 1.$$

Let $t_k \leq T$ be the largest value not bigger than T. We let the points $x_i = \gamma(t_i)$ and they obviously satisfy the properties that $r(x_i, x_{i+1}) = \beta^i + \beta^{i+1}$ and $r(x_k, x) < \beta^k + \beta^{k+1}$. Moreover, the closure of the set of geodesic balls $\{B_{x_i}(\beta^i)\}$ covers $\gamma\left(\left[0, 2\sum_{j=1}^{k} \beta^j - 1\right]\right)$ and the balls have disjoint interiors. We now claim that

$$V_{x_k}(\beta^k) \geq C_4 \left(\frac{\beta^m}{(\beta+2)^m - \beta^m}\right)^k V_p(1). \tag{29.2}$$

To see this, the volume comparison theorem implies that

$$V_{x_i}(\beta^i) \geq A_i (V_{x_i}(\beta^i + 2\beta^{i-1}) - V_{x_i}(\beta^i))$$
$$\geq A_i V_{x_{i-1}}(\beta^{i-1}),$$

where

$$A_i = \frac{\int_0^{\beta^i \sqrt{R(x_i,\beta^i+2\beta^{i-1})}} \sinh^{m-1}(t)\, dt}{\int_{\beta^i \sqrt{R(x_i,\beta^i+2\beta^{i-1})}}^{(\beta^i+2\beta^{i-1})\sqrt{R(x_i,\beta^i+2\beta^{i-1})}} \sinh^{m-1}(t)\, dt}$$

with
$$\mathcal{R}_{ij} \geq -(m-1)R(x_i, \beta^i + 2\beta^{i-1})$$

denotes the lower bound of the Ricci curvature on $B_{x_i}(\beta^i + 2\beta^{i-1})$. Iterating this inequality, we conclude that

$$V_{x_k}(\beta^k) \geq V_p(1) \prod_{j=1}^{k} A_j. \tag{29.3}$$

Since $r(x_i) = 2\sum_{j=0}^{i} \beta^j - 1 - \beta^i$, the curvature assumption implies that

$$R(x_i, \beta^i + 2\beta^{i-1}) \leq \delta(m) \left(2 \sum_{j=0}^{i-2} \beta^j\right)^{-2}$$

for $i \geq 2$. Therefore

$$\beta^i \sqrt{R(x_i, \beta^i + 2\beta^{i-1})} \leq \frac{\sqrt{\delta(m)}}{2} \frac{\beta^i}{\sum_{j=0}^{i-2} \beta^j}$$

$$\leq \frac{\sqrt{\delta(m)}}{2} \frac{\beta^2}{\sum_{j=0}^{i-2} \beta^{-j}}$$

$$\leq \frac{\sqrt{\delta(m)}}{2} \beta(\beta-1)(1-\beta^{-i+1})^{-1},$$

which can be made arbitrarily small for a fixed β by choosing $\delta(m)$ to be sufficiently small. In particular, simply by approximating $\sinh(t)$ by t, A_i can be approximated by

$$A_i \sim \frac{(\beta^i)^m}{(\beta^i + 2\beta^{i-1})^m - (\beta^i)^m}$$

$$= \frac{\beta^m}{(\beta+2)^m - \beta^m}.$$

Combining this with (29.3), we conclude (29.2).

To estimate $V_x(\beta^{k+1})$, we consider the following cases. First, let us assume that

$$r(x, x_k) \leq \beta^k(\beta - 1).$$

In this case, we see that

$$B_{x_k}(\beta^k) \subset B_x(\beta^{k+1}),$$

hence
$$V_{x_k}(\beta^k) \leq V_x(\beta^{k+1}).$$

In particular, we conclude that

$$V_x(\beta^{k+1}) \geq C_4 \left(\frac{\beta^m}{(\beta+2)^m - \beta^m}\right)^k V_p(1). \tag{29.4}$$

On the other hand, if
$$r(x, x_k) > \beta^k(\beta - 1),$$

then we have
$$B_{x_k}(\beta^k) \subset B_x(r(x, x_k) + \beta^k) \setminus B_x(r(x, x_k) - \beta^k).$$

Using the volume comparison theorem, we conclude that

$$V_x(\beta^k) \geq A\, (V_x(r(x, x_k) + \beta^k) - V_x(x(x, x_k) - \beta^k)$$
$$\geq A\, V_{x_k}(\beta^k),$$

where
$$A = \frac{\int_0^{\beta^k} \sqrt{R(x, r(x, x_k) + \beta^k)}\, \sinh^{m-1}(t)\, dt}{\int_{(r(x,x_k)-\beta^k)\sqrt{R(x,r(x,x_k)+\beta^k)}}^{(r(x,x_k)+\beta^k)\sqrt{R(x,r(x,x_k)+\beta^k)}} \sinh^{m-1}(t)\, dt}.$$

Since
$$(r(x, x_k) + \beta^k)\sqrt{R(x, r(x, x_k) + \beta^k)} \leq (\beta^{k+1} + 2\beta^k)\sqrt{R(x, r(x, x_k) + \beta^k)}$$
$$\leq \frac{\sqrt{\delta(m)}}{2}\beta(\beta - 1)$$

can be made sufficiently small, we can approximate A by
$$A \sim \frac{\beta^m}{(\beta+2)^m}.$$

Combining the value of A with (29.2), we again conclude that

$$V_x(\beta^{k+1}) \geq C_5 \left(\frac{\beta^m}{(\beta+2)^m - \beta^m}\right)^k V_p(1) \tag{29.5}$$

for some constant $C_5 > 0$.

29 L^q harmonic functions

Note that when $x \to \infty$, $k \to \infty$. Since the choice of β ensures that

$$\frac{\beta^m}{(\beta+2)^m - \beta^m} > 1,$$

the right-hand side of (29.5) tends to infinity.

Using the value $R = \beta^{k+1}$ in (29.1), we conclude that $f(x) \to 0$ as $x \to \infty$. The maximum principle now asserts that f must be identically 0. $\square$

Corollary 29.2 *Let M^m be a complete manifold. There exists a constant $\delta(m) > 0$, depending only on m, such that if the Ricci curvature of M satisfies the lower*

$$\mathcal{R}_{ij}(x) \geq -\delta(m)\, r^{-2}(x) \quad \text{for} \quad r(x) \to \infty,$$

then M does not admit any nontrivial L^q harmonic functions for $q \in (0, 1)$.

Proof Since the absolute value of a harmonic function is subharmonic, this corollary follows directly from Theorem 29.1. $\square$

Theorem 29.3 (Li) *Let M^m be a complete manifold. Suppose the Ricci curvature of M satisfies the lower*

$$\mathcal{R}_{ij}(x) \geq -C\,(1+r(x))^2$$

for some constant $C > 0$. Then any nonnegative, L^1 subharmonic function must be constant.

Proof Let f be a nonnegative, L^1 subharmonic function defined on M. By solving the heat equation with f as initial datum, we obtain

$$f(x,t) = \int_M H(x,y,t)\, f(y)\, dy$$

with

$$f(x,0) = f(x).$$

Recall that the heat semi-group is contractive in L^1 because of Theorem 12.4, hence

$$\int_M H(x,y,t)\, dy \leq 1. \tag{29.6}$$

We now claim that

$$\frac{\partial}{\partial t} f(x,t) \geq 0 \tag{29.7}$$

for all $x \in M$ and for all $t \in (0, \infty)$. To see this, we differentiate under the integral sign and obtain

$$\frac{\partial}{\partial t} f(x,t) = \int_M \frac{\partial}{\partial t} H(x, y, t) \, f(y) \, dy$$
$$= \int_M \Delta_y H(x, y, t) \, f(y) \, dy.$$

If we can justify integration by parts, namely,

$$\int_M \Delta H(x, y, t) \, f(y) \, dy = \int_M H(x, y, t) \, \Delta f(y) \, dy, \tag{29.8}$$

then the subharmonicity of f implies (29.7). Indeed, (29.8) follows from Green's identity

$$\left| \int_{B_p(\rho)} \Delta H(x, y, t) \, f(y) \, dy - \int_{B_p(\rho)} H(x, y, t) \, \Delta f(y) \, dy \right|$$
$$= \left| \int_{\partial B_p(\rho)} \frac{\partial}{\partial r} H(x, y, t) \, f(y) - \int_{\partial B_p(\rho)} H(x, y, t) \, \frac{\partial}{\partial r} f(y) \, dy \right|$$
$$= \int_{\partial B_p(\rho)} |\nabla H|(x, y, t)| \, f(y) \, dy + \int_{\partial B_p(\rho)} H(x, y, t) \, |\nabla f|(y) \, dy$$

and by showing that there exists a sequence $\{\rho_i\}$ with $\rho_i \to \infty$ such that the boundary terms on the right-hand side when setting $\rho = \rho_i$ tend to 0 as $\rho_i \to \infty$.

Using the mean value inequality (14.23) of Corollary 14.8, we obtain the growth estimate

$$\sup_{B_p(\rho)} f \leq C_1 \bar{V}(4\rho) \, \rho^{-m} \exp\left(C_2 \rho \sqrt{R}\right) V_p^{-1}(2\rho) \int_{B_p(2\rho)} f,$$

where $-(m-1)R$ is the lower bound of the Ricci curvature on $B_p(4\rho)$. However, the assumption on the curvature of M implies that

$$\bar{V}(4\rho) \leq C_6 \exp\left(4(m-1)\rho \sqrt{R}\right)$$
$$\leq C_6 \exp\left(C_7 \rho^2\right)$$

for some constants $C_6, C_7 > 0$. This implies that

$$\sup_{B_p(\rho)} f \leq C_8 V_p^{-1}(2\rho) \exp\left(C_9 \rho^2\right), \tag{29.9}$$

where C_8 depends also on the L^1-norm of f.

Let ϕ be a nonnegative cutoff function satisfying

$$\phi(x) = \begin{cases} 0 & \text{on} & B_p(\rho-1) \cup (M \setminus B_p(2\rho)), \\ r(x) - \rho + 1 & \text{on} & B_p(\rho) \setminus B_p(\rho-1), \\ 1 & \text{on} & B_p(\rho) \setminus B_p(\rho+1), \\ \dfrac{2\rho + 1 - r(x)}{\rho} & \text{on} & B_p(2\rho+1) \setminus B_p(\rho+1). \end{cases}$$

The subharmonicity of f implies that

$$0 \leq \int_M \phi^2 f \, \Delta f$$

$$= -2 \int_M \phi f \langle \nabla \phi, \nabla f \rangle - \int_M \phi^2 |\nabla f|^2$$

$$\leq 2 \int_M |\nabla \phi|^2 f^2 - \frac{1}{2} \int_M \phi^2 |\nabla f|^2.$$

Hence using the definition of ϕ, we obtain the estimate

$$\int_{B_p(\rho+1)\setminus B_p(\rho)} |\nabla f|^2 \leq 4 \int_{B_p(2\rho+1)} f^2$$

for $\rho \geq 1$. Combining the above inequality with (29.9) and applying Schwarz inequality, we obtain

$$\int_{B_p(\rho+1)\setminus B_p(\rho)} |\nabla f| \leq \left(\int_{B_p(\rho+1)\setminus B_p(\rho)} |\nabla f|^2 \right)^{1/2} V_p^{1/2}(\rho+1)$$

$$\leq C_8 \exp(C_{10} \rho^2). \tag{29.10}$$

To estimate $H(x, y, t)$ we recall Theorem 13.4 which asserts that

$$H(x, y, t) \leq C_{11} V_x^{-1/2}(\sqrt{t}) V_y^{-1/2}(\sqrt{t})$$

$$\times \exp\left(\frac{-r^2(x, y)}{(4+\epsilon)t} + C_{12}\sqrt{(\rho^{-2} + R)t} \right)$$

for all $x, y \in B_p(\rho)$. For a fixed $x \in M$ and for sufficiently large ρ, we may assume that $x \in B_p(\rho/4)$. In this case, if $y \in \partial B_p(\rho)$, then the curvature assumption implies that

$$H(x, y, t) \leq C_{11} V_x^{-1/2}(\sqrt{t}) V_y^{-1/2}(\sqrt{t}) \exp\left(\frac{-\alpha \rho^2}{t} + C_{12}\rho\sqrt{t} \right). \tag{29.11}$$

Using the volume comparison theorem, we have

$$V_x\left(\sqrt{t}\right) \leq V_y\left(r(x,y)+\sqrt{t}\right) - V_y\left(r(x,y)-\sqrt{t}\right)$$
$$\leq V_y\left(\sqrt{t}\right) \frac{\bar{V}\left(r+\sqrt{t}\right)}{\bar{V}\left(\sqrt{t}\right)}$$
$$\leq V_y\left(\sqrt{t}\right) C_{13} t^{\frac{m}{2}} \exp\left(C_{14}\rho\left(\rho+\sqrt{t}\right)\right).$$

Applying this to (29.11), we conclude that

$$H(x,y,t) \leq C_{15} V_x^{-1}\left(\sqrt{t}\right) \exp\left(\frac{-\alpha\rho^2}{t} + C_{16}\rho\left(\rho+\sqrt{t}\right)\right). \quad (29.12)$$

To estimate the gradient of H, we consider the integral

$$\int_M \phi^2(y) |\nabla H|^2(x,y,t)$$
$$= -2\int_M \langle H(x,y,t)\nabla\phi(y), \phi(y)\nabla H(x,y,t)\rangle$$
$$\quad - \int_M \phi^2(y) H(x,y,t) \Delta H(x,y,t)$$
$$\leq 2\int_M |\nabla\phi|^2(y) H^2(x,y,t) + \tfrac{1}{2}\int_M \phi^2(y) |\nabla H|^2(x,y,t)$$
$$\quad - \int_M \phi^2(y) H(x,y,t) \Delta H(x,y,t).$$

This implies that

$$\int_{B_p(\rho+1)\setminus B_p(\rho)} |\nabla H|^2$$
$$\leq \int_M \phi^2 |\nabla H|^2$$
$$\leq 4\int_M |\nabla\phi|^2 H^2 - 2\int_M \phi^2 H \Delta H$$
$$\leq 4\int_{B_p(\rho+1)\setminus B_p(\rho-1)} H^2 + 2\int_{B_p(2\rho+1)\setminus B_p(\rho-1)} H|\Delta H|$$
$$\leq 4\int_{B_p(2\rho+1)\setminus B_p(\rho-1)} H^2 + 2\left(\int_{B_p(2\rho+1)\setminus B_p(\rho-1)} H^2\right)^{1/2} \left(\int_M (\Delta H)^2\right)^{1/2}.$$
$$(29.13)$$

29 L^q harmonic functions

Using the upper bound (29.12) for H and (29.6), we obtain the estimate

$$\int_{B_p(2\rho+1)\setminus B_p(\rho-1)} H^2(x, y, t)$$

$$\leq \sup_{y\in B_p(2\rho+1)\setminus B_p(\rho-1)} H(x, y, t)$$

$$\leq C_{15} V_x^{-1}\left(\sqrt{t}\right) \exp\left(\frac{-\alpha\rho^2}{t} + C_{16}\rho\left(\rho + \sqrt{t}\right)\right). \tag{29.14}$$

We now claim that there exists a constant $C_{17} > 0$, such that

$$\int_M (\Delta H)^2(x, y, t) \leq C_{17} t^{-2} H(x, x, t). \tag{29.15}$$

To see this, we first prove the inequality for any Dirichlet heat kernel H defined on a compact subdomain Ω of M. Using the fact that the heat kernel on M can be obtained by taking the limits of Dirichlet heat kernels on a compact exhaustion of M, (29.15) follows. Indeed, if $H(x, y, t)$ is a Dirichlet heat kernel defined on a compact subdomain $\Omega \subset M$, we can write H using the eigenfunction expansion

$$H(x, y, t) = \sum_{i=1}^{\infty} e^{-\lambda_i t} \psi_i(x,)\,\psi_i(y),$$

where $\{\psi_i\}$ is the orthonormal basis of the space of L^2 functions with Dirichlet boundary value satisfying the equation

$$\Delta \psi_i = -\lambda_i \psi_i.$$

Differentiating with respect to the variable y, we have

$$\Delta H(x, y, t) = -\sum_{i=1}^{\infty} \lambda_i\, e^{-\lambda_i t} \psi_i(x)\,\psi_i(y).$$

Therefore, using the fact that

$$s^2 e^{-2s} \leq C_{17} e^{-s}$$

for all $0 \leq s < \infty$, we conclude that

$$\int_M (\Delta H)^2(x, y, t)\, dy \leq C_{17} t^{-2} \sum_{i=1}^{\infty} e^{-\lambda_i t} \psi_i^2(x)$$

$$= C_{17} t^{-2} H(x, x, t).$$

Combining (29.13), (29.14), and (29.15), we obtain

$$\int_{B_p(\rho+1)\setminus B_p(\rho)} |\nabla H|^2$$

$$\leq C_{18}\left(V_x^{-\frac{1}{2}}\left(\sqrt{t}\right) + t^{-1}H^{\frac{1}{2}}(x,x,t)\right)\exp\left(\frac{-\alpha\rho^2}{2t} + C_{19}\rho\left(\rho+\sqrt{t}\right)\right).$$

Applying Schwarz inequality and the volume comparison theorem, we conclude that

$$\int_{B_p(\rho+1)\setminus B_p(\rho)} |\nabla H| \leq C_{20}\left(V_x^{-1/2}\left(\sqrt{t}\right) + t^{-1}H^{1/2}(x,x,t)\right)^{1/2}$$

$$\times \exp\left(\frac{-\alpha\rho^2}{4t} + C_{21}\rho\left(\rho+\sqrt{t}\right)\right). \qquad (29.16)$$

Using (29.9), (29.10), (29.12), and (29.16), we conclude that for a fixed $x \in M$ and for any sufficiently small fixed $t > 0$

$$\int_{B_p(\rho+1)\setminus B_p(\rho)} |\nabla H|(x,y,t)\, f(y)\, dy$$

$$+ \int_{B_p(\rho+1)\setminus B_p(\rho)} H(x,y,t)\, |\nabla f|(y)\, dy \to 0$$

as $\rho \to \infty$. Hence, the mean value theorem implies that there is a sequence of $\{\rho_i\}$ with $\rho_i \to \infty$ such that

$$\int_{\partial B_p(\rho_i)} |\nabla H|(x,y,t)\, f(y)\, dy + \int_{\partial B_p(\rho_i)} H(x,y,t)\, |\nabla f|(y)\, dy \to 0,$$

and (29.8) is verified. In particular, we confirm the monotonicity of $f(x,t)$ for t sufficiently small. Monotonicity of $f(x,t)$ for all t follows from the semigroup property of the heat equation.

Let us also observe that since

$$\frac{\partial}{\partial t}\int_M H(x,y,t)\, dy = \int_M \Delta H(x,y,t)\, dy,$$

and because (29.16) implies that

$$\int_{B_p(\rho_i)} \Delta H(x,y,t)\, dy \leq \int_{\partial B_p(\rho_i)} |\nabla H|(x,y,t)\, dy$$

$$\to 0$$

as $\rho_i \to \infty$, we conclude that

$$\frac{\partial}{\partial t} \int_M H(x, y, t) \, dy = 0,$$

hence

$$\int_M H(x, y, t) \, dy = 1 \qquad (29.17)$$

for all $x \in M$ and for all t.

To finish the theorem, since (29.17) implies that

$$\int_M f(x, t) \, dx = \int_M \int_M H(x, y, t) \, f(y) \, dy \, dx$$
$$= \int_M f(y) \, dy$$

and by applying the monotonicity of $f(x, t)$, we conclude that $f(x, t) = f(x)$, hence $\Delta f(x) = 0$. On the other hand, for any arbitrary constant $a > 0$, let us we define the function

$$g(x) = \min\{f(x), a\}.$$

It must be nonnegative satisfying

$$g(x) \leq f(x),$$
$$|\nabla g|(x) \leq |\nabla f|(x),$$

and

$$\Delta g(x) \leq 0.$$

In particular, it will satisfy the same estimates, (29.9) and (29.10), as f. Hence we can show that

$$\frac{\partial}{\partial t} \int_M H(x, y, t) g(y) \, dy = \int_M H(x, y, t) \Delta g(y) \, dy$$
$$\leq 0.$$

Since g is also L^1, the same argument as before implies that $\Delta g(x) = 0$, hence regularity theory asserts that g must be smooth. Since a is arbitrary, this is impossible unless f is identically constant. □

Corollary 29.4 *Let M^m be a complete manifold. Suppose the Ricci curvature of M is bounded from below by*

$$\mathcal{R}_{ij}(x) \geq -C\left(1+r(x)\right)^2$$

for some constant $C > 0$. Then M does not admit any nonconstant L^1 harmonic functions.

30

Mean value constant, Liouville property, and minimal submanifolds

In this chapter, we will assume that M satisfies a slightly stronger version of the mean value property ($\mathcal{M}$) as in Definition 28.2.

Definition 30.1 A manifold is said to satisfy an L^1 mean valued inequality ($\mathcal{M}_1$) if there exists a constant $C_{\mathcal{M}_1} > 0$ such that for any $x \in M$, $\rho > 0$, and any nonnegative subharmonic function f defined on M, it must satisfy

$$f(x) \leq C_{\mathcal{M}_1} V_x^{-1}(\rho) \int_{B_x(\rho)} f(y)\, dy.$$

Note that by applying Schwarz's inequality, ($\mathcal{M}_1$) implies ($\mathcal{M}$), with $C_{\mathcal{M}} < C_{\mathcal{M}_1}^2$. It turns out that the value of $C_{\mathcal{M}_1}$ plays a significant role in both the regularity theory and the Liouville property of a manifold. It was proved by Li and Wang [LW2] that when $C_{\mathcal{M}_1}$ is sufficiently close to 1, the mean value inequality ($\mathcal{M}_1$) becomes more powerful.

Theorem 30.2 (Li–Wang) *Let M be a complete manifold satisfying the mean value inequality ($\mathcal{M}_1$). If $C_{\mathcal{M}_1} < 2$ and M has subexponential volume growth, then M has the Liouville property, i.e., M does not admit any nonconstant bounded harmonic functions.*

Proof Let f be a harmonic function defined on M. Let us define

$$s(r) = \sup_{\bar{B}_p(r)} f(x),$$

$$i(r) = \inf_{\bar{B}_p(r)} f(x),$$

and

$$\omega(r) = s(r) - i(r).$$

361

Assume that $x, y \in \bar{B}_p(r)$ such that $f(x) = s(r)$ and $f(y) = i(r)$. In fact, the maximum principle asserts that $x, y \in \partial B_p(r)$. For any $\rho > r$, the mean value inequality applying to the functions $f - i(\rho)$ and $s(\rho) - f$ gives

$$C_{\mathcal{M}_1} \int_{B_p(\rho)} (f - i(\rho)) \geq C_{\mathcal{M}_1} \int_{B_x(\rho-r)} (f - i(\rho))$$

$$\geq (s(r) - i(\rho)) V_x(\rho - r) \quad (30.1)$$

and

$$C_{\mathcal{M}_1} \int_{B_p(\rho)} (s(\rho) - f) \geq C_{\mathcal{M}_1} \int_{B_y(\rho-r)} (s(\rho) - f)$$

$$\geq (s(\rho) - i(r)) V_y(\rho - r). \quad (30.2)$$

By taking $\rho > 2r$, we have

$$B_p(\rho - 2r) \subset B_x(\rho - r)$$

and

$$B_p(\rho - 2r) \subset B_y(\rho - r),$$

hence we have

$$V_x(\rho - r) \geq V_p(\rho - 2r)$$

and

$$V_y(\rho - r) \geq V_p(\rho - 2r).$$

Combining these inequalities with (30.1) and (30.2), and adding the two estimates, we conclude that

$$C_{\mathcal{M}_1} \omega(\rho) V_p(\rho) \geq (\omega(r) + \omega(\rho)) V_p(\rho - 2r).$$

This implies that

$$\left(C_{\mathcal{M}_1} \frac{V_p(\rho)}{V_p(\rho - 2r)} - 1\right) \omega(\rho) \geq \omega(r). \quad (30.3)$$

Suppose now that f is a nonconstant bounded harmonic function. Then by scaling we may assume that the total oscillation $\sup f - \inf f = 1$, hence $\omega(\rho) \nearrow 1$ as $\rho \to \infty$. In particular, (30.3) becomes

$$C_{\mathcal{M}_1} \frac{V_p(\rho)}{V_p(\rho - 2r)} - 1 \geq \omega(r). \quad (30.4)$$

For any $r > 0$, we claim that the sequence $V_p(2(i+1)r)/V_p(2ir)$ must satisfy

$$\liminf_{i \to \infty} \frac{V_p(2(i+1)r)}{V_p(2ir)} = 1.$$

Indeed, if this is not the case, then there exists $i_0 > 0$ such that

$$\frac{V_p(2(i+1)r)}{V_p(2ir)} \geq 1 + \epsilon$$

for $i \geq i_0$. Iterating this inequality k times beginning from $i = i_0$, we have

$$V_p(2(k+i_0)r) \geq (1+\epsilon)^k V_p(2i_0 r).$$

Let $k \to \infty$, this implies that M has exponential volume growth, and hence contradicts the assumption. Therefore there is a subsequence i_j such that

$$\frac{V_p(2(i_j+1)r)}{V_p(2i_j r)} \to 1$$

as $j \to \infty$. Setting $\rho_j = 2(i_j + 1)r$, then

$$\frac{V_p(\rho_j)}{V_p(\rho_j - 2r)} \to 1.$$

Together with (30.4), this implies that

$$(C_{\mathcal{M}_1} - 1) \geq \omega(r)$$

for all $r > 0$. However, this contradicts the assumptions that $\omega(r) \nearrow 1$ and $C_{\mathcal{M}_1} < 2$, hence f must be identically constant. $\square$

Theorem 30.3 (Li–Wang) *Let M be a complete manifold satisfying the mean value inequality $(\mathcal{M}_1)$ and the volume comparison condition $\mathcal{V}_\mu$. If $C_\mathcal{V} C_{\mathcal{M}_1} < 2$, then there exists a constant $0 < \alpha < 1$ depending only on $C_\mathcal{V} C_{\mathcal{M}_1}$ and μ such that any harmonic function f defined on M satisfying*

$$|f(x)| = O(r^\alpha(x)),$$

where r is the distance to some fixed point $p \in M$, must be identically constant.

Proof Following the proof of Theorem 30.2, (30.1) and (30.2) are valid for any harmonic functions defined on M. On the other hand, using the volume growth condition, we have

$$C_\mathcal{V} \left(\frac{\rho - r}{\rho + r}\right)^\mu V_x(\rho - r) \geq V_x(\rho + r)$$

$$\geq V_p(\rho)$$

and similarly

$$C_V \left(\frac{\rho - r}{\rho + r}\right)^\mu V_y(\rho - r) \geq V_p(\rho).$$

Substituting the above inequalities into (30.1) and (30.2), and taking their sum, we conclude that

$$C_{\mathcal{M}_1} \omega(\rho) \geq C_V^{-1} (\omega(r) + \omega(\rho)) \left(\frac{\rho - r}{\rho + r}\right)^\mu.$$

This implies that

$$\left(C_V C_{\mathcal{M}_1} \left(\frac{\rho + r}{\rho - r}\right)^\mu - 1\right) \omega(\rho) \geq \omega(r). \tag{30.5}$$

Setting $\rho = (1 + \beta)r$, this becomes

$$\left(C_V C_{\mathcal{M}_1} \left(\frac{2 + \beta}{\beta}\right)^\mu - 1\right) \omega((1 + \beta)r) \geq \omega(r).$$

Iterating this inequality k times yields

$$\left(C_V C_{\mathcal{M}_1} \left(\frac{2 + \beta}{\beta}\right)^\mu - 1\right)^k \omega((1 + \beta)^k r) \geq \omega(r). \tag{30.6}$$

On the other hand, the assumption on the growth rate of f implies that

$$\omega((1 + \beta)^k r) \leq C (1 + \beta)^{\alpha k} r^\alpha,$$

and hence when combined with (30.6) gives

$$\left(\left(C_V C_{\mathcal{M}_1} \left(\frac{2}{\beta} + 1\right)^\mu - 1\right) (1 + \beta)^\alpha\right)^k \geq \frac{\omega(r)}{C r^\alpha}. \tag{30.7}$$

Obviously, by taking β sufficiently large depending on μ and using the assumption that $C_V C_{\mathcal{M}_1} < 2$, we may assume that

$$\left(\frac{2}{\beta} + 1\right)^\mu \leq \frac{C_V C_{\mathcal{M}_1} + 2}{2 C_V C_{\mathcal{M}_1}},$$

which implies

$$\left(C_V C_{\mathcal{M}_1} \left(\frac{2}{\beta} + 1\right)^\mu - 1\right) \leq \frac{C_V C_{\mathcal{M}_1}}{2}.$$

Therefore there exists α sufficiently small depending on β and $C_V C_{\mathcal{M}_1}$ such that

$$\left(C_V C_{\mathcal{M}_1} \left(\frac{2}{\beta}+1\right)^\mu - 1\right)(1+\beta)^\alpha < 1.$$

Under these choices, the left-hand side of (30.7) tends to 0 as $k \to \infty$, hence $\omega(r) = 0$ and f is identically constant. $\square$

In general, the proof of a growth estimate for $\rho \to \infty$ can also be modified to obtain a regularity estimate by considering $\rho \to 0$. In this spirit, one obtains a regularity type result for harmonic functions. Of course, when M is smooth, this does not yield any new insight since all harmonic functions are smooth. However, the argument in the next theorem can be applied to singular manifolds where the issue of regularity is not so apparent at a singular point.

Theorem 30.4 (Li–Wang) *Let M be a singular manifold satisfying the mean value inequality $\mathcal{M}_1$ and the weak volume growth condition $\mathcal{V}_\mu$ in a neighborhood $B_p(\rho)$ of a point $p \in M$. If $C_V C_{\mathcal{M}_1} < 2$, then any bounded harmonic function defined on $B_p(\rho)$ must be C^α at p for some $0 < \alpha < 1$ depending on $C_V C_{\mathcal{M}_1}$ and μ.*

Proof As in the proof of Theorem 30.3, by setting $r = \beta \rho$ for some $\beta < 1$ in (30.5), we have

$$\left(C_V C_{\mathcal{M}_1} \left(\frac{1-\beta}{1+\beta}\right)^\mu - 1\right)\omega(\rho) \geq \omega(\beta \rho).$$

Interating this k times, we obtain

$$\left(C_V C_{\mathcal{M}_1} \left(\frac{1-\beta}{1+\beta}\right)^\mu - 1\right)^k \omega(\rho) \geq \omega(\beta^k \rho).$$

Hölder continuity follows if we can find $\alpha > 0$ such that

$$\left(C_V C_{\mathcal{M}_1} \left(\frac{1-\beta}{1+\beta}\right)^\mu - 1\right)^k \leq C(\beta^k \rho)^\alpha.$$

This can be achieved by finding α such that

$$C_V C_{\mathcal{M}_1} \left(\frac{1-\beta}{1+\beta}\right)^\mu - 1 \leq \beta^\alpha.$$

To see this, we choose β sufficiently small that

$$\left(\frac{1-\beta}{1+\beta}\right)^\mu \le \frac{2 + C_V C_{\mathcal{M}_1}}{2 C_V C_{\mathcal{M}_1}},$$

which implies

$$C_V C_{\mathcal{M}_1} \left(\frac{1-\beta}{1+\beta}\right)^\mu - 1 \le \frac{C_V C_{\mathcal{M}_1}}{2}.$$

Therefore, there exists α sufficiently small that $\beta^\alpha \ge C_V C_{\mathcal{M}_1}/2$, and the theorem follows. $\square$

Lemma 30.5 *Let M^m be a minimal varifold in $\mathbb{R}^N$. Let $\bar{r}(x, y)$ be the extrinsic distance function between two points $x, y \in M$ obtained by the restriction of the Euclidean distance function to M, and $\bar{B}_x(\rho) = \{y \in M \mid \bar{r}(x, y) < \rho\}$ be the extrinsic ball centered at x of radius ρ. For any regular point $x \in M$ and any nonnegative subharmonic function h defined on M, it must satisfy the mean value inequality*

$$\omega_m \rho^m h(x) \le \int_{\bar{B}_x(\rho)} h(y)\, dy$$

for all $x \in M$ and $\rho \ge 0$.

Proof For a point $x \in M$, it is known that the extrinsic distance function $\bar{r}_x(y) = \bar{r}(x, y)$ satisfies

$$|\nabla \bar{r}_x| \le 1$$

and

$$\Delta \bar{r}_x = \frac{m-1}{\bar{r}_x}.$$

If we define

$$\bar{G}_x(y) = \frac{\bar{r}_x^{2-m}(y) - r^{2-m}}{m(m-2)\omega_m}$$

with $\omega_m = \bar{V}(1)$, then a direct computation shows that

$$\Delta \bar{G}_x = -\frac{(m-1)\bar{r}_x^{-m}}{m\,\omega_m} + \frac{(m-1)\bar{r}_x^{-m} |\nabla \bar{r}_x|^2}{m\,\omega_m}$$

$$\le 0.$$

Since x is a regular point, if we denote the Dirichlet Green's function on the extrinsic ball $\bar{B}_x(r)$ with pole at x by G_x, then $\bar{G}_x \sim G_x$ when $y \to x$.

Hence, together with the fact that $G_x(y) = 0 = \bar{G}_x(y)$ when $y \in \partial \bar{B}_x(r)$, the maximum principle implies that

$$\bar{G}_x(y) \geq G_x(y).$$

This also implies that

$$\frac{\partial G_x(y)}{\partial \mu} \geq \frac{\partial \bar{G}_x(y)}{\partial \nu}$$

$$= -\frac{\bar{r}_x^{1-m}(y)}{m\,\omega_m} \frac{\partial \bar{r}_x(y)}{\partial \nu}$$

$$\geq -\frac{\bar{r}_x^{1-m}(y)}{m\,\omega_m}$$

on $\partial \bar{B}_x(r)$ with ν being the outward pointing unit normal vector on $\partial \bar{B}_x(r)$.

Using the definition of the Green's function, if h is a nonnegative subharmonic function, we have

$$-h(x) = \int_{\bar{B}_x(r)} \Delta G_x(y)\, h(y)\, dy$$

$$= \int_{\bar{B}_x(r)} G_x(y)\, \Delta h(y) + \int_{\partial \bar{B}_x(r)} \frac{\partial G_x(y)}{\partial \nu}\, h(y)$$

$$\geq -\frac{r^{1-m}}{m\,\omega_m} \int_{\partial \bar{B}_x(r)} h(y).$$

Multiplying both sides by r^{m-1} and integrating over $[0, \rho]$, we obtain the mean value inequality

$$\int_{\bar{B}_x(\rho)} h = \int_0^\rho \int_{\bar{B}_x(r)} \frac{h}{|\nabla \bar{r}_x|}$$

$$\geq \int_0^\rho \int_{\bar{B}_x(r)} h$$

$$\geq \omega_m\, \rho^m\, h(x). \qquad \square$$

Theorem 30.6 (Li–Wang) *Let M^m be a minimal varifold in $\mathbb{R}^N$. Let $\bar{r}(x, y)$ be the extrinsic distance function between two points $x, y \in M$. For a fixed point $p \in M$, suppose the density function*

$$\theta_p(\rho) = \frac{\bar{V}_p(\rho)}{\omega_m \rho^m}$$

has an upper bound $\theta_p(\rho) \leq \bar{\theta} < 2$, where $\bar{V}_p(\rho)$ is the volume of the extrinsic ball of radius ρ with the center at p. Then there exists $0 < \alpha < 1$ depending only on $\bar{\theta}$ and m such that if f is a harmonic function satisfying the growth condition

$$|f(x)| = O(\bar{r}_p^\alpha(x)),$$

then f must be identically constant.

Proof If f is a harmonic function, then applying a similar argument to that in Theorem 30.3 and Lemma 30.5, we obtain

$$\int_{\bar{B}_p(\rho)} f - \bar{V}_p(\rho) i(\rho) \geq \bar{\omega}_m(\rho - r)^m f(x) - \omega_m(\rho - r)^m i(\rho)$$

and

$$\bar{V}_p(\rho) s(\rho) - \int_{\bar{B}_p(\rho)} f \geq \omega_m(\rho - r)^m s(\rho) - \omega_m(\rho - r)^m f(y)$$

for any smooth points $x, y \in \partial \bar{B}_p(r)$. In particular, since smooth points are dense in $\bar{B}_p(r)$, this implies that

$$\left(\theta_p(\rho)\left(\frac{\rho}{\rho - r}\right)^m - 1\right)\omega_p(\rho) \geq \omega_p(r).$$

The assumption that $\theta_p(\rho) \leq \bar{\theta} < 2$ and the argument of Theorem 30.3 yield the desired conclusion. □

Combining the arguments of Theorem 30.4 and Theorem 30.6, we have the following regularity theorem. However, we should point out that it was proved by Simon [Si] that under the assumption of Theorem 30.7, a Poincaré inequality is valid. In which case, the Moser iteration argument implies C^α regularity also.

Theorem 30.7 (Simon) *Let M^m be a stationary varifold in $\mathbb{R}^N$. Suppose the multiplicity at a point $p \in M$ defined by*

$$\theta_p = \lim_{\rho \to 0} \theta_p(\rho)$$

is strictly less than 2. Then any harmonic function defined in a neighborhood of p is C^α for some $0 < \alpha < 1$ depending only on θ_p and m.

Corollary 30.8 *Let M^m be a minimal stationary varifold in $\mathbb{R}^N$. Suppose the volume growth of M is Euclidean, i.e.,*

$$\bar{V}_x(\rho) \leq \bar{\theta}\,\omega_m\,\rho^m$$

for some constant $\bar{\theta} \geq \omega_m$. Then there exists a constant $C_1 > 0$ depending only on $\bar{\theta}$ and m such that

$$h^d(M) \leq C\,d^{m-1}$$

for all $d \geq 1$.

Proof Lemma 30.5 and the volume assumption assert that M satisfies the mean value inequality ($\mathcal{M}$) for extrinsic balls. On the other hand, since

$$\omega_m\,\rho^m \leq \bar{V}_x(\rho) \leq \bar{\theta}\,\omega_m\rho^m,$$

it satisfies the weak volume comparison condition ($\mathcal{V}_m$) with $C_\mathcal{V} = \bar{\theta}$. The theorem follows from the argument of Theorem 30.4. □

31
Massive sets

In this chapter, we will introduce the notion of d-massive sets. The notion of a 0-massive set was first introduced by Grigor'yan [G2] when he established the relationship between massive sets and harmonic functions. While it is not clear if such a relationship exists for $d > 0$, the notion of d-massive sets has important geometric and analytic implications.

Definition 31.1 For any real number $d \geq 0$, a d-massive set Ω is a subset of a manifold M that admits a nonnegative, subharmonic function f defined on Ω with the boundary condition

$$f = 0 \quad \text{on} \quad \partial\Omega,$$

and satisfying the growth property

$$f(x) \leq C r^d(x)$$

for all $x \in \Omega$ and for some constant $C > 0$. The function f is called a potential function of Ω. We also let $m^d(M)$ be the maximum number of disjoint d-massive sets admissible on M.

We will now give a proof of the Grigor'yan theorem [G2].

Theorem 31.2 (Grigor'yan) *Let M^m be a complete Riemannian manifold. The maximum number of disjoint 0-massive sets admissible on M is given by the dimension of the space of bounded harmonic functions on M, i.e.,*

$$m^0(M) = h^0(M).$$

Proof Note that $m^0(M) \geq 1$ because the constant function 1 is a nonnegative bounded harmonic function on M. Let Ω_i with $1 \leq i \leq k$ be a set of disjoint

0-massive sets in M. By scaling, we may assume that there are nonnegative subharmonic functions f_i on Ω_i such that

$$f_i = 0 \quad \text{on} \quad \partial \Omega_i$$

and

$$\sup_{\Omega_i} f_i = 1.$$

For a fixed $1 \leq i \leq k$, a point $p \in M$, and $\rho > 0$, let us solve the Dirichlet problem

$$\Delta u_\rho = 0 \quad \text{on} \quad B_p(\rho)$$

with boundary value

$$u_\rho = \begin{cases} f_i & \text{on} \quad \partial B_p(\rho) \cap \Omega_i, \\ 1 - f_\alpha & \text{on} \quad \partial B_p(\rho) \cap \Omega_\alpha, \text{ for } \alpha \neq i, \\ 0 & \text{on} \quad \partial B_p(\rho) \setminus (\cup_{j=1}^k \Omega_j). \end{cases}$$

Since both the functions f_i and $1 - f_\alpha$ are bounded between 0 and 1, the maximum principle implies that $0 \leq u_\rho \leq 1$ on $B_p(\rho)$. Applying the maximum principle on each Ω_j separately, we conclude that

$$u_\rho \geq f_i \quad \text{on} \quad B_p(\rho) \cap \Omega_i$$

and

$$u_\rho \leq 1 - f_\alpha \quad \text{on} \quad B_p(\rho) \cap \Omega_\alpha, \quad \text{for } \alpha \neq i.$$

The fact that u_ρ is bounded implies that there exists a sequence $\rho_j \to \infty$ such that

$$u_{\rho_j} \to u_i,$$

where u_i is a harmonic function defined on M satisfying $0 \leq u_i \leq 1$. Moreover,

$$u_i \geq f_i \quad \text{on} \quad \Omega_i$$

and

$$u_i \leq 1 - f_\alpha \quad \text{on} \quad \Omega_\alpha.$$

Clearly, if $\{x_j\}$ is a sequence of points in M such that $f_i(x_j) \to 1$, then $u_i(x_j) \to 1$. Similarly, if $\{y_j\}$ is a sequence of points such that $f_\alpha(y_j) \to 1$, then $u_i(y_j) \to 0$.

We claim that the set of functions $\{u_i\}_{i=1}^k$ forms a linearly independent set. Indeed, let

$$u = \sum_{i=1}^k a_i u_i$$

be a linear combination such that u is identically 0. For each $1 \leq i \leq k$, evaluating u on a sequence of points $\{x_j\}$ such that

$$f_i(x_j) \to 1$$

we conclude that

$$a_i = 0.$$

Hence $\{u_i\}$ are linearly independent and $k \leq h^0(M)$. Since k is arbitrary, this implies that $m^0(M) \leq h^0(M)$.

To prove the reverse inequality, we may assume that $m^0(M)$ is finite, otherwise the proposition is automatically true. Let $\hat{M}$ be the Stone–Čech compactification of M. Then every bounded continuous function on M can be continuously extended to $\hat{M}$. Let $\{\Omega_i\}_{i=1}^{m^0(M)}$ be a set of disjoint 0-massive sets in M. For each $i \in \{1, \ldots, m^0(M)\}$, let us define the set

$$S_i = \bigcap \{\hat{x} \in \hat{M} \mid f(\hat{x}) = \sup f\},$$

where the intersection is taken over all the potential functions f of Ω_i. The fact that f is subharmonic together with the maximum principle implies that $S_i \subset \hat{M} \setminus M$. We claim that $S_i \neq \emptyset$. In fact, for each potential function f of Ω_i, the set $\{\hat{x} \mid f(\hat{x}) = \sup f\}$ is a closed subset of $\hat{M} \setminus M$. By the compactness of $\hat{M} \setminus M$, we need only to show that for any finitely many potential functions $f_1, \ldots, f_l$ of Ω_i,

$$\bigcap_{j=1}^l \{\hat{x} \mid f_j(\hat{x}) = \sup f_j\} \neq \emptyset.$$

We will argue by induction on l. It is trivially true for one potential function. Let us assume that it is true for l potential functions that

$$\bigcap_{j=1}^l \{\hat{x} \mid f_j(\hat{x}) = \sup f_j\} \neq \emptyset.$$

After normalizing $\sup f_j = 1$, if we define the function $f = f_1 + \cdots + f_l$, then we have

$$\{\hat{x} \mid f(\hat{x}) = \sup f\} = \bigcap_{j=1}^{l} \{\hat{x} \mid f_j(\hat{x}) = \sup f_j\}.$$

Note that both f and f_{l+1} are potential functions of Ω_i. If

$$\{\hat{x} \mid f(\hat{x}) = \sup f\} \bigcap \{\hat{x} \mid f_{l+1}(\hat{x}) = \sup f_{l+1}\} = \emptyset,$$

then for sufficiently small ϵ, the sets

$$D_1 = \{x \in M \mid f(x) > \sup f - \epsilon\}$$

and

$$D_2 = \{x \in M \mid f_{l+1}(x) > \sup f_{l+1} - \epsilon\}$$

are disjoint. Clearly both D_1 and D_2 are subsets of Ω_i with the properties that $\partial D_1 \cap \partial \Omega_i = \emptyset$ and $\partial D_2 \cap \partial \Omega_i = \emptyset$ because $f = f_{l+1} = 0$ on $\partial \Omega_i$. Also, the functions

$$g_1 = f - \sup f + \epsilon$$

and

$$g_2 = f_{l+1} - \sup f_{l+1} + \epsilon$$

are potential functions of D_1 and D_2, respectively. In particular, this implies that M has $m^0(M) + 1$ disjoint massive sets given by

$$\{\Omega_1, \ldots, \Omega_{i-1}, D_1, D_2, \Omega_{i+1}, \ldots, \Omega_{k_0}\},$$

which is a contradiction. Therefore,

$$\bigcap_{j=1}^{l+1} \{\hat{x} \mid f_j(\hat{x}) = \sup f_j\} = \{\hat{x} \mid f(\hat{x}) = \sup f\} \cap \{\hat{x} \mid f_{l+1}(\hat{x}) = \sup f_{l+1}\}$$

$$\neq \emptyset,$$

and the claim that S_i is nonempty follows.

We now show that for each i there exists a potential function h_i of Ω_i such that

$$\{\hat{x} \mid h_i(\hat{x}) = \sup h_i\} = S_i.$$

The function h_i will be called a minimal potential function of Ω_i. For an arbitrary open set U in $\hat{M}$ such that $S_i \subset U$, note that

$$\hat{M} \setminus U \subset \hat{M} \setminus S_i = \bigcup \{\hat{x} \in \hat{M} \mid f(\hat{x}) < \sup f\},$$

where the union is over all potential functions f of Ω_i. The compactness of $\hat{M}\setminus U$ implies that there exist finitely many potential functions $f_1,\ldots,f_l$ of Ω_i such that

$$\hat{M}\setminus U \subset \bigcup_{j=1}^{l}\{\hat{x}\mid f_j(\hat{x}) < \sup f_j\}.$$

Let us define $g = f_1 + \cdots + f_l$, which has the property that $\{\hat{x}\mid g(\hat{x}) = \sup g\} \subset U$. One may assume by scaling g that $0 \leq g \leq 1$ on M and $\sup g = 1$. Now choose a sequence of open sets $U_n \subset \hat{M}$, $n = 1, 2, \ldots$, such that $U_n \subset U_{n+1}$ and $\cap_{n=1}^{\infty} U_n = S_i$. For each U_n, there exists a potential function g_n of Ω_i such that $0 \leq g_n \leq 1$, $\sup g_n = 1$ and

$$\{\hat{x}\mid g_n(\hat{x}) = \sup g_n\} \subset U_n.$$

By defining

$$h_i = \sum_{n=1}^{\infty} 2^{-n} g_n,$$

it is clear that h_i is a minimal potential function of Ω_i satisfying

$$\{\hat{x}\mid h_i(\hat{x}) = \sup h_i\} = S_i.$$

From now on, h_i will denote a minimal potential function of Ω_i.

For a bounded subharmonic function v on M, consider the set

$$S = \{\hat{x}\mid v(\hat{x}) = \sup v\}.$$

We claim that S must contain some S_i. Moreover, for each j, either $S \cap S_j = \emptyset$ or $S_j \subset S$. In fact, let us first argue that $S \cap S_i \neq \emptyset$ for some i. If this is not the case, then for $\epsilon > 0$ sufficiently small the sets

$$\Omega = \{x \in M \mid v(x) > \sup v - \epsilon\}$$

and

$$\tilde{\Omega}_i = \{x \in M \mid h_i(x) > \sup h_i - \epsilon\}$$

must satisfy $\Omega \cap \tilde{\Omega}_i = \emptyset$. Clearly $\tilde{\Omega}_i \subset \Omega_i$, and each $\tilde{\Omega}_i$ is a massive set with potential function $h_i - \sup h_i + \epsilon$. Also, Ω is a massive set with potential function $v - \sup v + \epsilon$. Therefore

$$\{\Omega, \tilde{\Omega}_1, \ldots, \tilde{\Omega}_{k_0}\}$$

31 Massive sets

are $m^0(M)+1$ disjoint massive sets of M, which is impossible, hence $S \cap S_i \neq \emptyset$ for some i. To see that $S_i \subset S$, let us consider the function $w = h_i + v$. Note that

$$\{\hat{x} \mid w(\hat{x}) = \sup w\} = S_i \cap S \subset S_i.$$

Thus, for sufficiently small $\epsilon > 0$, the massive set

$$W = \{x \mid w(x) > \sup w - \epsilon\} \subset \Omega_i,$$

has a potential function given by $f = w - \sup w + \epsilon$. In particular, by extending f to be zero outside W, f is a potential function of Ω_i with

$$\{\hat{x} \mid f(\hat{x}) = \sup f\} = \{\hat{x} \mid w(\hat{x}) = \sup w\}$$
$$= S_i \cap S.$$

The minimality of S_i implies that $S_i \subset S_i \cap S$, hence $S_i \subset S$. This argument also shows that for any j, either $S \cap S_j = \emptyset$ or $S_j \subset S$.

For each Ω_i, let h_i be a minimal potential function of Ω_i with

$$\{\hat{x} \mid h_i(\hat{x}) = \sup h_i\} = S_i.$$

After normalization, we may assume that $\sup h_i = 1$. Using the construction discussed in the first part of this proof, there exists $m^0(M)$ bounded harmonic functions $\{f_i\}$ with the properties that

$$0 \leq f_i \leq 1,$$
$$f_i = 1 \quad \text{on} \quad S_i,$$

and

$$f_i = 0 \quad \text{on} \quad S_\alpha \quad \text{for } \alpha \neq i.$$

The claim that $h^0(M) \leq m^0(M)$ follows if we show that any bounded harmonic function f can be written as linear combinations of the set $\{f_i\}$. Too see this, we first observe that the constant function is spanned by $\{f_i\}$. Indeed, if we let

$$g = \sum_{i=1}^{m^0(M)} f_i,$$

then

$$g \geq 0$$

with

$$g = 1 \text{ on } S_i \text{ for } 1 \leq i \leq m^0(M).$$

In fact, $\sup g = 1$ because otherwise the set

$$\{\hat{x} \mid g(\hat{x}) = \sup g\}$$

would be disjoint from $\bigcup S_i$. In particular, there exists a sufficiently small constant ϵ such that the set

$$\{x \mid g(x) \geq \sup g - \epsilon\}$$

is a 0-massive set disjoint from $\bigcup S_i$, hence contradicting the definition of $m^0(M)$.

Let us now consider h to be any nonconstant bounded harmonic function. Let

$$S_0(+) = \{\hat{x} \mid h(\hat{x}) = \sup h\}$$

and

$$S_0(-) = \{\hat{x} \mid h(\hat{x}) = \inf h\}.$$

The above argument showed that there is at least one S_i such that $S_i \subset \bar{S}_0(+)$ and at least one S_i such that $S_i \subset \bar{S}_0(-)$. Moreover $S_j \cap \bar{S}_0(+) \cup \bar{S}_0(-) = \emptyset$ if S_j is not contained in $\bar{S}_0(+) \cup \bar{S}_0(-)$. Let $I_0(+)$ and $I_0(-)$ be those is such that $S_i \subset \bar{S}_0(+)$ and $S_i \subset \bar{S}_0(-)$, respectively. If we define

$$g_0 = a_0 \left(\sum_{i \in I_0(+)} f_i \right) + +b_0 \left(\sum_{i \in I_0(-)} f_i \right)$$

with $a_0 = \sup h$ and $b_0 = \inf h$, then $g_0 = a_0$ on S_i for all $i \in I_0(+)$ and $g_0 = b_0$ on S_i for all $i \in I_0(+)$. In particular, the harmonic function

$$h_1 = h - g_0$$

satisfies the properties that

$$h_1 = 0 \text{ on } \bigcup_{i \in (I_0(+) \cup I_0(-))} S_i$$

and

$$h_1 = h \text{ on } S_j \text{ for } j \notin (I_0(+) \cup I_0(-)).$$

Applying the same process to h_1 taking combinations with the remaining f_j for $j \notin (I_0(+) \cup I_0(-))$, we obtain a harmonic function h_2 such that

$$h_2 = 0 \quad \text{on} \quad \bigcup_{i \in (I_1(+) \cup I_1(-))} S_i$$

and

$$h_2 = h_1 \quad \text{on} \quad S_i \quad \text{for } j \notin (I_1(+) \cup I_1(-)),$$

where $I_1(+)$ and $I_1(-)$ are those is such that $S_i \subset \{\hat{x} \mid h_1(\hat{x}) = \sup h_1\}$ and $S_i \subset \{\hat{x} \mid h_1(\hat{x}) = \inf h_1\}$, respectively. In particular,

$$h_2 = 0 \quad \text{on} \quad \bigcup_{i \in (I_0(+) \cup I_0(-) \cup I_1(+) \cup I_1(-))} S_i$$

and

$$h_2 = h \quad \text{on} \quad S_i \quad \text{for } j \notin (I_0(+) \cup I_0(-) \cup I_1(+) \cup I_1(-)).$$

Since $m^0(M) < \infty$, we can apply this inductive procedure finitely many times and end up with a function h_k that vanishes on S_i for all $1 \leq i \leq m^0(M)$. This function h_k must be identically 0. Indeed, if this is not the case, then either $\sup h_k$ or $\inf h_k$ is not 0. Let us assume that $\sup h_k > 0$. then for some $\epsilon > 0$ the set given by

$$\{\hat{x} \mid h_k(\hat{x}) = \sup h_k - \epsilon\}$$

is disjoint from $\bigcup S_i$ and produces another 0-massive set, which is a contradiction. This proves the inequality

$$h^0(M) \leq m^0(M),$$

and hence the theorem. $\square$

As pointed out earlier, there is no direct relationship between $h^d(M)$ and $m^d(M)$ for $d > 0$. However, there are parallel theories concerning the two numbers as demonstrated by the following theorem that mirrors Theorem 28.4 when applied to polynomial growth harmonic functions.

Theorem 31.3 (Li–Wang) *Let M^m be a complete manifold satisfying conditions $(\mathcal{V}_\mu)$ and $(\mathcal{M})$. For all $d \geq 1$, there exists a constant $C > 0$ depending only on μ such that*

$$m^d(M) \leq C\, C_\mathcal{M}\, d^{\mu-1}.$$

Proof The proof follows the proof of Theorem 28.4. Instead of using an orthonormal basis for a finite dimensional subspace of harmonic functions, we use a set of potential functions f_i with normalized L^2-norm. Since the functions have disjoint support, they are automatically perpendicular to each other. We also do not need to change bases and consider the space $K_x = \{u \mid u(x) = 0\}$ because for any $x \in M$ there is at most one f_i that does not vanish on x. The rest of the argument is exactly the same. □

Similarly, we also have the following finiteness theorem that mirrors Theorem 28.7.

Theorem 31.4 *Let M be a complete manifold satisfying the weak mean value property (WM). Suppose that the volume growth of M satisfies $V_p(\rho) = O(\rho^\mu)$ as $\rho \to \infty$ for some point $p \in M$. Then*

$$m^d(M) \leq C_{\mathcal{WM}} (2b+1)^{(2d+\mu)}.$$

The next theorem gives a sharp estimate of m^d on $\mathbb{R}^2$.

Theorem 31.5 *On $\mathbb{R}^2$,*

$$m^d(\mathbb{R}^2) \leq 2d$$

for all $d \geq 0$.

Proof Let $\{\Omega_1, \ldots, \Omega_k\}$ be disjoint d-massive sets and $\{u_1, \ldots, u_k\}$ be their corresponding potential functions. Note that since

$$\int_{\Omega_i \cap B(\rho)} |\nabla u_i|^2 \leq \int_{\Omega_i \cap \partial B(\rho)} u_i \frac{\partial u_i}{\partial r},$$

Schwarz inequality implies that

$$2\lambda_1^{\frac{1}{2}}(\Omega_i \cap \partial B(\rho)) \int_{\Omega_i \cap B(\rho)} |\nabla u_i|^2 \leq 2\lambda_1^{\frac{1}{2}}(\Omega_i \cap \partial B(\rho)) \int_{\Omega_i \cap \partial B(\rho)} u_i \frac{\partial u_i}{\partial r}$$

$$\leq \lambda_1(\Omega_i \cap \partial B(\rho)) \int_{\Omega_i \cap \partial B(\rho)} u_i^2$$

$$+ \int_{\Omega_i \cap \partial B(\rho)} \left(\frac{\partial u_i}{\partial r}\right)^2$$

$$\leq \int_{\Omega_i \cap \partial B(\rho)} |\bar\nabla u_i|^2 + \int_{\Omega_i \partial B(\rho)} \left(\frac{\partial u_i}{\partial r}\right)^2$$

$$= \int_{\Omega_i \cap \partial B(\rho)} |\nabla u_i|^2, \tag{31.1}$$

where $\lambda_1(\Omega_i \cap \partial B(\rho))$ denotes the first Dirichlet eigenvalue on $\Omega_i \cap \partial B_p(\rho)$. Using the fact that

$$\lambda_1(\Omega_i \cap \partial B(\rho)) \geq \frac{\pi^2}{A(\Omega_i \cap \partial B(\rho))^2},$$

we conclude that

$$2 \sum_{i=1}^{k} \lambda_1^{1/2}(\Omega_i \cap \partial B(\rho)) \geq 2\pi \sum_{i=1}^{k} \frac{1}{A(\Omega_i \cap \partial B(\rho))}. \tag{31.2}$$

On the other hand, when combined with the inequality

$$k^2 \leq \left(\sum_{i=1}^{k} A(\Omega_i \cap \partial B(\rho)) \right) \left(\sum_{i=1}^{k} \frac{1}{A(\Omega_i \cap \partial B(\rho))} \right)$$

$$\leq 2\pi\rho \left(\sum_{i=1}^{k} \frac{1}{A(\Omega_i \cap \partial B(\rho))} \right),$$

(31.1) and (31.2) imply that

$$\sum_{i=1}^{k} \frac{\int_{\Omega_i \cap \partial B(\rho)} |\nabla u_i|^2}{\int_{\Omega_i \cap B(\rho)} |\nabla u_i|^2} \geq \frac{k^2}{\rho}. \tag{31.3}$$

Observing that

$$\int_{\Omega_i \cap \partial B(r)} |\nabla u_i|^2 = \frac{\partial}{\partial r} \left(\int_{\Omega_i \cap B(\rho)} |\nabla u_i|^2 \right),$$

(31.3) can be written as

$$\frac{\partial}{\partial r} \ln \left(\prod_{i=1}^{k} \int_{\Omega_i \cap B(\rho)} |\nabla u_i|^2 \right) \geq \frac{k^2}{r}.$$

Integrating both sides from ρ_0 to ρ yields

$$\ln \left(\prod_{i=1}^{k} \left(\frac{\int_{\Omega_i \cap B(\rho)} |\nabla u_i|^2}{\int_{\Omega_i \cap B(\rho_0)} |\nabla u_i|^2} \right) \right) \geq k^2 \ln \left(\frac{\rho}{\rho_0} \right).$$

The growth assumption on the u_is asserts that

$$\prod_{i=1}^{k} \int_{\Omega_i \cap B(\rho)} |\nabla u_i|^2 \leq (C \, \rho^{2d-2} \, V(\rho))^k$$

$$\leq C^k \, \rho^{2kd},$$

hence we have

$$2kd \ln \rho + C_1 \geq k^2 \ln \rho - k^2 \ln \rho_0.$$

Letting $\rho \to \infty$ implies that

$$2d \geq k,$$

as was to be proven. □

32
The structure of harmonic maps into a Cartan–Hadamard manifold

In this chapter, we will apply the notion of massive sets to study the structure of the image of a harmonic map whose target is a hyperbolic space. In fact, Li and Wang [LW1] developed this theory which applies when the target is a strongly negatively curved Cartan–Hadamard manifold, or when it is a two-dimensional visibility manifold.

Throughout this chapter we shall assume that N is a Cartan–Hadamard manifold, namely, N is simply connected and has nonpositive sectional curvature. It is well known that N can be compactified by adding a sphere at infinity $S_\infty(N)$. The resulting compact space $\bar{N} = N \cup S_\infty(N)$ is homeomorphic to a closed Euclidean ball. Two geodesic rays γ_1 and γ_2 in N are called equivalent if their Hausdorff distance is finite. Then the geometric boundary $S_\infty(N)$ is simply given by the equivalence classes of geodesic rays in N. A sequence of points $\{x_i\}$ in $\bar{N}$ converges to $x \in \bar{N}$ if for some fixed point $p \in N$, the sequence of geodesic rays $\{\overline{px_i}\}$ converges to a geodesic ray $\gamma \in x$. In this case, we say γ is the geodesic segment $\overline{px}$ joining p to x. Recall that a subset C in N is strictly convex if any geodesic segment between any two points in C is also contained in C. For a subset K in N, the convex hull of K, denoted by $\mathcal{C}(K)$, is defined to be the smallest strictly convex subset C in N containing K. The convex hull can also be obtained by taking the intersection of all convex sets $C \subset N$ containing K. When N is a Cartan–Hadamard manifold, there is only one geodesic segment joining a pair of points in N. In this case, there is only one notion of convexity, and we will simply say a set is convex when it is a strictly convex set. For the purpose of this chapter, we will need a notion of convexity for $\bar{N}$. Since a geodesic line is a geodesic segment joining the two end points in $S_\infty(N)$, it still makes sense to talk about geodesics joining two points in $\bar{N}$. However, it is not true in general that any two points in $S_\infty(N)$ can always be joined by a geodesic segment given by a geodesic line, as indicated by two nonantipodal points in $S_\infty(\mathbb{R}^n)$. If every pair of points in $S_\infty(N)$ can be joined by a geodesic line in N, then N is said to be a visibility manifold. This class of manifolds was extensively studied in [EO]. A typical example of a visibility manifold is a Cartan–Hadamard manifold with sectional curvature bounded from above by a negative constant $-a < 0$.

To remedy the situation when N is not a visibility manifold, we define a generalized notion of geodesic segment joining two points at infinity.

Definition 32.1 A geodesic segment γ joining a pair of points x and y in $\bar{N}$ is the limiting set of a sequence of geodesic segments $\{\gamma_i\}$ in N with end points $\{x_i\}$ and $\{y_i\}$ such that $x_i \to x$ and $y_i \to y$. We will denote γ by $\overline{xy}$.

Observe that if $\overline{xy} \cap S_\infty(N) = \{x, y\}$, then $\overline{xy}$ must be a geodesic line in N and hence a geodesic segment in the traditional sense. For the case of two nonantipodal points in $S_\infty(\mathbb{R}^2)$, the shortest arc on $\mathbb{S}^1 = S_\infty(\mathbb{R}^2)$ joining the two points is the geodesic segment in the sense defined above. If the two points are antipodal in $S_\infty(\mathbb{R}^2)$, say the north pole and the south pole, then there are infinitely many geodesic segments joining them. Each vertical line is a geodesic segment in the genuine sense. Also, both arcs on $\mathbb{S}^1$ joining the two poles are geodesic segment joining them. Using this definition, for a pair of points in $S_\infty(N)$, it is possible to have more than one geodesic segment joining them. The convexity we will define will be in the sense of strictly convex.

Definition 32.2 A subset C of $\bar{N}$ is a convex set if for every pair of points in C, any geodesic segment joining them is also in C.

Definition 32.3 For a subset A in $\bar{N}$, we define its convex hull $\mathcal{C}(A)$ to be the smallest convex subset of $\bar{N}$ containing A.

In what follows, when we say that a subset is closed, we mean that it is closed in $\bar{N}$ unless otherwise noted. In general, we denote the closure for a subset A in $\bar{N}$ by $\bar{A}$. For a given sequence of closed subsets $\{A_i\}$ decreasing to A, it is natural to ask whether the convex hull of A_i in $\bar{N}$ decreases to the convex hull of A. For this purpose, we introduce the following definition.

Definition 32.4 A Cartan–Hadamard manifold N is said to satisfy the separation property if for every closed convex subset A in $\bar{N}$ and every point p not in A, there exists a closed convex set C properly containing A and separating p from A, i.e., $A \subset C$, $A \cap S_\infty(N)$ is contained in the interior of $C \cap S_\infty(N)$ and p is not in C.

For a two-dimensional visibility manifold or a Cartan–Hadamard manifold with constant negative curvature, it is easy to check that the separation property holds. In fact, for a point p not in the closed convex set A, pick up a point $q \in A$ such that $r(p, q) = r(p, A)$. Then the convexity of A and the first variation formula imply that for $z \in A$, $\angle(\overline{zq}, \overline{qp}) \geq \pi/2$. Let x be the midpoint of the geodesic segment between p and q, and

$$C = \{y \in \overline{N} : \angle(\overline{yx}, \overline{xp}) \geq \pi/2\}.$$

Then C is closed, convex as ∂C is evidently totally geodesic and C properly separates p from A.

Lemma 32.5 *A Cartan–Hadamard manifold N satisfies the separation property if and only if for every closed subset A and monotone decreasing sequence of closed subsets $\{A_i\}$ in $\bar{N}$ such that $\bigcap_{i=1}^{\infty} A_i = A$,*

$$\bigcap_{i=1}^{\infty} \overline{C(A_i)} = \overline{C(A)}.$$

Proof Suppose that N satisfies the separation property. Let $\{A_i\}$ be a decreasing sequence of closed subsets in $\bar{N}$ with $\bigcap_{i=1}^{\infty} A_i = A$. Obviously,

$$\overline{C(A)} \subset \bigcap_{i=1}^{\infty} \overline{C(A_i)}$$

from the definition of convex hull. Assume the contrary that

$$\bigcap_{i=1}^{\infty} \overline{C(A_i)} \neq \overline{C(A)}.$$

Then there exists a point $p \in \bigcap_{i=1}^{\infty} \overline{C(A_i)}$ but not in $\overline{C(A)}$. The separation property asserts that there is a closed convex subset C properly separating p from $\overline{C(A)}$. Let

$$C_\epsilon = \{x \in N : r(x, C) \leq \epsilon\}$$

be the ϵ-neighborhood of C. For sufficiently small $\epsilon > 0$, $\bar{C}_\epsilon$ also properly separates p from $\overline{C(A)}$. Since $A \cap S_\infty(N)$ is contained in the interior of $C \cap S_\infty(N)$ and A_i is decreasing to A, we conclude that for i sufficiently large, $A_i \subset \bar{C}_\epsilon$. Thus, $\overline{C(A_i)} \subset \bar{C}_\epsilon$ and $p \in \bar{C}_\epsilon$, which is a contradiction.

Conversely, to show that N satisfies the separation property, let A be a closed convex subset and p be a point not in A. We identify $\bar{N}$ with the closed unit ball of the Euclidean space endowed with the canonical metric. Let A_i be the tubular neighborhood of A of size $1/i$. It is then clear that A_i is a decreasing sequence of closed subsets with $\cap_{i=1}^{\infty} A_i = A$. Hence by the assumption,

$$\cap_{i=1}^{\infty} \overline{\mathcal{C}(A_i)} = \overline{\mathcal{C}(A)} = A.$$

The fact that $p \notin A$ implies that $p \notin \overline{\mathcal{C}(A_i)}$ for sufficiently large i. By choosing $C = \overline{\mathcal{C}(A_i)}$, it is clear that C properly separates p from A. □

According to our definition of a convex hull, it is possible that $\overline{\mathcal{C}(K)} \cap S_\infty(N)$ is a much bigger set than $K \cap S_\infty(N)$. In fact, if we consider K to be the y-axis in $\mathbb{R}^2$, then $K \cap S_\infty(\mathbb{R}^2)$ consists of the two poles in $\mathbb{S}^1$. However, $\mathcal{C}(K) = \mathbb{R}^2$ because every line given by $x = $ constant is a geodesic joining the two poles of $\mathbb{S}^1$. Hence, $\overline{\mathcal{C}(K)} \cap S_\infty(\mathbb{R}^2) = \mathbb{S}^1$. On the other hand, if we assume in addition that N satisfies the following separation property at infinity, then

$$\overline{\mathcal{C}(K)} \cap S_\infty(N) = K \cap S_\infty(N).$$

Definition 32.6 Let N be a Cartan–Hadamard manifold. N is said to satisfy the separation property at infinity if for any closed subset A of $S_\infty(N)$ and any point $p \in S_\infty(N) \setminus A$, there exists a closed convex subset C in $\bar{N}$ such that A is contained in the interior of $C \cap S_\infty(N)$ and p not in C.

It is easy to check that a two-dimensional visibility manifold alway satisfies the separation property at infinity. On the other hand, upon improving a result of Anderson [An], Borbély [Bo] has shown that the Cartan–Hadamard manifold N has the separation property at infinity provided that its sectional curvature satisfies $-Ce^{\alpha r(x)} \leq K_N(x) \leq -1$ for some constant $C > 0$ and $0 \leq \alpha < 1/3$, where $r(x)$ is the distance from point x to a fixed point $o \in N$. The interested reader should refer to [Bo] for a detailed proof. The following simple lemma gives a condition equivalent to the separation property at infinity.

Lemma 32.7 *Let N be a Cartan–Hadamard manifold. Then for every closed set K in $\bar{N}$,*

$$\overline{\mathcal{C}(K)} \cap S_\infty(N) = K \cap S_\infty(N)$$

if and only if N satisfies the separation property at infinity.

Proof Assume that N satisfies the separation property at infinity. For a given closed subset K, let $A = K \cap S_\infty(N)$. If $A = S_\infty(N)$, then there is nothing to prove. So we may assume this is not the case. The closeness of K implies that A is closed. Given $p \in S_\infty(N) \setminus A$, there is a closed convex subset C such that

A is contained in the interior of $C \cap S_\infty(N)$ and p is not in C. In particular, we conclude that

$$\sup_{x \in K} r(x, C) = R < \infty.$$

Let us consider the R-neighborhood,

$$\bar{C}_R = \{x \in N : r(x, C) \le R\},$$

of C. Then $\bar{C}_R$ is a closed convex subset and $K \subset \bar{C}_R$. Moreover,

$$C \cap S_\infty(N) = \bar{C}_R \cap S_\infty(N).$$

Therefore, $\overline{C(K)} \subset \bar{C}_R$ and p is not in $\bar{C}_R$. In particular, $p \notin \overline{C(K)} \cap S_\infty(N)$. This shows that $\overline{C(K)} \cap S_\infty(N) = A$.

Conversely, to show that N satisfies the separation property at infinity, let A be a closed subset of $S_\infty(N)$ and point $p \in S_\infty(N) \setminus A$. Then there exists a closed subset $K \subset S_\infty(N)$ such that A is in the interior of K and $p \notin K$. Let $C = \overline{C(K)}$ and by the assumption $C \cap S_\infty(N) = K$. Thus, $p \notin C$ and A is contained in the interior of $C \cap S_\infty(N)$. Thus, N satisfies the separation property at infinity and the lemma is proved. $\square$

Theorem 32.8 (Li–Wang) *Let M be a complete Riemannian manifold such that the dimension of the space of bounded harmonic functions $\mathcal{H}^0(M)$ is $h^0(M)$. Let $u : M \to N$ be a harmonic map from M into a Cartan–Hadamard manifold, N^n. Let $A = \overline{u(M)} \cap S_\infty(N)$, where $S_\infty(N) = \mathbb{S}^{n-1}$ is the geometric boundary of N. Then there exists a set of points $\{y_i\}_{i=1}^k \subset \overline{u(M)} \cap N$ with $k \le h^0(M)$ such that*

$$u(M) \subset \bigcap_j \overline{C\left(A_j \cup \{y_i\}_{i=1}^k\right)},$$

where $\{A_j\}$ is a monotonically decreasing sequence of closed subsets of $\bar{N}$ properly containing A and $\bigcap_j A_j = A$. In addition, if we assume that N has the separation property, then

$$u(M) \subset \overline{C\left(A \cup \{y_i\}_{i=1}^k\right)}.$$

Proof Let us pick a point $y_0 \in \overline{u(M)}$. If

$$u(M) \subset \bigcap_j \overline{C(A_j \cup \{y_0\})},$$

then we are done. Hence we may assume that there exists an A_j properly containing A such that the set

$$u(M)\setminus \overline{\mathcal{C}\left(A_j \cup \{y_0\}\right)} \neq \emptyset.$$

Moreover, one can easily check that $u(M)\setminus\overline{\mathcal{C}(A_j \cup \{y_0\})}$ is bounded in N. Since u is a harmonic map and the distance function $r(y, \overline{\mathcal{C}(A_j \cup \{y_0\})})$ to the set $\mathcal{C}(A_j \cup \{y_0\})$ is convex, the composition function

$$f(x) = r(u(x), \overline{\mathcal{C}\left(A_j \cup \{y_0\}\right)})$$

is a bounded nonconstant subharmonic function on M. Thus, f attains its maximum at every point of some supremum set S_1. In particular, for $\hat{x}_1 \in S_1$ and a net $\{x_\alpha\}$ in M converging to $\hat{x}_1$ in $\hat{M}$, a subnet of $\{u(x_\alpha)\}$ converges to $y_1 \in N$. Again, if

$$u(M) \subset \cap_j \overline{\mathcal{C}\left(A_j \cup \{y_0 \cup y_1\}\right)},$$

then the theorem is true, otherwise by choosing a larger j if necessary, the function

$$g(x) = r\left(u(x), \overline{\mathcal{C}\left(A_j \cup \{y_0 \cup y_1\}\right)}\right)$$

is a bounded nonconstant subharmonic function on M. If g achieves its maximum on S_1, then $g(\hat{x}) = \sup g$ for $\hat{x} \in S_1$. In particular, this implies

$$\sup g = g(\hat{x}_1)$$
$$= r\left(y_1, \overline{\mathcal{C}\left(A_j \cup \{y_0 \cup y_1\}\right)}\right)$$
$$= 0,$$

which is impossible. Therefore we may assume g achieves its maximum on S_2.

For a net $\{x_\alpha\}$ in M converging to a point $\hat{x}_2$ in S_2, there exists a subnet of $\{u(x_\alpha)\}$ that converges to $y_2 \in N$. Suppose that we have chosen l points $y_1, \ldots, y_l$ described in the above procedure. If

$$u(M) \subset \bigcap_j \overline{\mathcal{C}\left(A_j \cup \{y_i\}_{i=1}^l\right)},$$

then we are done, otherwise by choosing a larger j if necessary, we define the function

$$h(x) = r\left(u(x), \overline{\mathcal{C}\left(A_j \cup \{y_i\}_{i=1}^l\right)}\right).$$

which is a bounded nonconstant subharmonic function on M. We claim that h cannot achieve its maximum on $\bigcup_{i=1}^{l} S_i$. Indeed, if it does, then h must achieve its maximum at every point on S_i for some $1 \le i \le l$. Thus using an argument similar to that used before

$$h(\hat{x}_i) = r\left(y_i, \overline{\mathcal{C}\left(A_j \cup \{y_i\}_{i=1}^{l}\right)}\right) = 0,$$

which is a contradiction, hence h achieves its maximum on some S_i with $i > l$. We may assume that $i = l + 1$.

Let us pick a point $\hat{x}_{l+1} \in S_{l+1}$ and a net $\{x_\alpha\}$ converging to $\hat{x}_{l+1}$. Suppose y_{l+1} is an accumulation point of the net $\{u(x_\alpha)\}$. It is clear that this process must stop after at most $m_0(M)$ steps since there are only $m_0(M)$ massive sets. In particular, there exist k points $\{y_1, \ldots, y_k\}$ with $k \le m_0(M)$ such that

$$u(M) \subset \bigcap_j \overline{\mathcal{C}\left(A_j \cup \{y_i\}_{i=1}^{k}\right)}.$$

Moreover, $y_i \in \overline{u(M)}$, and the proof of the first statement is completed. The second statement follows from Lemma 32.5 □

Corollary 32.9 *If $h^0(M) = 1$ and N is a Cartan–Hadamard manifold, then every bounded harmonic map $M \to N$ must be constant.*

Theorem 32.10 *Let M be a complete manifold and we denote the space spanned by all positive harmonic functions on M by $\mathcal{H}_+(M)$. Suppose M has the property that*

$$h^0(M) = \dim \mathcal{H}_+(M) < \infty.$$

Assume that $u : M \to N$ is a harmonic map from M into a Cartan–Hadamard manifold N which either is a two-dimensional visibility manifold or has strongly negative sectional curvature, and that $A = \overline{u(M)} \cap S_\infty(N)$ consists of at most one point. Then the set A is necessarily empty, and there exists a set of k points $\{y_i\}_{i=1}^{k} \subset \overline{u(M)} \cap N$ with $k \le h^0(M)$ such that

$$u(M) \subset \overline{\mathcal{C}\left(\{y_i\}_{i=1}^{k}\right)}.$$

In particular, if M does not admit any nonconstant positive harmonic functions, then every such harmonic map must be a constant map.

Proof Theorem 32.8 implies that

$$u(M) \subset \overline{\mathcal{C}\left(A \cup \{y_i\}_{i=1}^{k}\right)}$$

for some set of k points $\{y_i\}_{i=1}^{k}$ in N with $k \leq h^0(M)$. If A contains exactly one point a, let γ be a geodesic line on $(-\infty, +\infty)$ such that its restriction to $(0, +\infty)$ represents a. For each y_i, there exists a unique point $\gamma(t_i)$ such that $r(y_i, \gamma) = r(y_i, \gamma(t_i))$. Choose a point $p = \gamma(t_0)$ with $t_0 < t_i$ for $i = 1, 2, \ldots, k$. Let δ be the geodesic ray given by the restriction of γ onto $(t_0, +\infty)$, and denote the Busemann function associated to δ by β. Recall that if δ is parametrized by arc-length, then

$$\beta(y) = \lim_{t \to \infty} (t - r(y, \delta(t))).$$

We claim that there exists a constant c such that

$$r(y, p) \leq \beta(y) + c$$

for $y \in \overline{\mathcal{C}\left(\{a\} \cup \{y_i\}_{i=1}^{k}\right)}$. In fact, by the convexity of the function $r(y, \gamma)$ and the choice of p, one easily checks that

$$r(y, \delta) = r(y, \gamma)$$
$$\leq \max_{1 \leq i \leq k} r(y_i, \gamma)$$
$$= c$$

for any $y \in \overline{\mathcal{C}\left(\{a\} \cup \{y_i\}_{i=1}^{k}\right)}$. Therefore, if we let $\bar{y} \in \delta$ be the point such that $r(y, \delta) = r(y, \bar{y})$, then

$$r(y, p) \leq r(y, \delta) + r(\bar{y}, p)$$
$$\leq c + \beta(\bar{y})$$
$$\leq 2c + \beta(y).$$

This justifies the claim that

$$r(u(x), p) \leq \beta(u(x)) + c$$

for all $x \in M$. Since u is a harmonic map and N is a Cartan–Hadamard manifold, the function $r(u(x), p)$ is subharmonic and the function $\beta(u(x)) + c$ is superharmonic. The sub–super solution method yields a harmonic function $f(x)$ on M such that

$$r(u(x), p) \leq f(x) \leq \beta(u(x)) + c.$$

Therefore f is an unbounded positive harmonic function on M, contradicting to our assumption that there is no such function. In conclusion, A must be empty and

$$\overline{\mathcal{C}(u(M))} = \overline{\mathcal{C}(\{y_i\}_{i=1}^k)}.$$

This proves our first statement. The second part of the theorem follows from the first part by taking $h^0(M) = 1$. □

Notice that the horoball of a visibility manifold intersects the geometric boundary at exactly one point (see [BGS]). Thus, we obtain the following Liouville type theorem which partially generalizes the results in [Sh] and [T].

Corollary 32.11 *Suppose M satisfies $\dim \mathcal{H}_+(M) = 1$. Assume that N is either a two-dimensional visibility manifold or has strongly negative sectional curvature. Then every harmonic map from M into a horoball of N must be constant.*

Recall that a manifold is parabolic if it does not admit a positive Green's function. Since a parabolic manifold has no massive subsets and every positive harmonic function must be constant, we can apply Theorem 32.8 to this case and obtain the following corollary.

Corollary 32.12 *Let u be a harmonic map from a parabolic manifold M into N^n. Assume that N is either a two-dimensional visibility manifold or has strongly negative sectional curvature. Then $u(M) \subset \overline{\mathcal{C}(A)}$, where $A = \overline{u(M)} \cap S_\infty(N)$.*

Proof In this case, we have $h^0(M) = 1$, hence Theorem 32.8 implies that

$$u(M) \subset \overline{\mathcal{C}(A \cup \{y\})}$$

for some $y \in \overline{u(M)} \cap N$. Let us assume the contrary that $u(M)$ is not contained in $\overline{\mathcal{C}(A)}$. In particular, the parabolicity of M implies that the function $r(u(x), \mathcal{C}(A))$ is unbounded. Lemma 32.5 then asserts that $u(M) \backslash \mathcal{C}(W)$ is nonempty for some open set $W \subset S_\infty(N)$ which properly contains A. Let us consider the function

$$f(x) = r(u(x), \overline{\mathcal{C}(W)}),$$

which is a nonconstant, nonnegative, bounded subharmonic function on M. However, the parabolicity assumption on M implies that such a function does not exist. This completes our proof. □

Theorem 32.13 *Let M be a complete manifold such that the maximum number of disjoint d-massive sets of M is $m^d(M)$. Suppose $u : M \to N$ is a harmonic map from M into N, and N satisfies the separation property at infinity. Assume* that there exists a point $o \in N$ such that $r(u(x), o) = O(r^d(x))$ as $x \to \infty$. Then

$$A = \overline{u(M)} \cap S_\infty(N) = \{a_i\}_{i=1}^{k'}$$

with $k' \le m^d(M) - m^0(M)$. If, in addition, N is either a two-dimensional visibility manifold or has strongly negative sectional curvature, then there exist k points $\{y_j\}_{j=1}^k \subset \overline{u(M)} \cap N$ with $k' + k \le m^d(M)$ such that

$$u(M) \subset \overline{\mathcal{C}\left(\{a_i\}_{i=1}^{k'} \cup \{y_j\}_{j=1}^k\right)}.$$

Proof Since a 0-massive set is always d-massive, we have $m^0(M) \le m^d(M)$. Theorem 32.8 implies that there exist k points $\{y_j\}_{j=1}^k \subset \overline{u(M)} \cap N$ with $k \le m^0(M)$, such that

$$u(M) \subset \bigcap_{\epsilon > 0} \overline{\mathcal{C}\left(A_\epsilon \cup \{y_j\}_{j=1}^k\right)}.$$

If A contains at least k' points, then there exist disjoint open sets $\{U_i\}_{i=1}^{k'}$ in $\bar{N}$ such that $U_i \cap A \ne \emptyset$ for $i = 1, 2, \ldots, k'$. Since N is assumed to satisfy the separation property at infinity, Lemma 32.7 yields that $u(M)$ is not a subset of $\overline{\mathcal{C}\left((\bar{N} \setminus U_i) \cup \{y_j\}_{j=1}^k\right)}$. In particular, the function

$$f_i(x) = r\left(u(x), \overline{\mathcal{C}\left((\bar{N} \setminus U_i) \cup \{y_j\}_{j=1}^k\right)}\right)$$

is not identically zero on $u^{-1}(U_i)$ and $\sup f_i = \infty$. Clearly, $f_i = 0$ on the boundary of $u^{-1}(U_i)$ and

$$f_i(x) = O(r^d(x)).$$

This implies that each set $u^{-1}(U_i)$ is d-massive but not massive. In particular, since they are disjoint, $k' \le m^d(M) - m^0(M)$. Thus A has at most $m^d(M) - m^0(M)$ points, and the first conclusion follows. If, in addition, N either is a two-dimensional visibility manifold or has strongly negative sectional curvature, then Theorem 32.8 yields that

$$u(M) \subset \overline{\mathcal{C}\left(\{a_i\}_{i=1}^{k'} \cup \{y_j\}_{j=1}^k\right)},$$

and the estimate $k' + k \le m^d(M)$ follows from the argument. This completes our proof. $\square$

Combining Theorem 32.13 with the estimates on $m^d(M)$ given by Theorem 31.3 and Theorem 31.4, we have the following corollary.

Corollary 32.14 *Let M be a complete manifold satisfying condition ($\mathcal{M}$) and its volume growth $V_p(R) = O(R^\mu)$ for some point $p \in M$. Suppose N is a Cartan–Hadamard manifold satisfying either of the following conditions:*

(1) it has strongly negative sectional curvature;
(2) it is a two-dimensional visibility manifold.

Let $u : M \to N$ be a harmonic map and suppose that there exists a point $o \in N$ such that $r(u(x), o) = O(r^d(x))$ as $x \to \infty$. Then there exist sets of k' points $\{a_i\}_{i=1}^{k'} = \overline{u(M)} \cap S_\infty(N)$ and k points $\{y_j\}_{j=1}^k \subset \overline{u(M)} \cap N$ with $k' + k \leq \lambda 3^{(2d+\mu)}$ such that

$$u(M) \subset \overline{\mathcal{C}\left(\{a_i\}_{i=1}^{k'} \cup \{y_j\}_{j=1}^k\right)}.$$

If M is further assumed to have property ($\mathcal{V}_\mu$), then we have $k' + k \leq C d^{\mu-1}$.

We would like to remark that though Lemma 32.5 states that separation property is necessary and sufficient to conclude

$$\cap_{j=1}^\infty \overline{\mathcal{C}(A_j)} = \overline{\mathcal{C}(A)},$$

after careful examination of the proof of Theorem 32.8, we only need to use the fact that $\cap_{j=1}^\infty \overline{(A_j)}$ is a bounded distance from $\overline{\mathcal{C}(A)}$. With this in mind, using a theorem of Anderson and Borbély, Theorem 32.8 and hence all consequential theorems of this chapter are valid when N is a strongly negatively curved manifold. A complete treatment can be found in [LW1].

Appendix A
Computation of warped product metrics

Let $M^m = \mathbb{R} \times N^{m-1}$ be the product manifold endowed with the warped product metric

$$ds_M^2 = dt^2 + f^2(t)\, ds_N^2,$$

where ds_N^2 is a given metric on N. Our purpose is to compute the curvature on M with respect to this warped product metric.

Let $\{\tilde{\omega}_2, \ldots, \tilde{\omega}_m\}$ be an orthonormal coframe on N with respect to ds_N^2. If we define $\omega_1 = dt$ and $\omega_\alpha = f(t)\,\tilde{\omega}_\alpha$ for $2 \leq \alpha \leq m$, then the set $\{\omega_i\}_{i=1}^m$ forms an orthonormal coframe of M with respect to ds_M^2. The first structural equations assert that

$$d\omega_i = \omega_{ij} \wedge \omega_j,$$

where ω_{ij} are the connection 1-forms with the property that

$$\omega_{ij} = -\omega_{ji}.$$

On the other hand, direct exterior differentiation yields

$$d\omega_1 = 0$$

and

$$\begin{aligned} d\omega_\alpha &= f'\,\omega_1 \wedge \tilde{\omega}_\alpha + f\,\tilde{\omega}_{\alpha\beta} \wedge \tilde{\omega}_\beta \\ &= -(\log f)'\,\omega_\alpha \wedge \omega_1 + \tilde{\omega}_{\alpha\beta}\wedge\beta, \end{aligned}$$

where $\tilde{\omega}_{\alpha\beta}$ are the connection 1-forms on N and f' is the derivative of f with respect to t. Hence we conclude that the connection 1-forms on M are

Appendix A Computation of warped product metrics

given by

$$\omega_{1\alpha} = -\omega_{\alpha 1}$$
$$= (\log f)' \omega_\alpha \qquad (A.1)$$

and

$$\omega_{\alpha\beta} = \tilde{\omega}_{\alpha\beta}. \qquad (A.2)$$

The second structural equations also assert that

$$d\omega_{ij} - \omega_{ik} \wedge \omega_{kj} = \tfrac{1}{2}\mathcal{R}_{ijkl}\,\omega_l \wedge \omega_k,$$

where $\mathcal{R}_{ijkl}$ is the curvature tensor on M. Exterior differentiation of (A.1) yields

$$d\omega_{1\alpha} = (\log f)'' \omega_1 \wedge \omega_\alpha + (\log f)'\left(-(\log f)' \omega_\alpha \wedge \omega_1 + \tilde{\omega}_{\alpha\beta} \wedge \omega_\beta\right).$$

Hence substituting (A.1) and (A.2) into the above identity yields

$$d\omega_{1\alpha} - \omega_{1\beta} \wedge \omega_{\beta\alpha} = \left((\log f)'' + ((\log f)')^2\right) \omega_1 \wedge \omega_\alpha.$$

Also, exterior differentiation of (A.2) gives

$$d\omega_{\alpha\beta} = d\tilde{\omega}_{\alpha\beta},$$

and

$$d\omega_{\alpha\beta} - \omega_{\alpha 1} \wedge \omega_{1\beta} + \omega_{\alpha\gamma} \wedge \omega_{\gamma\beta}$$
$$= d\tilde{\omega}_{\alpha\beta} - \tilde{\omega}_{\alpha\gamma} \wedge \tilde{\omega}_{\gamma\beta} + ((\log f)')^2 \omega_\alpha \wedge \omega_\beta$$
$$= \tfrac{1}{2}\tilde{\mathcal{R}}_{\alpha\beta\gamma\tau}\tilde{\omega}_\tau \wedge \tilde{\omega}_\gamma + ((\log f)')^2 \omega_\alpha \wedge \omega_\beta$$
$$= \tfrac{1}{2}\tilde{\mathcal{R}}_{\alpha\beta\gamma\tau}\,f^{-2}\,\omega_\tau \wedge \omega_\gamma + ((\log f)')^2 \omega_\alpha \wedge \omega_\beta,$$

where $\tilde{\mathcal{R}}_{\alpha\beta\gamma\tau}$ is the curvature tensor on N. In particular, the sectional curvature of the 2-plane section spanned by e_1 and e_α is given by

$$K(e_1, e_\alpha) = -\left((\log f)'' + ((\log f)')^2\right),$$

and the sectional curvature of the 2-plane section spanned by e_α and e_β is given by

$$K(e_\alpha, e_\beta) = f^{-2}\,\tilde{K}(e_\alpha, e_\beta) - ((\log f)')^2,$$

where $\tilde{K}$ is the sectional curvature of N. Moreover the curvature tensor is given by

$$\mathcal{R}_{1\alpha j k} = \begin{cases} (\log f)'' + ((\log f)')^2 & \text{if } j = \alpha, \ k = 1, \\ -(\log f)'' - ((\log f)')^2 & \text{if } j = 1, \ k = \alpha, \\ 0 & \text{otherwise} \end{cases}$$

and

$$\mathcal{R}_{\alpha\beta i j} = \begin{cases} f^{-2} \tilde{\mathcal{R}}_{\alpha\beta\gamma\tau} + ((\log f)')^2 (\delta_{\alpha\tau}\delta_{\beta\gamma} - \delta_{\alpha\gamma}\delta_{\beta\tau}) & \text{if } i = \gamma, j = \tau, \\ 0 & \text{otherwise.} \end{cases}$$

The Ricci curvature is then given by

$$\mathcal{R}_{1j} = \sum_{\alpha} \mathcal{R}_{1\alpha j\alpha}$$

$$= -(m-1)\left((\log f)'' + ((\log f)')^2\right)\delta_{1j} \qquad (A.3)$$

and

$$\mathcal{R}_{\alpha\beta} = \sum_{\gamma} \mathcal{R}_{\alpha\gamma\beta\gamma} + \mathcal{R}_{\alpha 1\beta 1}$$

$$= f^{-2}\tilde{\mathcal{R}}_{\alpha\beta} - \left((\log f)'' + (m-1)((\log f)')^2\right)\delta_{\alpha\beta}, \qquad (A.4)$$

where $\tilde{\mathcal{R}}_{\alpha\beta}$ is the Ricci tensor on N.

Appendix B
Polynomial growth harmonic functions on Euclidean space

In this appendix, we will determine all polynomial growth harmonic functions in $\mathbb{R}^m$. We will also compute their dimensions.

Recall that we denote the space of polynomial growth harmonic functions of order at most d on a complete manifold M by $\mathcal{H}^d(M)$ and its dimension is denoted by $h^d(M)$. Corollary 6.6 asserts that if M has nonnegative Ricci curvature, then $h^d(M) = 1$ for all $d < 1$. Moreover, if $M = \mathbb{R}^m$, with rectangular coordinates given by $\{x_1, \ldots, x^m\}$, and if $f \in \mathcal{H}^d(\mathbb{R}^m)$, then the function $\partial f/\partial x_i$ is also harmonic in $\mathcal{H}^{d-1}(\mathbb{R}^m)$ by Corollary 6.6. Hence if $d < 2$, then $\partial f/\partial x_i$ must be constant functions for all x_i. This implies that f must be a linear function spanned by the coordinate functions $\{x_i\}$ and the constant functions. Therefore we conclude that

$$h^d(\mathbb{R}^m) = m + 1 \quad \text{for} \quad 1 \leq d < 2.$$

We now claim that any $f \in \mathcal{H}^d(\mathbb{R}^m)$ must be a harmonic polynomial of degree at most d. To see this, we argue by induction and assume that this is true for some $d \geq 2$. To prove that this is also valid for $d + 1$, we consider $f \in \mathcal{H}^{d+1}(\mathbb{R}^m)$. The above observation asserts that $\partial f/\partial x_i \in \mathcal{H}^d(\mathbb{R}^m)$ for all $1 \leq i \leq m$. The induction hypothesis implies that each $\partial f/\partial x_i = p_i$, where p_i a harmonic polynomial of degree at most d. On the other hand,

$$f(x_1, 0, \ldots, 0) = \int_0^{x_1} p_1(t_1, 0, \ldots, 0)\, dt_1 + f(0, \ldots, 0)$$

implies that $f(x_1, 0, \ldots, 0)$ is a polynomial of at most degree $d+1$ in x_1. By using the formula

$$f(x_1, \ldots, x_k, 0, \ldots, 0)$$
$$= \int_0^{x_k} p_k(x_1, \ldots, x_{k-1}, t_k, 0 \ldots, 0) \, dt_k + f(x_1, \ldots, x_{k-1}, 0, \ldots, 0)$$

we can argue inductively that $f(x_1, \ldots x_m)$ is a polynomial of degree at most d and the claim is proved.

To determine $h^d(\mathbb{R}^m)$ for each $d \in \mathbb{Z}^+$, we let $x = x_1$ and $y = (x_2, \ldots, x_m)$. For any $f \in \mathcal{H}^d(\mathbb{R}^m)$, we can write

$$f(x, y) = \sum_{i=0}^{d} a_{d-i}(y) x^i,$$

where a_j are polynomials in the variables $y = (x_2, \ldots x_m)$ of degree at most j. Since f is harmonic, by separation of variables we have

$$0 = \Delta f$$
$$= \frac{\partial^2}{\partial x^2} f + \Delta_y f$$
$$= \sum_{i=2}^{d} a_{d-i} \, i(i-1) \, x^{i-2} + \sum_{i=0}^{d} \Delta_y a_{d-i} \, x^i$$
$$= \sum_{i=0}^{d-2} (a_{d-2-i} \, (i+2)(i+1) + \Delta_y a_{d-i}) \, x^i + \Delta_y a_1 \, x^{d-1} + \Delta_y a_0 \, x^d.$$

Note that since a_0 and a_1 are of degree 0 and 1, respectively, the last two terms vanish. Hence we conclude that

$$-a_{d-2-i} \, (i+2)(i+1) = \Delta_y a_{d-i}$$

for all $0 \leq i \leq d-2$. This gives an inductive formula for all the coefficients once a_d and a_{d-1} are fixed and arbitrary polynomials in y of degree at most d and $d-1$, respectively. Let $\mathcal{P}^d(\mathbb{R}^{m-1})$ be the space of all polynomial of degree at most d defined on $\mathbb{R}^{m-1}$, and $p^d(\mathbb{R}^{m-1}) = \dim \mathcal{P}^d(\mathbb{R}^{m-1})$. Clearly,

$$h^d(\mathbb{R}^m) = p^d(\mathbb{R}^{m-1}) + p^{d-1}(\mathbb{R}^{m-1}). \tag{B.1}$$

Appendix B Polynomial growth harmonic functions on Euclidean space

We now claim that
$$p^d(\mathbb{R}^m) = \binom{m+d}{d}.$$

Again, to see this we write any polynomial of degree at most d as
$$f(x, y) = \sum_{i=0}^{d} a_{d-i}(y) x^i.$$

Obviously, each $a_j \in \mathcal{P}^j(\mathbb{R}^m)$ and hence
$$p^d(\mathbb{R}^m) = \sum_{i=0}^{d} p^i(\mathbb{R}^{m-1}).$$

When $m = 1$, we see that
$$p^d(\mathbb{R}^1) = d + 1 = \binom{d+1}{d}.$$

Using induction on m we have
$$p^d(\mathbb{R}^m) = \sum_{i=0}^{d} \binom{m+i-1}{i}.$$

It remains to prove that
$$\binom{m+d}{d} = \sum_{i=0}^{d} \binom{m+i-1}{i}.$$

One can see this by induction on d. Obviously, the identity is true for $d = 1$ since
$$\binom{m+1}{1} = m + 1$$

and
$$\binom{m}{0} = 1$$

by convention. By the induction hypothesis, we have

$$\sum_{i=0}^{d}\binom{m+i-1}{i} = \binom{m+d-1}{d-1} + \binom{m+d-1}{d}$$

$$= \frac{(m+d-1)!}{m!\,(d-1)!} + \frac{(m+d-1)!}{(m-1)!\,d!}$$

$$= \frac{(m+d-1)!\,(d+m)}{m!\,d!}$$

$$= \binom{m+d}{d},$$

and the identity is proved.

Applying this to (B.1), we conclude that

$$h^d(\mathbb{R}^m) = \binom{m+d-1}{d} + \binom{m+d-2}{d-1}. \tag{B.2}$$

Note that by Sterling's formula

$$h^d(\mathbb{R}^m) \sim \frac{2}{(m-1)!} d^{m-1} \tag{B.3}$$

as $d \to \infty$. We also note that

$$\sum_{i=0}^{d} h^i(\mathbb{R}^m) = \sum_{i=0}^{d}\binom{m+i-1}{i} + \sum_{i=0}^{d}\binom{m+i-2}{i-1}$$

$$= h^d(\mathbb{R}^{m+1})$$

$$\sim \frac{2}{m!} d^m \tag{B.4}$$

as $d \to \infty$.

References

[A] F. Almgren, Jr., *Some interior regularity theorems for minimal surfaces and an extension of Bernstein's Theorem*, Ann. Math. **84** (1966), 277–292.

[An] M. Anderson, *The Dirichlet problem at infinity for manifolds of negative curvature*, J. Diff. Geom. **18** (1983), 701–721.

[B] S. Bochner, *Vector fields and Ricci curvature*, AMS Bull. **52** (1946), 776–797.

[BC] R. Bishop and R. Crittenden, *Geometry of manifolds*, Academic Press, New York, 1964.

[BDG] E. Bombieri, E. De Giorgi, and E. Guisti, *Minimal cones and the Bernstein problem*, Invent. Math. **7** (1969), 243–268.

[BGS] W. Ballmann, M. Gromov, and V. Schroeder, *Manifolds of nonpositive curvature*, vol. 61, Progr. Math, Birkhäuser, Berlin, 1985.

[Bm] E. Bombieri, *Theory of minimal surfaces and a counterexample to the Bernstein conjecture in high dimensions*, Unpublished lecture notes (1970).

[Bo] A. Borbély, *A note on the Dirichlet problem at infinity for manifolds of negative curvature*, Proc. Amer. Math. Soc. **114** (1992), 865–872.

[Br] R. Brooks, *A relation between growth and the spectrum of the Laplacian*, Math. Z. **178** (1981), 501–508.

[Bu] P. Buser, *On Cheeger's inequality $\lambda_1 \geq h^2/4$*, AMS Proc. Symp. Pure Math. **36** (1980), 29–77.

[C] J. Cheeger, *A lower bound for the smallest eigenvalue of the Laplacian*, Problems in analysis, a symposium in honor of S. Bochner, Princeton University Press, Princeton, 1970, pp. 195–199.

[CCM] J. Cheeger, T. Colding, and W. Minicozzi, *Linear growth harmonic functions on complete manifolds with nonnegative Ricci curvature*, Geom. Func. Anal. **5** (1995), 948–954.

[Cg1] S. Y. Cheng, *Eigenvalue comparison theorems and its geometric application*, Math. Z. **143** (1975), 279–297.

[Cg2] S. Y. Cheng, *Eigenfunctions and nodal sets*, Comment. Math. Helv. **51** (1976), 43–55.

[Cg3] S. Y. Cheng, *Liouville theorem for harmonic maps*, Proc. Symp. Pure Math. **36** (1980), 147–151.

[CG1] J. Cheeger and D. Gromoll, *On the structure of complete manifolds of nonnegative curvature*, Ann. Math. **92** (1972), 413–443.

[CG2] J. Cheeger and D. Gromoll, *The splitting theorem for manifolds of nonnegative Ricci curvature*, J. Diff. Geom. **6** (1971), 119–127.

[CgY] S. Y. Cheng and S. T. Yau, *Differential equations on Riemannian manifolds and their geometric applications*, Comm. Pure Appl. Math. **27**, (1975), 333–354.

References

[CM1] T. Colding and W. Minicozzi, *On function theory on spaces with a lower Ricci curvature bound*, Math. Res. Lett. **3** (1996), 241–246.

[CM2] T. Colding and W. Minicozzi, *Generalized Liouville properties of manifolds*, Math. Res. Lett. **3** (1996), 723–729.

[CM3] T. Colding and W. Minicozzi, *Harmonic functions with polynomial growth*, J. Diff. Geom. **46** (1997), 1–77.

[CM4] T. Colding and W. Minicozzi, *Harmonic functions on manifolds*, Ann. Math. **146** (1997), 725–747.

[CM5] T. Colding and W. Minicozzi, *Weyl type bounds for harmonic functions*, Invent. Math. **131** (1998), 257–298.

[CM6] T. Colding and W. Minicozzi, *Liouville theorems for harmonic sections and applications*, Comm. Pure Appl. Math. **51** (1998), 113–138.

[Cn] R. Chen, *Neumann eigenvalue estimate on a compact Riemannian manifold*, Proc. Amer. Math. Soc. **108** (1990), 961–970.

[CnL] R. Chen and P. Li, *On Poincaré type inequalities*, Trans. AMS **349** (1997), 1561–1585.

[CSZ] H. Cao, Y. Shen, and S. Zhu, *The structure of stable minimal hypersurfaces in $\mathbb{R}^{n+1}$*, Math. Res. Let. **4** (1997), 637–644.

[CV] S. Cohn-Vossen, *Kürzeste Wege and Totalkrümmung auf Flächen*, Compositio Math. **2** (1935), 69–133.

[CW] H. I. Choi and A. N. Wang, *A first eigenvalue estimate for minimal hypersurfaces*, J. Diff. Geom. **18** (1983), 559–562.

[CY] J. Cheeger and S. T. Yau, *A lower bound for the heat kernel*, Comm. Pure Appl. Math. **34** (1981), 465–480.

[dCP] M. do Carmo and C. K. Peng, *Stable complete minimal surfaces in $\mathbb{R}^3$ are planes*, Bull. AMS **1** (1979), 903–906.

[D] E. B. Davies, *Heat kernels and spectral theory*, Cambridge Tracts in Mathematics, vol. 92, Cambridge University Press, Cambridge, 1989.

[De] E. De Giorgi, *Una estensione del teorema di Bernstein*, Ann. Scuola Nor. Sup. Pisa **19** (1965), 79–85.

[E] J. Escobar, *Uniqueness theorems on conformal deformation of metrics, Sobolev inequalities and an eigenvalue estimate*, Comm. Pure Appl. Math. **43** (1990), 857–883.

[EO] P. Eberlein and B. O'Neill, *Visibility manifolds*, Pacific J. Math. **46** (1973), 45–110.

[F] C. Faber, *Beweiss, dass unter allen homogenen Membrane von gleicher Fläche und gleicher Spannung die kreisförmige die tiefsten Grundton gibt*, Sitzungsber.–Bayer. Akad. Wiss, Math.-Phys. München, 1923, 169–172.

[Fl] W. Fleming, *On the oriented plateau problem*, Rend. Circ. Mat. Palerino **11** (1962), 69–90.

[FC] D. Fischer-Colbrie, *On complete minimal surfaces with finite Morse index in three manifolds*, Invent. Math. **82** (1985), 121–132.

[FCS] D. Fischer-Colbrie and R. Schoen, *The structure of complete stable minimal surfaces in 3-manifolds of nonnegative scalar curvature*, Comm. Pure Appl. Math. **33** (1980), 199–211.

[FF] H. Federer and W. Fleming, *Normal and integral currents*, Ann. Math. **72** (1960), 458–520.

[G1] A. Grigor'yan, *On stochastically complete manifolds*, Soviet Math. Dokl. **34** (1987), 310–313.

[G2] A. Grigor'yan, *On the dimension of spaces of harmonic functions*, Math. Notes **48** (1990), 1114–1118.

[G3] A. Grigor'yan, *The heat equation on noncompact Riemannian manifolds*, Math. USSR Sbornik **72** (1992), 47–77.

[Ge] T. Gelander, *Homotopy type and volume of locally symmetric manifolds*, Duke Math. J. **124** (2004), 459–515.

[GM] S. Gallot and D. Meyer, *Operateur de courbure et Laplacien des formes différentielles d'une variété Riemannienne*, J. Math. Pures Appl. (9) **54** (1975), 259–274.

[Gu1] R. Gulliver, *Index and total curvature of complete minimal surfaces*, Geometric measure theory and the calculus of variations (Arcata, Calif., 1984), Proc. Sympos. Pure Math. 44, Amer. Math. Soc., Providence, RI., 1986, pp. 207–211.

[Gu2] R. Gulliver, *Minimal surfaces of finite index in manifolds of positive scalar curvature*, Lecture Notes in Mathematics: Calculus of variations and partial differential equations (Trento, 1986), vol. 1340, Springer, Berlin, 1988, pp. 115–122.

[HK] P. Hajtasz and P. Koskela, *Sobolev meets Poincaré*, C. R. Acad. Sci. Paris Sr. I Math. **320** (1995), 1211–1215.

[K] E. Krahn, *Über eine von Rayleigh formulierte Minimaleigenschaft des Kreises*, Math. Ann. **94** (1925), 97–100.

[Ka] A. Kasue, *A compactification of a manifold with asymptotically nonnegative curvature*, Ann. Sci. Ecole. Norm. Sup. **21** (1988), 593–622.

[KLZ] S. Kong, P. Li, and D. Zhou, *Spectrum of the Laplacian on quaternionic Kähler manifolds*, J. Diff. Geom. **78** (2008), 295–332.

[KL] L. Karp and P. Li, *The heat equation on complete Riemannian manifolds*, http://math.uci.edu/ pli/heat.pdf.

[L1] P. Li, *A lower bound for the first eigenvalue for the Laplacian on compact manifolds*, Indiana U. Math. J. **27** (1979), 1013–1019.

[L2] P. Li, *On the Sobolev constant and the p-spectrum of a compact Riemannian manifold*, Ann. Scient. Ecole. Norm. Sup. 4, T **13** (1980), 451–469.

[L3] P. Li, *Poincaré inequalities on Riemannian manifolds*, Seminar on Differential Geometry, Annals of Math. Studies. Edited by S. T. Yau, vol. 102, Princeton University Press, Princeton, 1982, pp. 73–83.

[L4] P. Li, *Uniqueness of L^1 solutions for the Laplace equation and the heat equation on Riemannian manifolds*, J. Diff. Geom. **20** (1984), 447–457.

[L5] P. Li, *Large time behavior of the heat equation on complete manifolds with nonnegative Ricci curvature*, Ann. Math. **124** (1986), 1–21.

[L6] P. Li, *Lecture notes on geometric analysis*, Lecture Notes Series No. 6 - Research Institute of Mathematics and Global Analysis Research Center, Seoul National University, Seoul, 1993.

[L7] P. Li, *Harmonic functions of linear growth on Kähler manifolds with nonnegative Ricci curvature*, Math. Res. Lett. **2** (1995), 79–94.

[L8] P. Li, *Harmonic sections of polynomial growth*, Math. Res. Lett. **4** (1997), 35–44.

[L9] P. Li, *Harmonic functions and applications to complete manifolds*, XIV Escola de Geometria Diferencial: Em homenagem a Shiing-Shen Chern, IMPA, Rio de Janeiro, 2006.

[LoT] G. Liao and L. F. Tam, *On the heat equation for harmonic maps from non-compact manifolds*, Pacific J. Math. **153** (1992), 129–145.

[LS] P. Li and R. Schoen, L^p *and mean value properties of subharmonic functions on Riemannian manifolds*, Acta Math. **153** (1984), 279–301.

[LT1] P. Li and L. F. Tam, *Positive harmonic functions on complete manifolds with nonnegative curvature outside a compact set.*, Ann. Math. **125** (1987), 171–207.

[LT2] P. Li and L. F. Tam, *Symmetric Green's functions on complete manifolds*, Amer. J. Math. **109** (1987), 1129–1154.

[LT3] P. Li and L. F. Tam, *Linear growth harmonic functions on a complete manifold*, J. Diff. Geom. **29** (1989), 421–425.

[LT4] P. Li and L. F. Tam, *Complete surfaces with finite total curvature*, J. Diff. Geom. **33** (1991), 139–168.

[LT5] P. Li and L. F. Tam, *The heat equation and harmonic maps of complete manifolds*, Invent. Math. **105** (1991), 1–46.

[LT6] P. Li and L. F. Tam, *Harmonic functions and the structure of complete manifolds*, J. Diff. Geom. **35** (1992), 359–383.

[LW1] P. Li and J. Wang, *Convex hull properties of harmonic maps*, J. Diff. Geom. **48** (1998), 497–530.

[LW2] P. Li and J. P. Wang, *Mean value inequalities*, Indiana Math. J. **48** (1999), 1257–1273.

[LW3] P. Li and J. P. Wang, *Counting massive sets and dimensions of harmonic functions*, J. Diff. Geom. **53** (1999), 237–278.

[LW4] P. Li and J. P. Wang, *Minimal hypersurfaces with finite index*, Math. Res. Lett. **9** (2002), 95–103.

[LW5] P. Li and J. P. Wang, *Complete manifolds with positive spectrum*, J. Diff. Geom. **58** (2001), 501–534.

[LW6] P. Li and J. Wang, *Complete manifolds with positive spectrum, II.*, J. Diff. Geom. **62** (2002), 143–162.

[LW7] P. Li and J. P. Wang, *Stable minimal hypersurfaces in a nonnegatively curved manifold*, J. Reine Angew. Math. (Crelles) **566** (2004), 215–230.

[LW8] P. Li and J. Wang, *Comparison theorem for Kähler manifolds and positivity of spectrum.*, J. Diff. Geom. **69** (2005), 43–74.

[LY1] P. Li and S. T. Yau, *Eigenvalues of a compact Riemannian manifold.*, AMS Proc. Symp. Pure Math. **36** (1980), 205–239.

[LY2] P. Li and S. T. Yau, *On the parabolic kernel of the Schrödinger operator*, Acta Math. **156** (1986), 153–201.

[Lz] A. Lichnerowicz, *Géometrie des groupes de transformations*, Dunod, Paris, 1958.

[M] B. Malgrange, *Existence et approximation des solutions der équations aux dérivées partielles et des équations de convolution*, Annales de l'Inst. Fourier **6** (1955), 271–355.

[MS] J. H. Michael and L. Simon, *Sobolev and mean-value inequalities on generalized submanifolds of $\mathbb{R}^n$*, Comm. Pure Appl. Math. **26** (1973), 361–379.

[NT] L. Ni and L. Tam, *Kähler–Ricci flow and the Poincaré–Lelong equation*, Comm. Anal. Geom. **12** (2004), 111–141.

[O] M. Obata, *Certain conditions for a Riemannian manifold to be isometric to the sphere*, J. Math. Soc. Japan **14** (1962), 333–340.

[R] R. Reilly, *Applications of the Hessian operator in a Riemannian manifold*, Indiana U. Math. J. **26** (1977), 459–472.

[Ro] H. Royden, *Harmonic functions on open Riemann surfaces*, Trans. AMS **73** (1952), 40–94.

[S] J. Simons, *Minimal varieties in Riemannian manifolds*, Ann. Math. **80** (1964), 1–21.

[SC] L. Saloff-Coste, *Uniformly elliptic operators on Riemannian manifolds*, J. Diff. Geom. **36** (1992), 417–450.

[Sh] Y. Shen, *A Liouville theorem for harmonic maps*, Amer. J. Math. **117** (1995), 773–785.

[Si] L. Simon, *Lectures on geometric measure theory*, Proc. Centre for Mathematical Analysis, Australian National University, Canberra, 1984.

[St] E. Stein, *Singular integrals and differentiability properties of functions*, Princeton mathematical series no. 30, Princeton University Press, Princeton, 1970.

[STW] C. J. Sung, L. F. Tam, and J. P. Wang, *Spaces of harmonic functions*, J. London Math. Soc. **61** (2000), 789–806.

[SY1] R. Schoen and S. T. Yau, *Harmonic maps and the topology of stable hypersurfaces and manifolds of nonnegative Ricci curvature*, Comm. Math. Helv. **39** (1976), 333–341.

[SY2] R. Schoen and S. T. Yau, *Lectures on differential geometry*, Conference Proceedings and Lecture Notes in Geometry and Topology, vol. I, International Press., Cambridge.
[T] L. F. Tam, *Liouville properties of harmonic maps*, Math. Res. Lett. **2** (1995), 719–735.
[V] N. Varopoulos, *Hardy–Littlewood theory for semigroups*, J. Funct. Anal. **63** (1985), 240–260.
[WZ] J. Wang and L. Zhou, *Gradient estimate for eigenforms of Hodge Laplacian*, preprint.
[Y1] S. T. Yau, *Harmonic functions on complete Riemannian manifolds*, Comm. Pure Appl. Math. **27** (1975), 201–228.
[Y2] S. T. Yau, *Some function-theoretic properties of complete Riemannian manifolds and their applications to geometry*, Indiana U. Math. J. **25** (1976), 659–670.
[Y3] S. T. Yau, *Isoperimetric constants and the first eigenvalue of a compact Riemannian manifold*, Ann. Scient. Ecole. Norm. Sup. **4** (1985), 487–507.
[ZY] J. Q. Zhong and H. C. Yang, *On the estimate of first eigenvalue of a compact Riemannian manifold*, Sci. Sinica Ser. A **27** (1984), 1265–1273.

Index

$(\mathcal{VD})$, 203
$C_{\mathcal{M}_1}$, 361
$C_{\mathcal{M}}$, 341
$C_{\mathcal{P},\phi}(B_p(\rho))$, 207
$C_{\mathcal{P}}(\Omega)$, 203
$C_{\mathcal{V}}$, 341
$C_{\mathcal{SD}}$, 98, 217
$C_{\mathcal{VD},\phi}(B_p(\rho))$, 207
$C_{\mathcal{VD}}(\Omega)$, 203
$C_{\mathcal{WM}}$, 345
$ID_\alpha(M)$, 86
$IN_\alpha(M)$, 86
L^1 mean valued inequality $(\mathcal{M}_1)$, 361
L^α subharmonic function, 68
L^q harmonic functions, 281, 349
$N'(M)$, 242
$N^0(M)$, 242
$N'_i(M)$, 242
$N^0_i(M)$, 242
$SD_\alpha(M)$, 87
$SN_\alpha(M)$, 87
$\mathcal{H}'(M)$, 73
$\mathcal{H}^d(M)$, 326
$\mathcal{H}_+(M)$, 387
$\mathcal{K}'(M)$, 262
$\mathcal{K}(M)$, 264
$\mathcal{K}^0(M)$, 262
$\mathcal{L}^q(E)$, 281
$\lambda_1(M)$, 91
$\mu_1(M)$, 91
Spec(Δ), 267
d-massive set, 370
kth Neumann eigenfunction, 116
kth Dirichlet eigenfunction, 113
kth Dirichlet eigenvalue, 113
kth Neumann eigenvalue, 116
$k'(M)$, 262

$k(M)$, 264
$k^0(M)$, 262
$m^d(M)$, 370
pth Betti number, 29

adjoint operator, 26
area, boundary of geodesic ball, 14

barrier function, 242
Betti number, 28
Bochner Laplacian, 26
Bochner technique, 19
Bochner–Weitzenböck formulas, 19
Buseman function, 35

C_{SN}, 115
Cartan's first structural equations, 19
Cartan's second structural equations, 19
Cartan–Hadamard manifold, 381
coarea formula, 87
Cohn–Vossen–Hartman formula, 300
commutation formula, 21
comparison theorem, heat equation, 142
compatible with metric, 1
conformally equivalent, 311
connection 1-forms, 19
constant mean curvature, 82
contractive on L^1, 170
contractive on L^∞, 169
contractive on L^p, 171
convex, 53, 131
convex hull $\mathcal{C}(A)$, 382
convex set, 382
covariant derivatives, 22
covariant differentiation, p-forms, 27
curvature operator, 29
curvature tensor, 2, 19
cut-locus, 10

Index

de-Rham decomposition, 29
diameter, 14
Dirichlet α-isoperimetric constant, 86
Dirichlet α-Sobolev constant, 87
Dirichlet eigenfunction, 37
Dirichlet eigenvalue, 37
Dirichlet heat kernel, 108
Dirichlet Laplacian, 36
distance function, 32
dual 1-form, 31
dual coframe, 19
Duhamel principle, 106

eigenfunctions, 97
eigenvalue comparison theorem, 36
eigenvalues, 97
end, 241
ends, 36
essential spectrum $\mu_e(M)$, 267
exterior p-form, 21

finite index, 321
first nonzero eigenvalue, 37
first variation of area, 3
fundamental solution, 96

Gauss lemma, 10
geodesic, 7
geodesically star-shaped, 157
geometric boundary, 381
gradient estimate, 57
greatest lower bound for the L^2-spectrum, 63
Green's function, 189

half-space model, 67
harmonic p-form, 29
harmonic 1-form, 28
harmonic function, polynomial growth, 395
harmonic functions, 57
harmonic functions, linear growth, 76
Harnack type inequality, 62
heat equation, 96
heat kernel, 96
Hessian, 20
Hodge decomposition, 29
homology group, 84
horoball, 389
hyperbolic m-space, 67

induced metric, 77
infimum of the spectrum, 63
infimum of the spectrum of Δ, 267
infinite volume, 17
infinitesimal generator, 30
injectivity radius, 110
isoperimetric, 86
isoperimetric inequalities, 86

kernel function, 96
Killing vector field, 30

Laplacian, 20
Laplacian, p-forms, 23
line, 34
Liouville property, 349

Marcinkiewicz interpolation theorem, 117
maximum principle method, 57
mean curvature vector, 2
mean value inequality, 68
mean value inequality ($\mathcal{M}$), 341
mean value inequality at infinity, 327
measured Neumann Poincaré inequality, 204, 207
measured Neumann Sobolev inequality, 208
metric, 1
minimal barrier function, 257
minimal Green's function, 248
minimal heat kernel, 128
minimal hypersurfaces, 306
Myers' theorem, 14

Nash–Mosers Harnack inequality, 216
Neumann α-isoperimetric constant, 86
Neumann α-Sobolev constant, 87
Neumann boundary condition, 48
Neumann heat equation, 105
Neumann heat kernel, 105
Neumann Poincaré inequality, 203
nonparabolic, 241
normal variations, 8
normalized Sobolev inequality, 217

orthonormal frame, 19
oscillation, 195

parabolic, 241
parabolic equations, 122
parabolic Schrödinger equation, 57
parabolicity, 247
parallel, 27
Poincaré inequality, 71
polynomial growth harmonic functions, 326
positive solutions, 134
potential function, 63, 370
pseudo-differential operator, 118

quasi-isometric, 239
quasi-isometric invariant, 247

ray, 34
Rayleigh principle, 38
Rayleigh quotient, 299
Ricci curvature, 2

Ricci identity, 21
Riemannian connection, 1
Riemannian manifold, 1

scaled local λ_1 bound, 203
second covariant derivative, 20
second fundamental form, 2
second variational formula, 4
second variational formula, length, 6
sectional curvature, 2
semi-group property, 112
separation property, 382
separation property at infinity, 384
Sobolev constants, 86
Sobolev inequality, 98
Sobolev type inequalities, 86
space form, 12
spherical harmonic, 65
stability inequality, 8, 306
stable, 8
stable minimal hypersurface, 306
star operator, 23
strictly convex, 381

subadditive operator, 117
subharmonic functions, 68

third covariant derivative, 21
torsion free, 1

unit $(m-1)$-sphere, 16
universal covering, 29

variational vector field, 4
visibility manifold, 382
Vitali covering, 212
volume doubling property, measure, 207
volume comparison condition $(\mathcal{V}_\mu)$, 341
volume comparison theorem, 10
volume doubling property, 203
volume entropy, 268
volume growth, 16
volume, geodesic ball, 14

warped product, 285, 286, 392
weak (p,q) type, 117
weak mean value inequality $(\mathcal{WM})$, 345